Human Variation
From the Laboratory
to the Field

Society for the Study of Human Biology Series

Numbers 1–9 were published by Pergamon Press, Headington Hill Hall, Headington, Oxford OX3 0BY. Numbers 10–24 were published by Taylor & Francis Ltd, 10–14 Macklin Street, London WC2B 5NF. Numbers 25–40 were published by Cambridge University Press, The Pitt Building, Trumpington Street Cambridge CB2 1RP. Further details and prices of back-list numbers are available from the Secretary of the Society for the Study of Human Biology.

Human Variation
From the Laboratory to the Field

Edited by
C. G. Nicholas Mascie-Taylor
Akira Yasukouchi
Stanley Ulijaszek

CRC Press
Taylor & Francis Group
Boca Raton London New York

CRC Press is an imprint of the
Taylor & Francis Group, an **informa** business

CRC Press
Taylor & Francis Group
6000 Broken Sound Parkway NW, Suite 300
Boca Raton, FL 33487-2742

© 2010 by Taylor and Francis Group, LLC
CRC Press is an imprint of Taylor & Francis Group, an Informa business

No claim to original U.S. Government works

Printed in the United States of America on acid-free paper
10 9 8 7 6 5 4 3 2 1

International Standard Book Number: 978-1-4200-8471-9 (Hardback)

Library of Congress Cataloging-in-Publication Data

Human variation : from the laboratory to the field / [editors] C. G. Nicholas
Mascie-Taylor, Akira Yasukouchi, Stanley Ulijaszek.
    p. cm. -- (Society for the study of human biology series)
  Includes bibliographical references and index.
  ISBN 978-1-4200-8471-9 (hardcover : alk. paper)
  1. Physical anthropology. 2. Anthropometry. 3. Human population genetics. I.
Mascie-Taylor, C. G. N. II. Yasukouchi, Akira. III. Ulijaszek, Stanley J. IV. Title. V. Series.

  GN62.8.H88 2010
  599.9--dc22

                     2009047375

Visit the Taylor & Francis Web site at
http://www.taylorandfrancis.com

and the CRC Press Web site at
http://www.crcpress.com

Contents

Acknowledgements

This volume arose out of a joint Society for the Study of Human Biology (SSHB) and Japan Society of Physiological Anthropology (JSPA) meeting held at the University of Cambridge, U.K., in 2007. Financial support is gratefully acknowledged from the Daiwa Foundation and SSHB. The editors would like to express special thanks to Dr. Rie Goto, who was responsible for most of the conference logistics and also helped in preparing the manuscript for publication.

Contributors

Alan H. Bittles
Centre for Comparative Genomics
Murdoch University
Edith Cowan University
Perth, Australia

Michael L. Black
Centre for Comparative Genomics
Murdoch University
Perth, Australia

Katherine Brooke-Wavell
Department of Human Sciences
Loughborough University
Leicestershire, United Kingdom

Anthony G. Comuzzie
Department of Genetics
Southwest Foundation for Biomedical
    Research
Southwest National Primate Research
    Center
San Antonio, Texas

Ralph M. Garruto
Department of Anthropology
Binghamton University
State University of New York
Binghamton, New York

Paul Higgins
Department of Genetics
Southwest Foundation for Biomedical
    Research
San Antonio, Texas

Shigekazu Higuchi
Kyushu University
Fukuoka, Japan

Koichi Iwanaga
Department of Design Science
Graduate School of Engineering
Chiba University
Chiba, Japan

John Komlos
Department of Economics
University of Munich
Munich, Germany

Susumu Kudo
Department of Mechanical
    Engineering
Shibaura Institute of Technology
Tokyo, Japan

Michael A. Little
Department of Anthropology
Binghamton University
State University of New York
Binghamton, New York

Takafumi Maeda
Laboratory of Environmental
    Ergonomics
Hokkaido University
Sapporo, Japan

Robert M. Malina
Department of Kinesiology and Health
    Education
University of Texas at Austin
Austin, Texas
Department of Health and Physical
    Education
Tarleton State University
Stephenville, Texas

Neil J. Mansfield
Department of Human Sciences
Loughborough University
Leicestershire, United Kingdom

C.G. Nicholas Mascie-Taylor
Department of Biological
Anthropology
University of Cambridge
Cambridge, United Kingdom

Lawrence M. Schell
Department of Anthropology
State University of New York
Albany, New York

Yoshiaki Sone
Graduate School of Human Life
Science
Osaka City University
Osaka, Japan

Kazuo Tanishita
Department of System Design
Engineering
Keio University
Yokohama, Japan

Stanley Ulijaszek
Institute of Social and Cultural
Anthropology
University of Oxford
Oxford, United Kingdom

V. Saroja Voruganti
Department of Genetics
Southwest Foundation for Biomedical
Research
San Antonio, Texas

Charles A. Weitz
Department of Anthropology
Temple University
Philadelphia, Pennsylvania

Akira Yasukouchi
Department of Physiological
Anthropology
Kyushu University
Fukuoka, Japan

1 Factors and Forces Influencing Human Variation

Alan H. Bittles[*,1,2] *and Michael L. Black*[1]
[1]Centre for Comparative Genomics,
Murdoch University, Perth, Australia
[2]Edith Cowan University, Perth, Australia

CONTENTS

---

[*] Address all correspondence to abittles@ccg.murdoch.edu.au.

1.1 INTRODUCTION

Just as tastes in fashion often are subject to abrupt change, so ideas and hypotheses in science can alter quite rapidly according to the methodologies available, the resultant data produced, and the interpretive skills of researchers. It is therefore not surprising that the major transition in anthropological and biomedical research methods over the course of the past fifty years, from anthropometric and craniometric measurements to current, large-scale microarray studies of gene structure and expression (Table 1.1) has resulted in continued revision of opinions and ideas relating to the factors and forces that drive human variation (Bittles et al. 2007).

At a more subtle level, scientific discourse can also be influenced by the adoption into everyday speech of terms which originally had quite precise scientific meanings. The word *mutation* is a prime example of this phenomenon, to the extent that most members of the general public and many scientists would find it difficult to accept that a mutation could be other than harmful. Although it is probable that a majority of mutations are indeed disadvantageous, the entire concept of evolutionary change driven by natural selection proposed by Charles Darwin in 1859 is dependent on mutational changes that better equip organisms to survive and reproduce in their respective environmental niches. In the absence of such changes across generations, it would have been extremely difficult, if not impossible, for small, separately evolving human groups to have migrated from their home environments in sub-Saharan Africa some 60,000–70,000 ybp (Behar et al. 2008) to successfully inhabit the temperate and even the Arctic regions of the globe.

1.1.1 SELECTION AND NEUTRAL MUTATION

While mutation is essential to the creation of genetic variants, the Darwinian concept of biological fitness is equally dependent on selection, a process which allows either for the proliferation or removal of variants from the gene pool. The concept of evolutionary change driven by selection was largely unchallenged until the 1960s and 1970s, when the Neutral Theory of Mutation was proposed to explain the vast numbers of apparently selection-independent genetic variants revealed by protein electrophoresis (Kimura 1968, 1969), followed by the modified 'Nearly Neutral' theory (Ohta 1973). More recently, the influence of the Human Genome Project has

---

TABLE 1.1
Methods for the Assessment of Human Genetic Variation

Anthropometric and craniometric analysis
Blood group antigen analysis
Serum protein electrophoresis
Gene-based studies
Microarray studies of gene structure and expression

---

been apparent in attempts to bridge the gap between theories solely or primarily based on selection as opposed to neutral variation—for example, the Neoselectionist Theory of Genome Evolution (Bernardi 2007), which as its name suggests concentrates on genomic and epigenomic events, and the proposal that neutral mutations act as predecessors of subsequent evolutionary adaptation (Wagner 2008).

1.1.2　THE ABO BLOOD GROUP SYSTEM AND DISEASE ASSOCIATIONS

Just as the introduction of new methodologies provided enhanced insights into human genetic variation, a changing understanding of the demographic structure of human populations has resulted in quite dramatic revisions in the conclusions being drawn from disease association studies. The initial description of the ABO blood group system in the 1920s was followed by the development of laboratory protocols that would allow blood transfusions to become routine, life-saving procedures. Further progress in blood group serology led researchers to conduct association studies between ABO blood groups and organic and infectious diseases, fertility, and constructs such as IQ and socioeconomic status, resulting in the publication of numerous scientific papers and reviews in major medical and scientific journals from the late 1950s to the 1990s.

Interest in an association between the ABO blood groups and disease was first aroused by a report that stomach cancer was significantly more common in people with blood group A (Aird et al. 1953), interpreted by the authors as resulting from the action of natural selection on the ABO blood group system. As indicated in Table 1.2, these findings were closely followed by the report of an apparent predisposition to pernicious anaemia in persons with blood group A (Aird et al. 1956), to peptic and duodenal ulceration in those with blood group O (Aird et al. 1954; Clarke et al. 1955), and to a wide range of other organic disorders which showed a higher prevalence and/or more severe symptoms in people with one or other of the common ABO blood groups (Dawood et al. 1971; Mourant et al. 1971; Whincup et al. 1990; Juvonen and Niemelä 1992).

TABLE 1.2
ABO Blood Groups and Organic Diseases

Blood Group	Disease	Study
Group A	Stomach cancer	Aird et al., 1953
	Pernicious anaemia	Aird et al., 1956
	Choriocarcinoma	Dawood et al., 1971
	Ischaemic heart disease	Whincup et al., 1990
	Gall stones	Juvonen and Niemelä, 1992
Group O	Peptic ulcer	Aird et al., 1954
	Duodenal ulcer	Clarke et al., 1955
	Thromboembolic disease	Mourant et al., 1971

TABLE 1.3
ABO Blood Groups and Infectious Diseases

Blood Group	Disease	Study
Groups A and AB	Smallpox	Vogel and Chakravartii, 1966
Group A	Pulmonary TB	Jain, 1970
Groups B and AB	Gram-ve enterobacteria	Robinson et al., 1971
Group B	Neonatal streptococcus B	Regan et al., 1978
Group O	Cholera	Chaudhuri and De, 1977

A similar picture emerged with respect to major bacterial and viral diseases (Table 1.3), with a significant positive association between both susceptibility and the severity of smallpox infection in people with blood groups A and AB (Vogel and Chakravartii 1966). Conversely, cholera was more common among those with blood group O (Chaudhuri and De 1977), major enteric diseases due to *Escherischia coli* and *Salmonella* infections were associated with blood groups B and AB (Robinson et al. 1971), and neonatal streptococcus B infections were more prevalent in infants with blood group B (Regan et al. 1978).

Additional proof of the concept that the ABO blood group system was a stable polymorphism (Sheppard 1959) appeared to be offered by the differential role of the ABO blood groups with respect to fertility. Groups A and B mothers showed higher infertility (Bennett and Walker 1956), whereas ABO-incompatible pregnancies resulted in more abortions (Matsunaga and Itoh 1958) and stillbirths (Reed and Kelly 1958), and the pregnancies of group O mothers were more likely to be adversely affected by Rhesus incompatibility (Rosenfield 1955).

From a less obviously biological perspective, it also was reported that in the Otmoor villages of Oxfordshire, mean IQ scores varied according to ABO phenotype (Gibson et al. 1973), and among English blood donors Group A individuals were over-represented in the top two social classes, I and II (Beardmore and Karimi-Booshehri 1983), although the findings of the latter study were strongly queried, using data on the parents of children enrolled in the UK National Child Development Study of 1958 (Mascie-Taylor and McManus 1984).

1.1.3 THE INFLUENCE OF POPULATION STRATIFICATION ON ASSOCIATION STUDIES

Overall, these diverse findings added up to a quite compelling narrative indicating that the ABO blood group phenotypes were subject to strong but balancing selectional effects. That is, until a short commentary on the claimed findings of the association between the ABO blood group system and IQ (Thomson and Bodmer 1975), which pointed out that the results obtained could be an outcome of population stratification, since autochthonous and nonlocal residents had significant differences in their ABO phenotype frequencies, a possibility acknowledged by the authors of the original study (Gibson et al. 1975).

Papers claiming ABO blood group associations with different disease states continue to be published, such as one report that blood group O individuals are more resistant to severe falciparum malaria due to their reduced binding of erythrocytes infected with the *Plasmodium falciparum* to uninfected red cells (Rowe et al. 2007). Since the spontaneous binding of infected and uninfected red cells contributes to the pathogenesis of severe malaria by obstructing microvascular blood flow (Kaul et al. 1991), the reduced levels observed in blood group O persons are advantageous and restrict the development of cerebral malaria (Uneke 2007).

Nevertheless, ABO blood group investigations largely were replaced by association studies based on the highly polymorphic Human Leukocyte Antigen (HLA) system, initially using antigenic analysis but with more specific DNA-based methods gradually introduced for HLA haplotype identification (Table 1.1). The pattern of change is ongoing, with virtually all current disease association investigations now relying on genome-wide microarrays to identify disease susceptibility loci (Wellcome Trust Consortium 2007). But whatever the investigative method used, population stratification remains an underlying problem in case-control comparisons, especially in populations that are known to be multiply subdivided, for example, on geographical, ethnic, tribal, language, religious, and social grounds.

1.2 THE ROLE OF ETHNICITY IN THE DEFINITION OF POPULATION GENETIC STRUCTURE

A germ cell mutation in an individual can either be lost via selection due to infertility or pre-reproductive lethality, or be transmitted. If transmitted, then through time the mutation may be passed on to future generations and potentially affect ever-larger numbers of individuals. The transmission of gene mutations can be significantly influenced by a number of well-documented population attributes, including founder effects, bottlenecks, effective size, and random drift (Table 1.4), all of which in turn are influenced by total population numbers and internal subdivisions. Although the roles of nongenetic variables in determining population genetic structure are now generally accepted, they largely have remained underinvestigated. Where such studies have been undertaken, the influence of population variables on the gene pools of individual, co-resident communities is readily apparent, for example, between co-resident *biraderi* (literally, brotherhoods) living in the Rawalpindi district of northern Pakistan (Wang et al. 2000).

TABLE 1.4
Population Factors Influencing Human Genetic Variation

Founder effect
Bottlenecks
Random drift
Effective population size, including sex ratios
Language and regional dialects/accents
Religion
Socioeconomic status

1.2.1 THE INVESTIGATION OF ETHNIC DIFFERENCES IN PR CHINA

The importance of population subdivisions to human variation applies in both small and large populations, but it becomes of special global importance in the People's Republic of China with an estimated population of 1,325 million, spread across thirty-four provinces, 656 cities, and 48,000 districts (PRB 2008). Some 93% of the population are of the majority Han ethnicity; however, there are another fifty-five officially recognised ethnic minorities (*minzu*), and while Mandarin is the uniform written language, the spoken language can differ from province to province with eight recognised major dialects in everyday use: Mandarin, Wu, Gan, Xiang, North Min, South Min, Hakka, and Yue (Moser 1985). Therefore, in PR China it is inappropriate to refer to human genetic variation in 'the Chinese population', and instead it is necessary to specifically identify which Chinese *minzu* is under investigation (Wang et al. 2003; Black et al. 2006).

As discussed in Sections 1.2.4 and 1.2.5, there also is convincing evidence of population subdivisions within many *minzu*. In the case of the Han who number over 1,200 million with ten recognised branches, genomic analysis of the Pinghua Han subpopulation in Guangxi province has indicated their probable male and female descent from southern Chinese ethnic minorities, with subsequent adoption of the Han language and culture and eventual assimilation into the Han *minzu* (Gan et al. 2008).

1.2.2 CHANGING PATTERNS OF ETHNIC SELF-IDENTIFICATION IN PR CHINA

Given the extended, complex history of human migration into, within, and out of China (Black et al. 2007), even ethnic identity may be insufficient to ensure the appropriate identification of specific subpopulations. Besides the Pinghua Han, there is documented evidence that communities have been prepared to change their ethnic self-identification for political and social advantage. Thus through time the northern Jurchen ('barbarian') tribes of Mongolia adopted Chinese customs and culture, which resulted in the formation of the Man (Manchu) nationality in Manchuria (present-day Liaoning province) and later the establishment of the Chinese Qing Dynasty (1644–1911 AD) (Fairbank and Reischer 1985).

So many Han migrated into Manchuria that they outnumbered the local Man inhabitants within a generation, to become the dominant ethnic population in the region (Campbell et al. 2002). During the middle periods of the Qing dynasty (1700–1800 AD), an indeterminate number of Han migrants and their descendants changed their family names to Man surnames, thus altering their apparent ethnic status. However, as the power of the Qing dynasty declined, many individuals and families with Man family names adopted Han surnames in order to gain economic and political advantage. While readily understandable from a contemporary economic perspective, these historical exchanges of family names create major difficulties in planning studies on genetic variation, for example, in deciding whether an individual should actually be classified as being of Han or Man ancestry (Black et al. 2007).

1.2.3 THE *MINZU* SYSTEM OF ETHNIC CLASSIFICATION IN **PR** CHINA

Major ethnic misidentification and subpopulation agglomeration appears to have occurred with the establishment of the *minzu* system of ethnic classification introduced by the newly established People's Republic of China in 1949. Although fifty-six specific ethnic populations (*minzu*) received official governmental recognition, many other anthropologically discrete communities were excluded from the system, or in some instances were included within a *minzu* alongside communities with whom they had no obvious social, religious, language, or ancestry connections (Black et al. 2006, 2007).

1.2.4 THE AGGLOMERATION OF ETHNIC SUBCOMMUNITIES WITHIN THE YAO *MINZU*

Twenty-five of the fifty-six officially recognised *minzu* live in Yunnan province, on the southern borders of PR China with Myanmar and Laos and close to Tibet. The Yao *minzu* numbers some 8.9 million and besides Yunnan, they are resident in five other mainly mountainous regions in the south and southwest of the country (National Bureau of Statistics 2002). Historical and anthropological evidence suggest that individual Yao communities had very different ancestral backgrounds, with over thirty different 'subgroups' or *zhixi* within the official Yao *minzu* structure (Yang et al. 2005). From a linguistic perspective, the Yao language comprises multiple dialects; in terms of their spiritual beliefs, separate deities are worshipped in different Yao communities, with community-specific traditions in terms of their mythological origins and patterns of religious festivities (Litzinger 1995).

Not surprisingly the genomic data published to date confirm that the Yao are genetically heterogeneous, with a number of discrete ethnic groups apparently coalesced into a single *minzu*. As indicated in Table 1.5, Y-chromosome analysis showed that one Yao community closely matched the Y-chromosome profile of Tibetans, both having high frequencies of Y-chromosome haplogroup D (Black et al. 2006; Bittles et al. 2007). But an earlier study of the Yao reported a complete absence of Y-chromosome haplogroup D with instead a largely northern Chinese

---

TABLE 1.5
Y-Chromosome Haplogroup Profiles (%) in Selected Chinese *Minzu*

Haplogroup	Global Distribution	Han	Yao	Miao	Tibetan
C	North Asia/Mongolia	5	–	5	–
D	Tibet/Japan/Andaman Islands	–	57	–	55
F*(xK)	Southern China/SE Asia	15	7	13	10
O	China	23	7	28	25
03	China/SE Asia/Pacific	48	29	54	–
Minor haplogroups ($n = 3$)	Various locations	9	0	0	10

TABLE 1.6

mtDNA Haplogroup Profiles (%) in Selected Chinese *Minzu*

Haplogroup	Global Distribution	Han	Yao	Miao	Tibetan
A	East Asia/Americas	–	–	–	13
B	Southern China/SE Asia	–	10	22	7
C	East Asia/Americas	8	–	–	–
D	East Asia/Americas	50	27	11	53
F	Southern China/SE Asia	8	18	33	–
R*	Southern China/SE Asia	16	27	22	7
Minor haplogroups ($n = 4$)	China and/or SE Asia	17	18	12	20

paternal ancestry of Y-chromosome haplogroup O (Qian et al. 2001). By comparison, mitochondrial DNA studies indicated that Yao founder females were of Southeast Asian origin, mainly comprising mtDNA haplogroups B, D, and F plus other similar minor haplogroups (Table 1.6), thus suggesting a historically sex-biased admixture of northern Chinese male immigrants with local southern Chinese females (Hammer and Horai 1995; Wen et al. 2004).

1.2.5   ETHNIC SUBDIVISION AND OVERLAP WITHIN THE MIAO *MINZU*

An anthropological study of the Miao *minzu*, also mainly resident in southwestern China and numbering 2.64 million (National Bureau of Statistics 2002), suggested similar problems in their classification as a single ethnic group (Diamond 1995). The Miao can be subdivided into three broad cultural and linguistic groups—the Hmong in Yunnan, the Hmu in Guizhou, and the Xioob in Guizhou and Hunan provinces—with intergroup differences mostly observed at village level. At least three major Miao languages have been described, each of which is subdivided into multiple dialects (Diamond 1995; Black et al. 2006).

Genetic diversity studies in the Miao have demonstrated very diverse male genetic ancestries, with one investigation suggesting a northern Chinese origin (Black et al. 2006), while others indicated multiple origins in different communities (Qian et al. 2001; Wen et al. 2005). Further, Qian et al. (2001) proposed that the genetic ancestries of different Yao and Miao communities were intermixed and difficult to distinguish without prior classification into linguistic and geographical groupings. As with the Yao, the mtDNA profiles of Miao females very clearly demonstrated their Southeast Asian ancestry, with haplogroups B, D, F, and R* predominating (Table 1.6).

1.2.6   COMMUNITY IDENTIFICATION IN PR CHINA

The overall conclusion from these and other studies, most of which are still at an early stage and often reliant on small sample sizes, is that due diligence needs to be exercised in the recruitment of volunteer subjects for genomic investigations (Black et al. 2006, 2007). As indicated in Table 1.7, there is a wide range of basic

---

TABLE 1.7

Human Genetic Variation in PR China

Demographic determinants
    Population structure
    Marriage customs
    Family size
    Internal migration
Ethnological and sociopolitical divides
    Ethnicity
    Language and dialect boundaries
    Religion
    Political opinions
    Urbanisation

---

demographic determinants, and ethnological, sociopolitical, and religious attributes, each of which can separately influence the male and female genomic profiles of Chinese ethnic populations and therefore potentially determine the results of studies into human variation.

A cautious approach to the concept of ethnicity in China is therefore needed, with reliance on genomic data alone almost certainly incapable of revealing the complex ancestral structures of the various ethnic populations. For this reason, to accurately assess genetic diversity in the many Chinese ethnic populations, and hence identify the forces and factors that have influenced this diversity, access to archaeological, historical, anthropological, and demographic sources is needed, as opposed to tacit acceptance of the officially prescribed *minzu* classificatory system (Black et al. 2006, 2007).

1.3 GENOMIC MARKER SYSTEMS IN THE ASSESSMENT OF HUMAN GENETIC VARIATION

The Human Genome Project profoundly changed our knowledge and understanding of human genetic variation, and during the past five years the rate at which new findings and insights have been reported has sharply accelerated. Much of this increase can be ascribed to the new technologies and methods of genomic analysis now widely available at moderate cost. The scope of the analyses available also has increased significantly, allowing multiple complementary marker systems to be investigated and providing more comprehensive, multifaceted perspectives of the human genome (Table 1.8). The findings and conclusions of some representative studies with respect to human genetic variation are summarised in Sections 1.3.1 to 1.3.6.

1.3.1 Evidence of Recent Positive Selection on the Human Genome

Comparative single nucleotide polymorphism (SNP) studies on DNA samples from volunteers of East Asian, North and West European, and sub-Saharan (Yoruba)

TABLE 1.8

Genetic Marker Systems in the Assessment of Human Genetic Variation

Autosomes and X-chromosome

Y-chromosome

Mitochondrial DNA

Alu repeats and nuclear mitochondrial pseudogenes

Copy number variants

RNA variants

Microarray expression studies

ancestry have detected recent positive selection at a diverse range of gene loci. In the first such study, the investigation of ~800,000 single nucleotide polymorphisms (SNPs) in 309 individuals identified interethnic differences mainly occurring in the Holocene, and favouring mutations that have not yet reached fixation (Voight et al. 2006). The genes studied included variants affecting fertility and reproduction, morphology, including European skin colour and African hair type, skeletal development, brain development and function, major histocompatibility complex (MHC)–mediated immunity, and components of the electron transport chain.

A second study employed ~1.2 million SNPs to investigate the genomes of African-American, European-American, and Chinese individuals and identified 101 genomic regions, including olfactory receptors, pigmentation pathways, nervous system development and function, the immune system, the dystrophin protein complex, and heat shock genes, which had undergone a recent selective sweep, that is, with adaptive evolution at the molecular level (Williamson et al. 2007). Fixation had occurred at gene loci located on all autosomes from 1 to 20, with 10% of the identified changes in African-Americans, 26% in European-Americans, 55% in Chinese, and 21% in the combined sample.

However, as illustrated in a study based on 2.8 million Phase II HapMap SNPs, negative as well as positive selection has occurred in human populations. While positive selection assisted 'local' population adaptation via increased differentiation at specific gene regions, especially in nonsynonymous variants, negative selection served to globally reduce population differentiation in amino acid mutations, exerting particular effect on the incidence of disease genes (Barreiro et al. 2008).

1.3.2 COPY NUMBER VARIANTS IN HUMANS

Copy number variants (CNV) are DNA structural variants comprising deletions, insertions, duplications, and complex multisite variants, and are usually defined in terms of segments of approximately 1 Kb or less to several megabases in size (Redon et al. 2006). CNVs can significantly influence gene expression, phenotypic variation, and adaptation by disrupting genes and altering gene dosage effects (Redon et al. 2006). To identify the prevalence and distribution of CNVs in the genome, samples in the HapMap collection from individuals of four different ethnicities/ancestries were screened by SNP genotyping arrays and clone-based comparative genomic hybridisation. The study identified 1,447 CNV regions covering 360 megabases (Mb) of the genome, that is, representing approximately 12% of the total (Redon et al. 2006).

It has been proposed that CNVs could be a major source of interindividual genetic variation, by variously affecting the penetrance of single gene disorders, the phenotypic expression of aneuploidies and sporadic traits, and the aetiology of complex multifactorial traits (Beckmann et al. 2007). To date, most attention has focussed on disease associations involving CNVs which are millions of bases in length, but with improved technologies the possible equivalent role of common small CNVs should become more apparent.

1.3.3   ENCODE Project

The Encyclopaedia of DNA Elements (ENCODE) project involves thirty-five international research groups with the initial aim of examining approximately 30 Mb (~1%) of the human genome, divided into forty-four genomic regions. In this phase some 400 million experimental data points were generated, comprising information on DNA, RNA, chromatin, histones, and sequence-specific factors (Encode Project Consortium 2007). Among the conclusions reached were: (i) The human genome is 'pervasively transcribed', with the majority of bases found in primary transcripts, including noncoding transcripts and those exhibiting extensive overlap; (ii) approximately 5% of the bases examined were under evolutionary constraint, with mutations at specific loci effectively suppressed; and (iii) many of the functional elements in the genome appeared to have been unconstrained across mammalian evolution, suggesting the existence of a large pool of biochemically active but selectively neutral elements conferring no specific identifiable benefit on the organism.

1.3.4   Gene Expression as a Quantitative Phenotype

By comparison with gene mapping, gene expression studies are still largely in their infancy. To compare gene expression in individuals of 'European' (Utah, U.S.A.), Han Chinese (Beijing), and Japanese (Tokyo) ancestries, ~4,200 lymphoblastoid cell line genes were analysed by microarray (Spielman et al. 2007). While the 'European' ($n = 60$) and Chinese samples ($n = 41$) differed in gene expression at 22.3% of loci, and European samples differed from the Japanese ($n = 41$) at 18.0%, the Chinese and Japanese samples differed at only 0.6% of loci, emphasising the relevance of information on gene expression in comparative human studies.

1.3.5   Ancient Population Bottlenecks and Human Phenotypic Variation

To determine whether a relationship could be determined between anthropological data and DNA-derived information, 4,666 male skulls from 105 populations were examined for thirty-seven morphometric characteristics and matched with microsatellite analysis of DNA samples held in the Human Genome Diversity Project (HGDP)–Centre d'Étude du Polymorphisme Humain (CEPH) collection (Manica et al. 2007). There was no evidence to support a multiregional basis of human origins, and the study indicated that loss of genetic variability was mirrored by a reduction in phenotypic variability, with distance from Africa accounting for 19–25% of the heritable variation in the craniometric measurements. From a wider anthropological viewpoint, the demonstration of parallel genetic and craniometric changes is a promising development in drawing together quite disparate analytical systems in the study of human evolution.

1.3.6   INDIVIDUAL HUMAN GENOME SEQUENCES

The early studies on the structure and function of the human genome utilised samples from different individuals. To determine the level of variation in the diploid genome sequence of an individual of European origin (Craig Venter), 2,810 million bases of contiguous DNA sequence were examined (Levy et al. 2007). A remarkable overall level of individual genetic variability was found, with 4.1 million DNA variants over 12.3 Mb of the genome, including 3.2 million SNPs, 53,823 block substitutions, 292,102 heterozygous and 559,473 insertion/deletion (indel) events, 90 inversions, and numerous segmental duplications and copy number variants. Of the 4.1 million DNA variants detected, 1.29 million (~31.5%) were unique, and although non-SNP variation accounted for 22% of all observed events, they involved 74% of all the variant bases. An estimated 44% of genes were heterozygous for one or more variants.

Since this initial report, three other individuals have been sequenced, most notably James Watson (NatureNews 2007), together with an unidentified Han Chinese male (Wang et al. 2008) and an anonymous Yoruba African (Bentley et al. 2008), with future plans both to sequence 1,000 human genomes and to offer commercial sequencing services to persons who might wish to investigate their genomic individuality.

1.4   DISCUSSION

From the above studies it is clear that with the advent of advanced genomic analysis, 'Big Science' is the order of the day, which in turn necessitates large budgets and probably makes the growth of collaborative, multinational research teams an inevitability. The increasingly sophisticated analytical tools employed have consistently revealed previously unsuspected levels and types of genomic variation. Studies based on the HapMap SNP dataset have indicated that human demographic growth has been an important contributory factor in the acceleration of selection over the past 40,000 years (Hawks et al. 2007), a finding in keeping with the prediction by Darwin (1859) that population size would be important in determining the rate of natural selection. Since the most dramatic growth in human population size has occurred within the past four to six generations and, given the continuing increase in global population numbers, it seems highly probable that this trend will continue into the foreseeable future.

As yet, theoreticians have found it difficult to keep abreast of the speed with which new experimental findings have emerged, and the evolutionary significance of many of these findings is far from clear. As indicated in examples in Sections 1.3.1 to

1.3.5, it also is important that a large majority of the reported results have been based on small and possibly unrepresentative sample sizes, and with rather broad ancestral descriptions applied, for example, 'European-American' or 'Chinese'. Whether the interethnic differences so far determined will be replicated, or even extended, remains to be seen, but for the moment their interpretation merits caution, especially in evolutionary terms.

Genomic studies often produce statistically significant but biologically enigmatic results. In many of these genome-based studies, little consideration was given to the demographic and social backgrounds of the study subjects. As the transmission of genes predominantly occurs within societal norms, it is important that future laboratory-based studies in disciplines such as human biology and physiological anthropology should be complemented by a detailed understanding of the history, demography, and social structures of study populations, including their historically preferred marriage patterns (Bittles 2008). This contention is supported by recent reports which, for example, have indicated a higher effective number of women in patrilineal herder groups than among bilineal agricultural communities (Ségurel et al. 2008), and the importance of lower male versus female effective population size in shaping patterns of human genomic variability (Hammer et al. 2008).

To some extent the selection versus neutral mutation debates of the 1960s and 1970s have reappeared in recent controversies within evolutionary developmental biology, that is, whether mutations in the coding regions of developmental genes (Hoekstra and Coyne 2007) or noncoding *cis* regulatory elements in the DNA (Carroll 2005, 2008) are primarily responsible for evolutionary changes in morphology. As yet, these disagreements have largely centred on species such as *Drosophila*, but a recent study on *cis* element-driven developmental changes in mice (Cretekos et al. 2008) suggests that the role of gene expression in human evolutionary changes will increasingly be examined.

Despite the major investigative advances summarised in Sections 1.3.1 to 1.3.6, our knowledge of the forces and factors that influence human genetic variation remains far from complete. Even with the introduction of new, increasingly sensitive and precise analytical methods, the very high levels of personal genetic variation so far determined, together with increasing recognition of the role(s) played by epigenetic mechanisms, suggest that a resolution to this complex issue is some years distant.

REFERENCES

Aird, I., Bentall, H.H. and J.A. Fraser Roberts. 1953. A relationship between cancer of stomach and the ABO blood groups. *Brit. Med. J.* 1:799–801.

Aird, I., Bentall, H.H., Mehigan, J.A. and J.A. Fraser Roberts. 1954. The blood groups in relation to peptic ulceration and carcinoma of colon, rectum, breast, and bronchus. *Brit. Med. J.* 2:315–21.

Aird, I., Bentall, H.H., Bingham, J. et al. 1956. An association between blood group A and pernicious anaemia. *Brit. Med. J.* 2:723–4.

Barreiro, L.B., Laval, G., Quach, H., Patin, E. and L. Quintana-Murci. 2008. Natural selection has driven population differentiation in modern humans. *Nat. Genet.* 40:340–5.

Beardmore, J.A. and F. Karimi-Booshehri. 1983. ABO genes are differentially distributed in socio-economic groups in England. *Nature* 303:522–4.

Beckmann, J.S., Estivill, X. and S.E. Antonarakis. 2007. Copy number variants and genetic traits: Closer to the resolution of phenotypic to genotypic variability. *Nat. Rev. Genet.* 8:639–46.

Behar, D.M., Villems, R., Soodyal, H.M. et al. 2008. The dawn of human matrilineal diversity. *Am. J. Hum. Genet.* 82:1130–40.

Bennett, J.H. and C.B.V. Walker. 1956. Fertility and blood groups of some East Anglian blood donors. *Ann. Hum. Genet.* 20:299–308.

Bentley, D.R., Balusubramanian, S., Swerdlow, H.P. et al. 2008. Accurate whole genome sequencing using reversible terminator chemistry. *Nature* 456:53–9.

Bernardi, G. 2007. The neoselectionist theory of genome evolution. *Proc. Natl. Acad. Sci. U.S.A.* 104:8385–90.

Bittles, A.H. 2008. A Community Genetics perspective on consanguineous marriage. *Community Genet.* 11:324–30.

Bittles, A.H., Black, M.L. and W. Wang. 2007. Physical anthropology and ethnicity in Asia: The transition from anthropometry to genome-based studies. *J. Physiol. Anthropol.* 26:77–82.

Black, M.L., Wang, W. and A.H. Bittles. 2006. Genetics and population history in the study of ethnic diversity in PR China. *Hum. Biol.* 78:277–93.

Black, M.L., Wang, W. and A.H. Bittles. 2007. Unity and diversity: Human genetic studies on the population of China. In *Recent Advances in Molecular Biology and Evolution: Applications to Biological Anthropology*, ed. C. Santos and M. Lima, 349–71. Thiruvananthapuram: Research Signpost.

Campbell, C., Lee, J.E. and M. Elliot. 2002. Identity construction and reconstruction: Naming and Manchu ethnicity in North East China, 1749–1909. *Hist. Methods* 35:101–15.

Carroll, S.B. 2005. Evolution at two levels: On genes and form. *PLoS Biol.* 3:e245.

Carroll, S.B. 2008. Evo-devo and an expanding evolutionary synthesis: A genetic theory of morphological evolution. *Cell* 134:25–36.

Chaudhuri, A. and S. De. 1977. Cholera and blood-groups. *Lancet* 2:404.

Clarke, C.A., Cowan, W.K., Edwards, E.W., Howel-Evans, A.W., McConnell, R.B. and P.M. Sheppard. 1955. The relationship of the ABO blood groups to duodenal and gastric ulceration. *Br. Med. J.* 2:643–6.

Cretekos, C.J., Wang, Y., Green, E.D., Martin, J.F., Rasweiler, J.J. and R.R. Behringer. 2008. Regulatory divergence modifies limb length between mammals. *Genes Dev.* 22:141–51.

Darwin, C. 1859. *The Origin of Species*. London: John Murray.

Dawood, M.Y., Teoh, E.S. and S.S. Ratnam. 1971. ABO blood group in trophoblastic disease. *J. Obstet. Gynaecol. Br. Commonw.* 78:918–23.

Diamond, N. 1995. Defining the Miao: Ming, Qing and contemporary views. In *Cultural Encounters on China's Ethnic Frontiers*, ed. S. Harrell, 92–116. Seattle: University of Washington Press.

Encode Project Consortium. 2007. Identification and analysis of functional elements in 1% of the human genome by the ENCODE pilot project. *Nature* 447:799–816.

Fairbank, J.K. and E.O. Reischer. 1985. *China, Tradition & Transformation*. Sydney: Allen & Unwin.

Gan, R.J., Pan, S.L., Mustavich, L.F. et al. 2008. Pinghua population as an exception of Han Chinese's coherent genetic structure. *J. Hum. Genet.* 53:301–13.

Gibson, J.B., Harrison, G.A., Clarke, V.A. and R.W. Hiorns. 1973. IQ and ABO blood groups. *Nature* 246:498–500.

Gibson, J.B., Harrison, G.A. and R.W. Hiorns. 1975. Population stratification as an explanation of IQ and ABO association. *Nature* 254:363–4.

Hammer, M.F. and S. Horai. 1995. Y chromosomal DNA variation and the peopling of Japan. *Am. J. Hum. Genet.* 56:951–62.

Hammer, M.F., Mendez, F.L., Cox, M.P., Woerner, A.E. and J.D. Wall. 2008. Sex-biased evolutionary forces shape genomic patterns of human diversity. *PLoS Genet.* 4:e1000202.

Hawks, J., Wang, E.T., Cochran, G.M., Harpending, H.C. and R.K. Moyzis. 2007. Recent acceleration of human adaptive evolution. *Proc. Natl. Acad. Sci. U.S.A.* 104:20753–8.

Hoekstra, H.E. and J.A. Coyne. 2007. The locus of evolution: Evo devo and the genetics of adaptation. *Evolution* 61:995–1016.

Jain, R.C. 1970. ABO blood groups and pulmonary tuberculosis. *Tubercle* 51:322–3.

Juvonen, T. and O. Niemelä. 1992. ABO blood group disease and gall stone disease. *Br. Med. J.* 305:26–7.

Kaul, D.K., Roth, E.F.J., Nagel, R.L., Howard, R.J. and S.M. Handunnetti. 1991. Rosetting of *Plasmodium falciparum*-infected red blood cells with uninfected red blood cells enhances microvascular obstruction under flow conditions. *Blood* 78:812–9.

Kimura, M. 1968. Evolutionary rate at the molecular level. *Nature* 217:624–6.

Kimura, M. 1969. Genetic variability maintained in a finite population due to mutational production of neutral and nearly neutral mutations. *Genet. Res.* 11:247–69.

Levy, S., Sutton, G., Ng, P.C. et al. 2007. The diploid genome sequence of an individual human. *PLoS Biol.* 5:e254.

Litzinger, R.A. 1995. Making histories: Contending conceptions of the Yao past. In *Cultural Encounters on China's Ethnic Frontiers,* ed. S. Harrell, 117–39. Seattle: University of Washington Press.

Manica, A., Amos, W., Balloux, F. and T. Hanihara. 2007. The effect of ancient population bottlenecks on human phenotypic variation. *Nature* 448:346–9.

Mascie-Taylor, C.G.N. and I.C. McManus. 1984. Blood group and socio-economic class. *Nature* 309:395–6.

Matsunaga, E. and Itoh, S. 1958. Blood groups and fertility in a Japanese population, with special reference to intra-uterine selection due to maternal-foetal incompatibility. *Ann. Hum. Genet.* 22:111–31.

Moser, L.J. 1985. *The Chinese Mosaic: The Peoples and Provinces of China.* Colorado: Westview Press.

Mourant, A.E., Kopec, A.C. and K. Domaniewska-Sobczak. 1971, Blood-groups and blood-clotting. *Lancet* 1:223–7.

National Bureau of Statistics of China. 2002. *China Population Handbook.* Beijing: China Statistics Press.

Nature News. 2007. James Watson's genome sequenced. http://www.nature.com/news/2007/070528/full/070528-10.html. http://www.jimwatsonsequence.cshl.edu/cgi-perl/browse/jwsequence/.

Ohta, T. 1973. Slightly deleterious mutant substitutions in evolution. *Nature* 246:96–8.

PRB. 2008. *World Population Data Sheet.* Washington: Population Reference Bureau.

Qian, Y.P., Chu, Z.T., Dai, Q. et al. 2001. Mitochondrial DNA polymorphisms in Yunnan nationalities in China. *J. Hum. Genet.* 46:211–20.

Redon, R., Ishikawa, S., Fitch, K.R. et al. 2006. Global variation in copy number in the human genome. *Nature* 444:444–54.

Reed, T.E. and E.L. Kelly. 1958. The completed reproductive performances of 161 couples selected before marriage and classified by ABO blood group. *Ann. Hum. Genet.* 22:165–81.

Regan, J.A., Chao, S. and L.S. James. 1978. Maternal ABO blood group type B: A risk factor in the development of neonatal group B streptococcal disease. *Pediatrics* 62:504–9.

Robinson, M.G., Tolchin D. and C. Halpern. 1971. Enteric bacterial agents and the ABO blood groups. *Am. J. Hum. Genet.* 23:135–45.

Rosenfield, R.E. 1955. A-B haemolytic disease of the newborn. *Blood* 10:17–28.

Rowe, J.A., Handel, I.G., Thera, M.A. et al. 2007. Blood group O protects against severe *Plasmodium falciparum* malaria through the mechanism of reduced resetting. *Proc. Natl. Acad. Sci. U.S.A.* 104:17471–6.

Ségurel, L., Martínez-Cruz, B., Quintana-Murci, L. et al. 2008. Sex-specific genetic structure and social organization in Central Asia: Insights from a multi-locus study. *PLoS Genet.* 4:e1000200.

Sheppard, P.M. 1959. Blood groups and natural selection. *Br. Med. Bull.* 15:134–9.

Spielman, R.S., Bastone, L.A., Burdick, J.T., Morley, M., Ewens, W.J and V.G. Cheung. 2007. Common genetic variants account for differences in gene expression among ethnic groups. *Nat. Genet.* 39:226–31.

Thomson, G. and W.F. Bodmer. 1975. Population stratification as an explanation of IQ and ABO association. *Nature* 254:363.

Uneke, C.J. 2007. *Plasmodium falciparum* malaria and ABO blood group: Is there any relationship? *Parasitol. Res.* 100:759–65.

Vogel, F. and M.R. Chakravartii. 1966. ABO blood groups and smallpox in a rural population of West Bengal and Bihar (India). *Humangenetik* 3:168–80.

Voight, B.F., Kudaravalli, S., Wen, X. and J.K. Pritchard. 2006. A map of recent positive selection in the human genome. *PLoS Biol.* 4:e72.

Wagner, A. 2008. Neutralism and selectionism: A network-based reconciliation. *Nat. Rev. Genet.* 9:965–74.

Wang, W., Sullivan, S.G., Ahmed, S., Chandler, D., Zhivotovsky, L.A. and A.H. Bittles. 2000. A genome-based study of consanguinity in three co-resident endogamous Pakistan communities. *Ann. Hum. Genet.* 64:41–9.

Wang, W., Wise, C., Baric, T., Black, M.L. and A.H. Bittles. 2003. The origins and genetic structure of three co-resident Chinese Muslim populations: The Salar, Bo'an and Dongxiang. *Hum. Genet.* 113:244–52.

Wang, J., Wang, W., Li, R. et al. 2008. The diploid genome sequence of an Asian individual. *Nature* 456:60–5.

Wellcome Trust Case Control Consortium. 2007. Genome-wide association study of 14,000 cases of seven common diseases and 3,000 shared controls. *Nature* 447:661–84.

Wen, B., Li, H., Gao, S. et al. 2005. Genetic structure of Hmong-Mien speaking populations in East Asia as revealed by mtDNA lineages. *Mol. Biol. Evol.* 22:725–34.

Wen, B., Li, H., Lu, D. et al. 2004. Genetic evidence supports demic diffusion of Han culture. *Nature* 431:302–5.

Whincup, P.H., Cook, D.G., Phillips, A.N. and A.G. Shaper. 1990. ABO blood groups and ischaemic heart disease in British men. *Br. Med. J.* 300:1679–82.

Williamson, S.H., Hubisz, M.J., Clark, A.G., Payseur, B.A., Bustamente, C.D. and R. Nielsen. 2007. Localizing recent adaptive evolution in the human genome. *PLoS Genet.* 3:e90.

Yang, Z., Yang, Z., Dong, Y. et al. 2005. The distribution of Y chromosome haplogroups in the nationalities from Yunnan Province of China. *Ann. Hum. Biol.* 32:80–7.

2 Genetics of Body Weight and Obesity

Paul Higgins[1], V. Saroja Voruganti[1], and Anthony G. Comuzzie[*1,2]
[1]Department of Genetics, Southwest Foundation for Biomedical Research, San Antonio, Texas
[2]Southwest National Primate Research Center, San Antonio, Texas

CONTENTS

2.1 INTRODUCTION

Complex phenotypes, such as those represented by anthropometric characters, exhibit quantitative variation in their expression, resulting from the combined contribution of multiple genes acting cumulatively to produce their physical manifestation. While environmental factors such as diet also make a significant contribution to the observed variation in these traits, it is the individual's unique genetic background that determines the response to these environmental factors. Recent advances in molecular and statistical genetics have provided a variety of tools that allow us to elucidate the genetic architecture underlying such complex phenotypes as anthropometric traits. Three general approaches have been used to date in the search for genes

---

* Address all correspondence to tony@sfbrgenetics.org.

underlying common, complex phenotypes. The first approach focuses on a priori selected candidate genes believed to have some plausible role in the trait of interest (for example, obesity) on the basis of their known or presumed biological function. This approach has had limited success in identifying genes involved in the development of disease at the population level. An alternative approach attempts to localise genes and requires no presumptions on the function of the gene, and is based on the detection of unique patterns of segregation among related individuals. Chief among this type of approach has been linkage analysis. Recent advances in the ability to evaluate linkage analysis data from large family pedigrees has shown great promise in identifying genomic regions associated with the development of complex phenotypes such as obesity, but the identification of the specific causal genetic variants has remained somewhat elusive. In addition to genome-wide linkage analyses, the last couple of years have also seen the widespread application of genome-wide association analyses. This approach differs from the linkage approach by its primary use of large samples of unrelated individuals. Recently, RNA-based technologies have also started to be very useful in the identification of genes differentially expressed in tissues of healthy and diseased individuals. This chapter reviews current knowledge of the genetic contribution to variation in anthropometric traits with a particular focus in those measures associated with obesity.

The World Health Organisation's most recent global estimates indicate that approximately 1.6 billion individuals over the age of fifteen are overweight (BMI ≥ 25 kg/m^2) and approximately 400 million are obese (BMI ≥ 30 kg/m^2). They have also estimated that approximately 2.3 billion adults will be overweight and more than 700 million will be obese by 2015 (World Health Organisation 2007). Worldwide, approximately 20 million children under the age of five years were considered overweight in 2005. Given the negative health outcomes associated with obesity, the current and future societal impact of these rates cannot be understated. Obesity ultimately results from an imbalance of energy intake and energy expenditure and is largely influenced by the environment. However, individuals exposed to similar environments can exhibit large variation in body weight, and studies conducted in the past three decades have clearly demonstrated a genetic contribution to body weight and adiposity.

A complex and highly regulated biologic system is responsible for the control of body weight, and estimates suggest that between 40 and 70% of the variation in body weight is heritable (Comuzzie and Allison 1998). However, body weight is a complex trait that can be influenced by many genes. Studies in lower organisms such as *Drosophila melanogaster* suggest that there are several hundred genes that control body weight (Dohrmann 2004). In support, genome-wide searches in mice point to over fifty chromosomal regions containing genes influencing adiposity (Wuschke et al. 2007). A number of rare monogenic forms of obesity have been identified in human populations and the genetic variation underlying them has been successfully identified. However, these allelic variants appear to have little or no influence on the common form of human obesity. The genetic influence on the common form of human obesity is complex. Nonetheless, efforts to unravel the precise allelic variation underlying this genetic component have yielded important insights and are summarised in Sections 2.2 through 2.6.

Initially, we discuss the principal methods for quantifying and defining overweight and obese phenotypes. Then, the aetiology of rare monogenic forms of obesity is described. In Section 2.4, the analytic approaches used in genetic studies of common complex phenotypes such as body weight and obesity are detailed, followed by descriptions of investigations of the genetics of human obesity.

2.2 BODY FAT PHENOTYPES

The quantification of adiposity and the definition of the obese or overweight phenotype are critical considerations in the study of the genetics of obesity and body weight regulation. Many technically advanced methodologies such as magnetic resonance imaging (MRI), dual-energy X-ray absorbtiometry (DXA), and air displacement plethysmography are now available for the accurate and precise quantification of total body fat and fat distribution. Unfortunately, given the expense, technical demands, and lack of portability, these techniques are seldom employed in large-scale human genetic studies. More common are techniques such as anthropometry and bioelectrical impedance analysis (BIA). Anthropometric measures typically include weight corrected for height, circumferences, and skinfold thicknesses.

The simplest and most commonly used obesity-related phenotype in genetic studies of human obesity is the body mass index (BMI) expressed in kg/m^2. Although BMI is a better index of body fat than simple body weight, it has several limitations. In particular, the relation between BMI and body fat is not uniform across age, gender, and ethnicity. For example, Asians have more body fat and more adverse metabolic profiles at lower BMI values than do Europeans (Chung et al. 2005; Jafar et al. 2006; Kim et al. 2004). These observations have led to suggestions that obesity should be defined at lower BMI thresholds in Asian populations (Gallagher 2004). Another example is the use of BMI in children. Because children are growing, the relationship between BMI and adiposity varies greatly by stage of maturation; increases in BMI in children may represent increases in fat-free tissue mass rather than fat mass (Horlick 2001; Ellis 2000, 2001). These findings can lead to spurious results when BMI is used to define obesity in case-control genetic studies, this measure being more robust when used as a continuous variable.

Waist circumference assessment is a useful and easily attainable measure of body fat. The relation between waist circumference and total body fat is strong in most cases (Higgins et al. 2001). Moreover, there is a strong relationship between central fat deposition and metabolic perturbations. As waist circumference is highly related to central fat deposition relative to total body fat, it may better predict disease risk associated with obesity (Shen et al. 2006; McCarthy 2006). However, gender differences may also exist (Shen et al. 2006). Therefore, waist circumference is a strong surrogate measure of body fat and may provide additional information on body fat distribution and metabolic disease risk (Klein et al. 2007). Although not completely free of limitations, waist circumference may be one of the most robust and practical anthropometric techniques for use in large human genetic studies.

Skinfold thicknesses have a long history of use in field and clinical genetic studies. These are based on measurements of the thickness of the subcutaneous fat layer and are readily derived using mechanical callipers. Measures are taken at a number of sites including the iliac crest, biceps, triceps, and subscapular region, and can be extrapolated to estimates of total body fat by means of regression-based prediction equations. A plethora of prediction equations exist, perhaps suggesting the limitations of this approach. When using skinfold as a surrogate for fat mass in population genetic studies, the problem of inaccurate prediction equations can be bypassed in part by using raw thickness values in a site-specific manner. However, the utility of this approach is limited by inconsistency in the relation between site-specific skinfold measurements. High interobserver variability and inaccurate extrapolation from prediction equations suggest that skinfold thickness assessment is limited in the estimation of total body fat.

Bioelectrical impedance analysis (BIA) is a useful and noninvasive field technique. A very weak electrical current is passed through the body between a set of electrodes; the resultant voltage drop between the electrodes is proportional of the volume of total body water. As only the fat-free mass portion of the body contains water, an estimate of total body water can be translated into an estimate of fat-free mass, and hence fat mass. BIA results are generally reproducible and accurate but care must be exercised in conditions in which the hydration status of the fat-free mass may vary.

2.3   MONOGENIC HUMAN OBESITY

Several forms of human obesity result from rare allelic variants that impact the regulation of energy balance and are inherited in a simple Mendelian fashion (Farooqi and O'Rahilly 2005). The genes and in most cases the allelic variation(s) underlying these conditions have been determined. An example is a rare defect in the gene that produces the adipocyte-derived peptide leptin. Leptin is secreted primarily from adipose tissue in proportion to fat mass. It crosses the blood brain barrier and interacts with neurons to reduce appetite and increase energy expenditure. A number of children have been identified with a frameshift mutation in the leptin (ob/ob) gene that resulted in circulating leptin deficiency and caused severe hyperphagia and obesity (Montague et al. 1997). Exogenous leptin administration significantly reduced food intake and normalised body weight in these individuals (Farooqi and O'Rahilly 2005).

Melanocortin 4 receptor is a 332-amino acid protein encoded by *MC4R* on chromosome 18q22 and is expressed widely in the brain, including the cortex, thalamus, hypothalamus, brainstem, and spinal cord. In human *MC4R* deficiency, multiple, rare loss-of-function variants are associated with hyperphagia, early-onset obesity, increased fat mass, and hyperinsulinemia. *MC4R* variants constitute the most common form of human monogenic obesity, with prevalence rates ranging from 0.5% to 5.8% (Farooqi and O'Rahilly 2005). In addition to *MC4R* variants contribution to severe monogenic forms of obesity, recent work in several populations suggest that they may also play a significant role in normal variation in phenotypes associated with body weight and energy expenditure as well (Cai et al. 2006).

Despite considerably increasing our understanding of the biologic pathways involved in body weight regulation, candidate gene studies utilising the information garnered from the study of monogenic obesity has had little impact on the understanding of the aetiology of the more common forms of obesity.

2.4 ANALYTICAL APPROACHES

2.4.1 CANDIDATE GENE STUDIES

For many years the candidate gene approach was used to identify genes underlying variation in a wide variety of complex phenotypes including obesity. Candidate genes are selected based on a priori information regarding their potential role in biologic pathways involved in the regulation of the physical expression of the phenotype of interest. Specific genetic variation at these loci is then tested for association with the phenotype under study, typically in unrelated individuals. However, this approach has been relatively fruitless in identifying major human obesity genes. Consequently, alternative approaches are now routinely adopted to study the genetics of obesity as well as most other complex phenotypes.

2.4.2 GENOME-WIDE LINKAGE ANALYSIS

Quantitative genetic analysis can be used to estimate the heritability of a trait and is a crucial primary step in gene identification. Heritability in this case is defined in the narrow sense, which is the proportion of phenotypic variance attributable to additive genetic variation (Comuzzie et al. 2001). From this, the total phenotypic variance can be decomposed into two components: the additive genetic and environmental (including nonadditive genetic) components (Rogers et al. 1999). These basic calculations can then be expanded to identify specific chromosomal regions influencing the trait in question, and these are typically referred to as quantitative trait loci (QTLs). QTLs are mapped using whole genome linkage analysis. Linkage analysis is undertaken in extended family pedigrees using information from polymorphisms (typically microsatellite markers) scattered at various intervals throughout the genome. Thus, QTLs can be identified that have significant cosegregation with the phenotype under study. In contrast to the candidate gene approach, linkage analysis makes no a priori assumptions regarding the potential role of a gene in the expression of a trait. Each family member is genotyped for the markers and the identity-by-descent (IBD) matrix is calculated; this matrix contains the probability that two alleles at a given locus have been inherited from a common ancestor (Rogers et al. 1999) for each set of relative pairs represented in the analysis. The correlation of the phenotype between related pairs is calculated to estimate the proportion of variance that is explained by these shared genetic markers. Thus, regions that show significant 'linkage' to the phenotype in question are identified for additional molecular analysis. QTLs can contain numerous genes, and fine-mapping or sequencing approaches are required to isolate the specific functional variants.

2.4.3 GENE EXPRESSION QTL MAPPING

A recently adopted approach involves the integrated use of gene expression information acquired from quantitative measurement of mRNA abundance in conjunction with linkage analysis described above. This approach has been referred to as genetical genomics or expression genetics (Broman 2005; Cheung and Spielman 2002).

Large-scale gene expression assessments are undertaken using microarray technology, and QTL mapping approaches (as described in Section 2.4.2) are applied to each measure of transcript (mRNA) abundance. Hence, transcript abundance becomes the phenotype and QTLs influencing gene expression can be isolated. These expression QTLs will be either *cis*-acting (mapped to the gene transcripts own genomic location) or *trans*-acting (mapped to a different genomic location). A particular advantage of this approach is that the identification of *trans*-acting loci can lead to the isolation of networks of co-regulated genes and aid in the understanding of complex phenotypes such as obesity. Moreover, this approach can also be used to more rapidly isolate the genes underlying identified QTLs. The efficacy of this approach for isolating human obesity genes remains to be determined; nonetheless, recent studies have shown considerable promise (Goring et al. 2007; Schadt et al. 2003).

2.4.4 WHOLE GENOME ASSOCIATION STUDIES

A new approach for detecting genes underlying risk for complex diseases has emerged from the Human Genome Project's characterisation of single nucleotide polymorphisms (SNPs) with high heterozygosity across unrelated human populations. This approach, commonly referred to as whole genome association, hinges on the hypothesis that common diseases result from common genetic variants (Reich and Lander 2001) and is aided by the availability of high-throughput SNP genotyping technology. SNPs occur on average about every 300 bases across the human genome. Cassettes of linkage disequilibrium or haplotypes are then identified among tightly-linked SNPs across populations and are used to conduct association studies with phenotypes of interest and determine the haplotypes that occur most frequently in individuals with a given disease (Terwilliger et al. 2002). Disease-related SNPs may lie within or near the disease-causing gene. While there is still some debate as to the ultimate utility of this approach (Couzin 2002), recent whole genome association studies have identified potential new obesity genes (Frayling et al. 2007).

2.5 FINDINGS FROM GENETIC STUDIES OF OBESITY

Heritabilities for anthropometric measurements have been estimated in various populations across multiple age categories. For anthropometric measurements, heritability estimates have ranged from 0.20 to 0.80. Whether from twin, sib-pair, or family-based studies, heritabilities for body weight and related traits have been found to be highly significant. In a study in a Belgian population, heritability estimates for waist-to-hip ratio, sum of four skinfold thicknesses, and lean body mass were 70, 74, and 81%, respectively (Souren et al. 2007). Similarly, several other twin studies have reported significant heritabilities for weight-related traits. The heritability of BMI was estimated to be 0.46 for males aged 46–59 years, 0.61 for males aged 60–76 years, 0.77 for females aged 46–59 years, and 0.75 for females aged 60–76 years in a study in Danish twin pairs (Herskind et al. 1996). Besides twin studies, family-based studies have reported similar heritability estimates. Studies in children have shown heritabilities to be in the range of 0.3 to 0.41 in a Hispanic population in Houston from the Viva la Familia study (Voruganti et al. 2007), 0.6 to 0.7 in a population from West Bengal in India (Salces et al. 2007), 0.25 to 0.61 in a population from Vishakhapatnam, India (Arya et al. 2002), and 0.39 for BMI and waist circumference each in a population of Canadians from the Quebec family study (Rice et al. 1999; Hunt et al. 2002). Moreover, a study by Hunt et al. (2002) showed a significant heritability for change in BMI in children over a period of seven years. In a longitudinal study where childhood growth patterns were compared to the adult health outcomes, heritabilities for growth parameters were quite high and ranged between 0.65 and 0.98 (Czerwinski et al. 2007).

Studies conducted in adults have found similar results for anthropometric traits. These studies have been conducted all around the world, including the United States, Canada, and Mexico (Rice et al. 1999; Hunt et al. 2002; Voruganti et al. 2007; Bastarrachea et al. 2007); Europe (Hanisch et al. 2004; Souren et al. 2007); and Asia (Wu, Cooper et al. 2002; Arya et al. 2002; Wu et al. 2003; Salces et al. 2007; Bayoumi et al. 2007), and across several ethnicities and communities (North et al. 2003; Comuzzie, Mitchell et al. 2003; Juo et al. 2004; Sale et al. 2005; Voruganti et al. 2006; Butte et al. 2006). In addition to humans, studies in animals have also shown significant heritabilities for morphometric traits (Comuzzie, Cole et al. 2003; Johnson and Nugent 2003; Cai et al. 2004; Voruganti et al. 2007; Kavanagh et al. 2007). Once heritability of a trait has been determined, the next step then becomes to map the genes responsible for the observed variation in the focal phenotype. Linkage is an important tool for mapping these genetic loci.

2.5.1 CANDIDATE GENE STUDIES

One of the first approaches to mapping susceptibility loci was to test association between a genetic variant and a particular phenotype. Variants to be tested were selected from candidate genes that had been established as playing a role in the regulation of the phenotype of interest and presumed to have biological effects on the phenotype. Although several gene variants have been reported to be associated with anthropometric traits, specifically BMI, there have been few gene variants whose association with BMI and related traits has been consistent across several studies, populations, and ethnicities. Polymorphisms in beta-adrenergic receptor (*ADRB*), candidate genes for energy regulation and lipolysis, have been associated with BMI, waist-hip ratio (WHR), and waist circumference. This has been consistent across different study populations. In Caucasian women, Dionne et al. (2002) found an association between variants in *ADRB1* and body weight and BMI. Similarly, variants in *ADRB2* and *ADRB3* were also associated with body weight (Pereira et al. 2003; Park et al. 2005). In particular, TRP64Arg mutation in the *ADRB3* gene has been associated with body weight and BMI (Large et al. 1997; Mitchell et al. 1998; Endo et al. 2000; Marti et al. 2002). In other studies, the same polymorphism was not associated with body weight or related traits (Terra et al. 2005; Nagano et al. 2005).

In a study in African-American and White families, interaction between polymorphisms in *ADRB*s showed differential patterns in men and women. In men, interaction between variants in *ADRB1* and *ADRB2* were associated with weight gain in men whereas interaction between variants in *ADRB1* and *ADRB3* were associated with weight gain in women (Ellsworth et al. 2005). Another group of genes whose

TABLE 2.1
Genes Associated with Anthropometric Measurements in Humans

Gene (Symbol)	Chromosomal Location	Traits	Reference
Adiponectin (*ACDC*)		BMI, BSA, height	Iacobellis et al., 2006
		BMI, waist circumference	Owecki et al., 2007
Neuropeptide Y (*NPY*)		BMI	Mattevi et al., 2002
			Ding et al., 2005
Canabinoid receptor 1 (*CNB1*)		BMI	Gazzerro et al., 2007
Interleukin 6 receptor (*IL6R*)		BMI	Bustamante et al., 2007
Tumor necrosis factor alpha (*TNF alpha*)	6p21	WHR	Um et al., 2004
Melanocortin 4 receptor (*MC4R*)	18q21–q22	BMI	Dempfle et al., 2004
			Hainerova et al., 2007
		Stature	Santoro et al., 2005

polymorphisms have been associated with BMI, WHR, or waist circumference are those coding for uncoupling proteins (UCP), which are involved in thermogenesis (Schrauwen et al. 1999) and thus play an important role in obesity. Polymorphisms in the *UCP1* (Evans et al. 2000; Herrmann et al. 2003), *UCP2* (Evans et al. 2000; Evans et al. 2001), and *UCP3* (Halsall et al. 2001; Herrmann et al. 2003) genes have been associated with BMI, WHR, skinfolds, and change in body weight. Leptin has an important role in energy metabolism and adipose tissue mass regulation. Its role is mediated through its receptor (*LEPR*), indicating that polymorphism in the *LEPR* gene might have effects on obesity. A single nucleotide polymorphism (SNP) in the *LEPR* gene was found to be associated with BMI and body weight (Quinton et al. 2001; Mattevi et al. 2002; Park et al. 2006) and hip circumference (Wauters et al. 2001). The Pro12Ala polymorphism in the peroxisome proliferators-activated receptor gamma (*PPARG*) gene (a transcription factor with an important role in adipogenesis) has been associated with BMI and waist circumference (Cole et al. 2000; Franks et al. 2007). Other genes whose polymorphisms have been commonly associated with BMI and related anthropometric traits are shown in Table 2.1.

2.5.2 GENOME-WIDE LINKAGE STUDIES

In order to identify genetic locations that might contain susceptibility loci for a particular disease, population studies utilise the linkage mapping approach. Genome-wide scans are conducted using polymorphic markers such as microsatellites positioned at equal distances on chromosomes. Linkage mapping not only depends upon the sample size of the study population but also on the type of pedigree structure and relative pairs (Comuzzie and Allison 1998). Therefore, these results are population- and sample-specific, and generalisability might be restricted. To date, several univariate genome-wide scans for anthropometric traits have been published. These include body weight, height, BMI, WHR, waist circumference, and skinfold measures as phenotypes. Quantitative trait loci (QTLs) for anthropometric traits have been identified all over the genome but some regions have been consistent. For example, 2p22 region has been linked to BMI in several studies. This region contains the *POMC* gene whose polymorphisms have also been associated with BMI (Buono et al. 2005; Baker et al. 2005). It has been reported to be linked with BMI in study populations of Mexican-Americans (Comuzzie et al. 1997), Europeans (Moslehi et al. 2003), and African-Americans (Rotimi et al. 1999). Likewise, marker *D7S1875* near the leptin (*LEP*) gene on chromosome 7 has been associated with BMI and WHR in German and Mexican-American families (Roth et al. 1997; Bray et al. 1999), while the regions of 3q26–27 (Kissebah et al. 2000; Wu, Cooper et al. 2002; Moslehi et al. 2003; Luke et al. 2003), 4p15 (Stone et al. 2002; Arya et al. 2004), and 6q22–23 (Geller et al. 2003; Meyre et al. 2004) have been associated with BMI. In a meta-analysis conducted by Johnson et al. (2005), the p12–23 region of chromosome 8 was linked to BMI in five different populations, namely Nigerians, Whites, Old-Order Amish, Mexican-Americans, and Finnish. Other regions that have been associated with BMI or related traits such as waist circumference, skinfold measures, and height are shown in Table 2.2. Although

TABLE 2.2
Genome-Wide Scan Results for Anthropometric Measurements

Trait	Chromosome	LOD Score	Reference
BMI	2q35	2.4	Tang et al., 2003
	10p	4.23	Atwood et al., 2006
	4q35	5.17	Almasy et al., 2007
Birth weight	6q	4.3	Arya et al., 2006
	10q22	3.09	Cai et al., 2007
Height	12q11	3.18	Dempfle et al., 2006
	2q21	3.21	Liu et al., 2006
Waist	1q22–24	3.7	Ng et al., 2004
	6q22–23	3.77	Fox et al., 2004
Waist-hip ratio	2q35	1.72	Tang et al., 2003
	5q34	1.56	Li et al., 2006
Body surface area	2q37	3.12	Liu et al., 2006

genome-wide scans for anthropometric traits have not been successful in identifying genes, they have been very useful as hypothesis-generating models and provide crucial preliminary data for genome-wide association studies.

2.5.3 GENOME-WIDE ASSOCIATION STUDIES

Exploring variants in a gene associated with a particular disease in a population on a genome-wide scale is yet another approach in the quest to identify genes that play an important role in the development of complex disease. With recent advances in molecular and analytical tools, it is possible to analyse precisely whole genome samples for genetic variations associated with a particular disease. So far, genome-wide association studies have been able to identify association between SNPs in genes with disease-related traits. A major gene whose variants have been associated with BMI is the *FTO* (fat mass and obesity associated) gene (Frayling et al. 2007). Similar observations were made by Scuteri et al. (2007), Dina et al. (2007) and Peeters et al. (2007), who reported significant associations between SNPs in the *FTO* gene and increased BMI, weight, and hip circumference. Thus the *FTO* gene is a potential candidate gene for increasing BMI or obesity, although its role is not yet clear. In the Framingham Heart Study, 310 families have been genotyped over a period of 30 years. Association analysis conducted in this population revealed a strong association of BMI and waist circumference with SNPs in genes such as *PPARG* and *ADIPOQ* (Fox et al. 2007). Also, they were able to replicate an association of a SNP in the *INSIG2* gene and BMI reported for an earlier study by Herbert et al. (2006). Other genome-wide associations related to anthropometric measurements have mainly focused on height/stature (Weedon et al. 2007; Dempfle et al. 2006).

2.6 TRANSLATION OF FINDINGS

Implicit in the effective translation of genetic findings to clinical treatments or preventive measures is the determination of the functional effects of the identified allelic variants. Obesity must result from an imbalance in energy intake and/or energy expenditure, and therefore genes conferring risk must work through one or both of these factors. The disorder is increasingly recognised as resulting from defects in the systems that govern these processes. Energy intake is known to be regulated by nuclei located in the hypothalamus and in the nucleus tractus solitarius of the brain stem. Studies have demonstrated that energy expenditure can also be regulated from these sites. Importantly, many of the identified genetic factors influencing monogenic forms of obesity are known to be expressed and have been shown to exert effects in the central nervous system (Farooqi and O'Rahilly 2005). Taken together, these findings suggest that alterations in the brains regulation of both central and potentially peripheral metabolic pathways may contribute largely to obesity. Clearly, findings from genetic studies will yield further insight into the biologic regulation of body weight and adiposity. Advances in genotyping methodologies and in vivo gene manipulation technologies will permit swift translation of these findings to the clinic.

REFERENCES

Almasy, L., Göring, H.H., Diego, V. et al. 2007. A novel obesity locus on chromosome 4q: The Strong Heart Family Study. *Obesity* 15:1741–8.

Arya, R., Duggirala, R., Comuzzie, A.G. et al. 2002. Heritability of anthropometric phenotypes in caste populations of Visakhapatnam, India. *Hum. Biol.* 74(3):325–44.

Arya, R., Duggirala, R., Jenkinson, C.P. et al. 2004. Evidence of a novel quantitative-trait locus for obesity on chromosome 4p in Mexican Americans. *Am. J. Hum. Genet.* 74:272–82.

Arya, R., Demerath, E., Jenkinson, C.P. et al. 2006. A quantitative trait locus (QTL) on chromosome 6q influences birth weight in two independent family studies. *Hum. Mol. Genet.* 15:1569–79.

Atwood, L.D., Heard-Costa, N.L., Fox, C.S., Jaquish, C.E. and L.A. Cupples. 2006. Sex and age specific effects of chromosomal regions linked to body mass index in the Framingham Study. *BMC Genet.* 26(7):7.

Baker, M., Gaukrodger, N., Mayosi, B.M. et al. 2005. Association between common polymorphisms of the proopiomelanocortin gene and body fat distribution: A family study. *Diabetes* 54:2492–6.

Bastarrachea, R.A., Kent, J.W. Jr., Rozada, G. et al. 2007. Heritability and genetic correlations of metabolic disease-related phenotypes in Mexico: Preliminary report from the GEMM Family Study. *Hum. Biol.* 79:121–9.

Bayoumi, R.A., Al-Yahyaee, S.A., Albarwani, S.A. et al. 2007. Heritability of determinants of the metabolic syndrome among healthy Arabs of the Oman family study. *Obesity* 15:551–6.

Bray, M.S., Boerwinkle, E. and C.L. Hanis. 1999. Linkage analysis of candidate obesity genes among the Mexican-American population of Starr County, Texas. *Genet. Epidemiol.* 16:397–411.

Broman, K.W. 2005. Mapping expression in randomized rodent genomes. *Nat. Genet.* 37:209–10.

Buono, P., Pasanisi, F., Nardelli, C. et al. 2005. Six novel mutations in the proopiomelanocortin and melanocortin receptor 4 genes in severely obese adults living in southern Italy. *Clin. Chem.* 51:1358–64.

Bustamante, M., Nogués, X., Mellibovsky, L. et al. 2007. Polymorphisms in the interleukin-6 receptor gene are associated with bone mineral density and body mass index in Spanish postmenopausal women. *Eur. J. Endocrinol.* 157(5):677–84.

Butte, N.F., Cai, G., Cole, S.A. and A.G. Comuzzie. 2006. Viva la Familia Study: Genetic and environmental contributions to childhood obesity and its comorbidities in the Hispanic population. *Am. J. Clin. Nutr.* 84:646–54.

Cai, G., Cole, S.A., Tejero, M.E. et al. 2004. Pleiotropic effects of genes for insulin resistance on adiposity in baboons. *Obes. Res.* 12:1766–72.

Cai, G., Cole, S.A., Butte, N.F. et al. 2006. A quantitative trait locus on chromosome 18q for physical activity and dietary intake in Hispanic children. *Obesity* 14:1596–604

Cai, G., Cole, S.A., Haack, K., Butte, N.F. and A.G. Comuzzie. 2007. Bivariate linkage confirms genetic contribution to fetal origins of childhood growth and cardiovascular disease risk in Hispanic children. *Hum. Genet.* 121:737–44.

Cheung, V.G. and R.S. Spielman. 2002. The genetics of variation in gene expression. *Nat. Genet.* 32:522–5 (Suppl).

Chung, S., Song, M.Y., Shin, H.D. et al. 2005. Korean and Caucasian overweight premenopausal women have different relationship of body mass index to percent body fat with age. *J. Appl. Physiol.* 99:103–7.

Cole, S.A., Mitchell, B.D., Hsueh, W.C. et al. 2000. The Pro012Ala variant of peroxisome proliferator-activated receptor-gamma2 (PPAR-gamma2) is associated with measures of obesity in Mexican Americans. *Int. J. Obes. Relat. Metab. Disord.* 24:522–4.

Comuzzie, A.G. and D.B. Allison. 1998. The search for human obesity genes. *Science* 280(5368):1374–7.

Comuzzie, A.G., Hixson, J.E., Almasy, L. et al. 1997. A major quantitative trait locus determining serum leptin levels and fat mass is located on human chromosome 2. *Nat. Genet.* 15:273–6.

Comuzzie, A.G., Williams, J.T., Martin, L.J. and J. Blangero. 2001. Searching for genes underlying normal variation in human adiposity. *J. Mol. Med.* 79:57–70.

Comuzzie, A.G., Cole, S.A., Martin, L. et al. 2003. The baboon as a nonhuman primate model for the study of the genetics of obesity. *Obes. Res.* 11:75–80.

Comuzzie, A.G., Mitchell, B.D., Cole, S. et al. 2003. The genetics of obesity in Mexican Americans: The evidence from genome scanning efforts in the San Antonio family heart study. *Hum. Biol.* 75:635–46.

Couzin, J. 2002. New mapping project splits the community. *Science* 296:1391–2.

Czerwinski, S.A., Lee, M., Choh, A.C. et al. 2007. Genetic factors in physical growth and development and their relationship to subsequent health outcomes. *Am. J. Hum. Biol.* 19:684–91.

Dempfle, A., Hinney, A., Heinzel-Gutenbrunner, M. et al. 2004. Large quantitative effect of melanocortin-4 receptor gene mutations on body mass index. *J. Med. Genet.* 41(10):795–800.

Dempfle, A., Wudy, S.A., Saar, K. et al. 2006. Evidence for involvement of the vitamin D receptor gene in idiopathic short stature via a genome-wide linkage study and subsequent association studies. *Hum. Mol. Genet.* 15(18):2772–83.

Dina, C., Meyre, D., Gallina, S. et al. 2007. Variation in FTO contributes to childhood obesity and severe adult obesity. *Nat. Genet.* 39(6):724–6.

Ding, B., Kull, B., Liu, Z. et al. 2005. Human neuropeptide Y signal peptide gain-of-function polymorphism is associated with increased body mass index: Possible mode of function. *Regul. Pept.* 127:45–53.

Dionne, I.J., Garant, M.J., Nolan, A.A. et al. 2002. Association between obesity and a polymorphism in the beta(1)-adrenoceptor gene (Gly389Arg ADRB1) in Caucasian women. *Int. J. Obes. Relat. Metab. Disord.* 26:633–9.

Dohrmann, C.E. 2004. Target discovery in metabolic disease. *Drug Discov. Today* 9:785–794.

Ellis, K.J. 2000. Human body composition: In vivo methods. *Physiol. Rev.* 80:649–80.

Ellis, K.J. 2001. Selected body composition methods can be used in field studies. *J. Nutr.* 131:S 1589–95.

Ellsworth, D.L., Coady, S.A., Chen, W., Srinivasan, S.R., Boerwinkle, E. and G.S. Berenson. 2005. Interactive effects between polymorphisms in the beta-adrenergic receptors and longitudinal changes in obesity. *Obes. Res.* 13:519–526.

Endo, K., Yanagi, H., Hirano, C., Hamaguchi, H., Tsuchiya, S. and S. Tomura. 2000. Association of trp63Arg polymorphism of the beta-3 adrenergic receptor gene and no association of Gln223Arg polymorphism of the leptin receptor gene in Japanese schoolchildren with obesity. *Int. J. Obes. Relat. Metab. Disord.* 24:443–9.

Evans, D., Minouchehr, S., Hagemann, G. et al. 2000. Frequency of and interaction between polymorphisms in the beta3-adrenergic receptor and in uncoupling proteins 1 and 2 and obesity in Germans. *Int. J. Obes. Relat. Metab. Disord.* 24:1239–45.

Evans, D., Wolf, A.M., Nellessen, U. et al. 2001. Association between polymorphisms in candidate genes and morbid obesity. *Int. J. Obes. Relat. Metab. Disord.* 25:S19–21.

Farooqi, S. and S. O'Rahilly. 2005. Monogenic obesity in humans. *Annu. Rev. Med.* 56:443–58.

Fox, C.S., Heard-Costa, N.L., Wilson, P.W., Levy, D., D'Agostino, R.B. Sr. and L.D. Atwood. 2004. Genome-wide linkage to chromosome 6 for waist circumference in the Framingham Heart Study. *Diabetes* 53(5):1399–402.

Fox, C.S., Heard-Costa, N., Cupples, L.A., Dupuis, J., Vasan, R.S. and L.D. Atwood. 2007. Genome-wide association to body mass index and waist circumference: The Framingham Heart Study 100K project. *BMC Med. Genet.* 8(Suppl 1):S18.

Franks, P.W., Jablonski, K.A., Delahanty, L. et al. 2007. The Pr012Ala variant at the peroxisome proliferator-activated receptor gamma gene and change in obesity-related traits in the Diabetes Prevention Program. *Diabetologia* 50:2451–60.

Frayling, T.M., Timpson, N.J., Weedon, M.N. et al. 2007. A common variant in the FTO gene is associated with body mass index and predisposes to childhood and adult obesity. *Science* 316(5826):889–94.

Gallagher, D. 2004. Overweight and obesity BMI cut-offs and their relation to metabolic disorders in Korean/Asians. *Obes. Res.* 12:440–1.

Gazzerro, P., Caruso, M.G., Natornicola, M. et al. 2007. Association between cannabinoid receptor type-1 receptor polymorphism and body mass index in a southern Italian population. *Int. J. Obes.* 31:908–12.

Geller, F., Dempfle, A. and T. Gorg. 2003. Framingham Heart Study. Genome scan for body mass index and height in the Framingham Heart Study. *BMC Genet.* 31(Suppl 1):S91.

Goring, H.H.H., Curran, J.E., Johnson, M.P. et al. 2007. Discovery of expression QTLs using large scale transcriptional profiling in human lymphocytes. *Nat. Genet.* 39:1208–16.

Hainerová, I., Larsen, L.H., Holst, B. et al. 2007. Melanocortin 4 receptor mutations in obese Czech children: Studies of prevalence, phenotype development, weight reduction response, and functional analysis. *J. Clin. Endocrinol. Metab.* 92:3689–96.

Halsall, D.J., Luan, J., Saker, P. et al. 2001. Uncoupling protein 3 genetic variants in human obesity: The c-55t promoter polymorphism is negatively correlated with body mass index in a UK Caucasian population. *Int. J. Obes. Relat. Metab. Disord.* 25(4):472–7.

Hanisch, D., Dittmar, M., Höhler, T. and K.W. Alt. 2004. Contribution of genetic and environmental factors to variation in body compartments: A twin study in adults. *Anthropol. Anz.* 62:51–60.

Herbert, A., Gerry, N.P., McQueen, M.B. et al. 2006. A common genetic variant is associated with adult and childhood obesity. *Science* 312(5771):279–83.

Herrmann, S.M., Wang, J.G., Staessen, J.A. et al. 2003. Uncoupling protein 1 and 3 polymorphisms are associated with waist-to-hip ratio. *J. Mol. Med.* 81(5):327–32.

Herskind, A.M., McGue, M., Sorensen, T.I. and B. Harvald. 1996. Sex and age specific assessment of genetic and environmental influences on body mass index in twins. *Int. J. Obes. Relat. Metab. Disord.* 20:106–13

Higgins, P.B., Gower, B.A., Hunter, G.R. and M.I. Goran. 2001. Defining health-related obesity in prepubertal children. *Obes. Res.* 9:233–40.

Horlick, M. 2001. Editorial. Measuring body mass index in childhood: Measuring a moving target. *J. Clin. Endocrinol. Metab.* 86:4059–60.

Hunt, M.S., Katzmarzyk, P.T., Pérusse, L., Rice, T., Rao, D.C. and C. Bouchard. 2002. Familial resemblance of 7-year changes in body mass and adiposity. *Obes. Res.* 10:507–17.

Iacobellis, G., Petrone, A., Leonetti, F. and R. Buzzetti. 2006. Left ventricular mass and +276 G/G single nucleotide polymorphism of the adiponectin gene in uncomplicated obesity. *Obesity (Silver Spring)* 14:368–72.

Jafar, T.H., Chaturvedi, N. and G. Pappas. 2006. Prevalence of overweight and obesity and their association with hypertension and diabetes mellitus in an Indo-Asian population. *CMAJ* 175:1071–7.

Johnson, Z.B. and R.A. Nugent III. 2003. Heritability of body length and measures of body density and their relationship to backfat thickness and loin muscle area in swine. *J. Anim. Sci.* 81(8):1943–9.

Johnson, L., Luke, A., Adeyemo, A. et al. 2005. Meta-analysis of five genome-wide linkage studies for body mass index reveals significant evidence for linkage to chromosome 8p. *Int. J. Obes. (Lond).* 29(4):413–9.

Juo, S.H., Lin, H.F., Rundek, T. et al. 2004. Genetic and environmental contributions to carotid intima-media thickness and obesity phenotypes in the Northern Manhattan Family Study. *Stroke* 35:2243–7.

Kavanagh, K., Fairbanks, L.A., Bailey, J.N. et al. 2007. Characterization and heritability of obesity and associated risk factors in vervet monkeys. *Obesity (Silver Spring)* 15:1666–74.

Kim, Y., Suh, Y.K. and H. Choi. 2004. BMI and metabolic disorders in South Korean adults: 1998 Korea National Health and Nutrition Study. *Obes. Res.* 12:445–53.

Kissebah, A.H., Sonnenberg, G.E., Myklebust, J. et al. 2000. Quantitative trait loci on chromosomes 3 and 17 influence phenotypes of the metabolic syndrome. *Proc. Natl. Acad. Sci. U.S.A.* 97:14478–83.

Klein, S., Allison, D.B., Heymsfield, S.B. et al. 2007. Waist circumference and cardiometabolic risk: A consensus statement from shaping America's health: Association for weight management and obesity prevention; NAASO, the Obesity Society; the American Society for Nutrition; and the American Diabetes Association. *Obesity* 15:1061–7.

Large, V., Hellström, L., Reynisdottir, S. et al. 1997. Human beta-2 adrenoceptor gene polymorphisms are highly frequent in obesity and associate with altered adipocyte beta-2 adrenoceptor function. *J. Clin. Invest.* 100:3005–13.

Li, H., Wu, Y., Loos, R.J. et al. 2007. Variants in FTO gene are not associated with obesity in a Chinese Han population. *Diabetes* [E-pub ahead of print].

Liu, Y.Z., Guo, Y.F., Xiao, P. et al. 2006. Epistasis between loci on chromosomes 2 and 6 influences human height. *J. Clin. Endocrinol. Metab.* 91(10):3821–5.

Luke, A., Wu, X., Zhu, X., Kan, D., Su, Y. and R. Cooper. 2003. Linkage for BMI at 3q27 region confirmed in an African-American population. *Diabetes* 52:1284–7.

Marti, A., Corbalán, M.S., Martínez-Gonzalez, M.A. and J.A. Martinez. 2002. TRP64ARG polymorphism of the beta 3-adrenergic receptor gene and obesity risk: Effect modification by a sedentary lifestyle. *Diabetes Obes. Metab.* 4:428–30.

Mattevi, V.S., Zembrzuski, V.M. and M.H. Hutz. 2002. Association analysis of genes involved in the leptin-signaling pathway with obesity in Brazil. *Int. J. Obes. Relat. Metab. Disord.* 26:1179–85.

McCarthy, H.D. 2006. Body fat measurements in children as predictors for the metabolic syndrome: Focus on waist circumference. *Proc. Nutr. Soc.* 65:385–92.

Meyre, D., Lecoeur, C., Delplanque, J. et al. 2004. Genome-wide scan for childhood obesity-associated traits in French families shows significant linkage on chromosome 6q22.31-q23.2. *Diabetes* 53(3):803–11.

Mitchell, B.D., Blangero, J., Comuzzie, A.G. et al. 1998. A paired sibling analysis of the beta-3 adrenergic receptor and obesity in Mexican Americans. *J. Clin. Invest.* 101:584–7.

Montague, C.T., Farooqi, I.S., Whitehead, J.P. et al. 1997. Congenital leptin deficiency is associated with severe early-onset obesity in humans. *Nature* 387:903–8.

Moslehi, R., Goldstein, A.M., Beerman, M., Goldin, L., Bergen, A.W. and Framingham Heart Study. 2003. A genome-wide linkage scan for body mass index on Framingham Heart Study families. *BMC Genet.* 4(Suppl 1):S97.

Nagano, T., Matsuda, Y., Tanioka, T. et al. 2005. No association of the Trp64Arg mutation of the beta3-adrenergic receptor gene with obesity, type 2 diabetes mellitus, hyperlipidemia, and hypertension in Japanese patients with schizophrenia. *J. Med. Invest.* 52:57–64.

Ng, M.C., So, W.Y., Lam, V.K. et al. 2004. Genome-wide scan for metabolic syndrome and related quantitative traits in Hong Kong Chinese and confirmation of a susceptibility locus on chromosome 1q21-q25. *Diabetes* 53(10):2676–83.

North, K.E., MacCluer, J.W., Williams, J.T. et al. 2003. Evidence for distinct genetic effects on obesity and lipid-related CVD risk factors in diabetic compared to nondiabetic American Indians: The Strong Heart Family Study. *Diabetes Metab. Res. Rev.* 19:140–7.

Owecki, M., Miczke, A., Kaczmarek, M. et al. 2007. The Y111 H (T415C) Polymorphism in exon 3 of the gene encoding adiponectin is uncommon in Polish obese patients. *Horm. Metab. Res.* 39(11):797–800.

Park, H.S., Kim, Y. and C. Lee. 2005. Single nucleotide variants in the beta2-adrenergic and beta3-adrenergic receptor genes explained 18.3% of adolescent obesity variation. *J. Hum. Genet.* 50:365–9.

Park, K.S., Shin, H.D., Park, B.L. et al. 2006. Polymorphisms in the leptin receptor (LEPR)-putative association with obesity and T2DM. *J. Hum. Genet.* 51:85–91.

Peeters, A., Beckers, S., Verrijken, A. et al. 2007. Variants in the FTO gene are associated with common obesity in the Belgian population. *Mol. Genet. Metab.* [E-pub ahead of print]

Pereira, A.C., Floriano, M.S., Mota, G.F. et al. 2003. Beta2 adrenoceptor functional gene variants, obesity, and blood pressure level interactions in the general population. *Hypertension* 42:685–92.

Quinton, N.D., Lee, A.J., Ross, R.J., Eastell, R. and A.I. Blakemore. 2001. A single nucleotide polymorphism (SNP) in the leptin receptor is associated with BMI, fat mass and leptin levels in postmenopausal Caucasian women. *Hum. Genet.* 108:233–6.

Reich, D.E. and E.S. Lander. 2001. On the allelic spectrum of human disease. *Trends Genet.* 17:502–10.

Rice, T., Pérusse, L., Bouchard, C. and D.C. Rao. 1999. Familial aggregation of body mass index and subcutaneous fat measures in the longitudinal Québec family study. *Genet. Epidemiol.* 16:316–34.

Rogers, J., Mahaney, M.C., Almasy, L., Comuzzie, A.G. and J. Blangero. 1999. Quantitative trait linkage mapping in anthropology. *Yearb. Phys. Anthropol.* 42:127–51.

Roth, H., Hinney, A., Ziegler, A. et al. 1997. Further support for linkage of extreme obesity to the obese gene in a study group of obese children and adolescents. *Exp. Clin. Endocrinol. Diabetes* 105:341–4.

Rotimi, C.N., Comuzzie, A.G., Lowe, W.L., Luke, A., Blangero, J. and R.S. Cooper. 1999. The quantitative trait locus on chromosome 2 for serum leptin levels is confirmed in African-Americans. *Diabetes* 48(3):643–4.

Salces, I., Rebato, E., Susanne, C., Hauspie, R.C., Saha, R. and P. Dasgupta. 2007. Heritability variations of morphometric traits in West Bengal (India) children aged 4–19 years: A mixed-longitudinal growth study. *Ann. Hum. Biol.* 34:226–39.

Sale, M.M., Freedman, B.I., Hicks, P.J. et al. 2005. Loci contributing to adult height and body mass index in African American families ascertained for type 2 diabetes. *Ann. Hum. Genet.* 69:517–27.

Santoro, N., Rankinen, T., Pérusse, L., Loos, R.J. and C. Bouchard. 2005. MC4R marker associated with stature in children and young adults: A longitudinal study. *J. Pediatr. Endocrinol. Metab.* 18:859–63.

Schadt, E.E., Monks, S.A., Drake, T.A. et al. 2003. Genetics of gene expression surveyed in maize, mice, and men. *Nature* 422:297–302.

Schrauwen, P., Walder, K. and E. Ravussin. 1999. Human uncoupling proteins and obesity. *Obes. Res.* 7(1):97–105.

Scuteri, A., Sanna, S., Chen, W.M. et al. 2007. Genome-wide association scan shows genetic variants in the FTO gene are associated with obesity-related traits. *PLoS Genet.* 3(7):e115.

Shen, W., Punyanitya, M., Chen, J. et al. 2006. Waist circumference correlates with metabolic syndrome indicators better than percentage fat. *Obesity* 14:727–36.

Souren, N.Y., Paulussen, A.D., Loos, R.J. et al. 2007. Anthropometry, carbohydrate and lipid metabolism in the East Flanders Prospective Twin Survey: Heritabilities. *Diabetologia* 50:2107–16.

Stone, S., Abkevich, V., Hunt, S.C. et al. 2002. A major predisposition locus for severe obesity, at 4p15-p14. *Am. J. Hum. Genet.* 70(6):1459–68.

Tang, W., Miller, M.B., Rich, S.S. et al. 2003. Linkage analysis of a composite factor for the multiple metabolic syndrome: The National Heart, Lung, and Blood Institute Family Heart Study. *Diabetes* 52(11):2840–7.

Terra, S.D., McGorray, S.P., Wu, R. et al. 2005. Association of β-adrenergic receptor polymorphisms and their G-protein-coupled receptors with body mass index and obesity in women: A report from the NHLBI-sponsored WISE study. *Int. J. Obes.* 29:746–54.

Terwilliger, J.D., Haghighi, F., Hiekkalinna, T.S. and H.H.H. Goring. 2002. A biased assessment of the use of SNPs in human complex traits. *Curr. Opin. Genet. Dev.* 12:726–34.

Um, J.Y., Kang, B.K., Lee, S.H., Shin, J.Y., Hong, S.H. and H.M. Kim. 2004. Polymorphism of the tumor necrosis factor alpha gene and waist-hip ratio in obese Korean women. *Mol. Cells* 18(3):340–5.

Voruganti, V.S., Cai, G., Cole, S.A. et al. 2006. Common set of genes regulates low-density lipoprotein size and obesity-related factors in Alaskan Eskimos: Results from the GOCADAN study. *Am. J. Hum. Biol.* 18:525–31.

Voruganti, V.S., Göring, H.H., Diego, V.P. et al. 2007. Genome-wide scan for serum ghrelin detects linkage on chromosome 1p36 in Hispanic children: Results from the Viva La Familia study. *Pediatr. Res.* 62:445–50.

Voruganti, V.S., Tejero, M.E., Proffitt, J.M., Cole, S.A., Freeland-Graves, J.H. and A.G. Comuzzie. 2007. Genome-wide scan of plasma cholecystokinin in baboons shows linkage to human chromosome 17. *Obesity (Silver Spring)* 15:2043–50.

Wauters, M., Mertens, I., Chagnon, M. et al. 2001. Polymorphisms in the leptin receptor gene, body composition and fat distribution in overweight and obese women. *Int. J. Obes. Relat. Metab. Disord.* 25(5):714–20.

Weedon, M.N., Lettre, G., Freathy, R.M. et al. 2007. A common variant of HMGA2 is associated with adult and childhood height in the general population. *Nat. Genet.* 39(10):1245–50.

World Health Organization. 2007. http://www.who.int/mediacentre/factsheets/fs311/en/index. html (accessed September 11, 2007).

Wu, K.D., Hsiao, C.F., Ho, L.T. et al. 2002. Clustering and heritability of insulin resistance in Chinese and Japanese hypertensive families: A Stanford-Asian Pacific Program in Hypertension and Insulin Resistance sibling study. *Hypertens. Res.* 25:529–36.

Wu, X., Cooper, R.S., Borecki, I. et al. 2002. A combined analysis of genomewide linkage scans for body mass index from the National Heart, Lung, and Blood Institute Family Blood Pressure Program. *Am. J. Hum. Genet.* 70(5):1247–56.

Wu, D.M., Hong, Y., Sun, C.A., Sung, P.K., Rao, D.C. and N.F. Chu. 2003. Familial resemblance of adiposity-related parameters: Results from a health check-up population in Taiwan. *Eur. J. Epidemiol.* 18:221–6.

Wuschke, S., Dahm, S., Schmidt, C., Joost, H.G. and H. Al-Hasani. 2007. A meta-analysis of quantitative trait loci associated with body weight and adiposity in mice. *Int. J. Obes.* 31:829–41.

3 Diversity of Human Adaptability to Environments
A Physiological Anthropology Perspective

*Akira Yasukouchi**
Department of Physiological Anthropology,
Kyushu University, Fukuoka, Japan

CONTENTS

---

* Address all correspondence to yasukouc@design.kyushu-u.ac.jp.

3.1 INTRODUCTION

There is always a danger when analysing data that results are presented in terms of some central tendency and the individual variation underlying this central tendency is ignored. In addition, apart from anthropometric data, issues of measurement precision and instability of individual physiological functions have contributed to the delay in understanding the fuller perspectives of environmental adaptability. However, significant individual differences in physiological functions have recently been detected due to advances in measuring devices and high technology in controlling physical environmental factors in climatic chambers.

Given such a favourable technical background, it is therefore possible to investigate concretely the relationships among specific environmental factors and/or special features of an individual or group with their environmental adaptability by examining physiological polytypism, functional potentiality, and whole-body coordination. These three terms serve as the basic parameters in the pursuit of a systemic approach to understanding the mechanisms of diversity in environmental adaptability. The three basic terms (physiological polytypism, functional potentiality, and whole-body coordination) may be briefly defined as follows. There is the maximum functional capacity composed of manifest and latent elements, where changes in the boundary between both these components are dependent on the type, frequency, and intensity (behavioural history of individuals) of stress encountered by an individual intention; viz., physiological polytypism is a characteristic phenotype of accompanying manifested component, depicting the specific difference manifested by the individual. Although physiological polytypism is influenced by gene- and environment-related factors, the effect on the latter is especially more of a plasticising nature. Moreover, functional potentiality defines the superficially nonapparent component of the maximum functional capacity. The function subject to these two factors incorporates the component as a coordinated response of the functional adjustment system to keep homeostasis against certain stresses. In cases where the coordination system is systemic, it is known as the whole-body coordination.

The elements affecting phenotype in polytypism include gene-, physical environment-, and culture-related factors, where the effects of the respective factor are subject to mutual interactions (Baker 1997). In cases of the lived environment, which is a subject of adaptation, lifestyle and environment are consistently shifting and changing in tandem with technological change. Therefore, the factors influencing the phenotype are more likely to be physical and cultural ones than genetic ones. Selections of resources, which are obtained from various living environments supported by the advent of technology, are dependent on individual preference and are largely homeostatic because of the innately comfort-seeking behaviour of humans (Eccles 1989). Contemporary humans in industrialised societies are therefore more likely to live in comfortable environments at lower stress levels than ever before. Tolerance of differences in individual physiological responses is extended in humans living in comfortable environments with low stress levels. In addition, the tolerance capacity to stress in individuals is dependent on the type, frequency, and intensity of the stress experienced, whereas selection of resources obtained from the living environment is dependent on individuals and their behavioural histories. It might be expected that low-stress environments might lead to deterioration of adaptability and tolerance capacity to stress. For example, abnormal weather due to global warming and transient nonavailability of environmental resources due to electrical blackouts could trigger unprecedented exacerbation of environmental stress.

In this chapter, physiological anthropology is discussed in relation to adaptability to modern comfortable living environments based on viewpoints derived from the terms of physiological anthropology. An overview of the history of anthropology and physiological anthropology in Japan is first presented before proceeding to the main theme.

3.2 HISTORY OF PHYSIOLOGICAL ANTHROPOLOGY IN JAPAN

Anthropology in Japan started in 1877 when the Department of Animal Studies of Tokyo Imperial University invited the late Professor Edward S. Morse (1838–1925) and he excavated the Shell Mounds of Omori. In 1884, Shogoro Tsuboi, a student at Tokyo Imperial University, founded an organisation known as the Group of Anthropologists (Jinruigaku no Tomo), which in 1886 became the Anthropological Society of Japan. This took place a mere twenty-five years after the foundation by Paul Broca (1824–1880) of the Paris Anthropological Society, the first such society anywhere.

In 1939, the Department of Anthropology was formally inaugurated in the Faculty of Science at Tokyo University. The first to assume the chair of the Department of Anthropology was Professor Kotondo Hasebe (1882–1969), and he established a course of physiological anthropology in addition to usual courses of morphological anthropology and prehistory (Terada 1981). Professor Toshihiko Tokizane (1909–1973), a neurophysiologist from the Department of Physiology, delivered the first lecture on physiological anthropology in early 1950s. One of Professor Tokizane's students was Masahiko Sato, who later served as professor at Kyushu Institute of Design in 1968 to assume the chair as professor of physiological anthropology. Professor Sato later developed the world-renowned Biotron model, where six climatic chambers were integrated as a single-environment chamber. While reinforcing the groundwork of physiological anthropology in Japan, Professor Sato continued to train and educate numerous disciples.

The mainstream of anthropological studies in Japan up to the 1970s has focussed on morphology, while research activities using physiological methods were undertaken by the Anthropological Society of Nippon. In 1978, the Organisation of Physiological Anthropology became independent, and formed the Conversazione of Physiology Anthropology, later (in 1987) changing its name to the Japan Society of Physiological Anthropology.

Physiological anthropology in Japan has its own research history. From the 1950s until today, the research trend may be divided into four stages (Sato et al. 1983; Sato 1995).

3.2.1 STAGE 1: HUMAN MOVEMENT STUDIES FROM THE 1950s TO THE LATE 1960s

Studies during this period were pioneered by Professor Tokizane; electromyography (EMG) was developed as a method to study and evaluate human vocalisation

(Tokizane and Kondo 1954), and the Research Society of Electromyography was established to support relevant research activities. EMG was then employed as one of the most popular research tools in the field of physiological anthropology. Contemporary studies of particular emphasis that appropriated the use of EMG include muscle fatigue, neuromuscular mechanisms of skilled movement, locomotive activity, posture, and so forth (Sato 1959, 1963, 1966; Tomita 1963). Other relevant studies worthy of mention in the early 1950s include measurements related to motor activity, such as reaction time and the manipulation skill test.

In the 1960s, there were two approaches to the study of fatigue: (i) functional differentiation of skeletal muscles associated with various movements (Tokizane and Shimazu 1964), and (ii) human adaptability to modern living involving different work types. The main issue for physiological anthropology in the 1960s was the evaluation of muscle fatigue using frequency analysis with EMG. In addition, studies using electrocardiography (ECG) and electroencephalography (EEG) were exploited in this stage (Sato and Tsuruma 1967; Hayami et al. 1968).

An event worthy of mention was the initiation of the Livelihood Fatigue Research Group supported by grants from the Ministry of Education, Science, and Culture of Japan in 1966. Research findings were disseminated at a session of the eighth International Congress of Anthropology and Ethnological Sciences at Tokyo in 1968. Monitoring of cardiorespiratory function and metabolism were also undertaken from this time and were extensively exploited in studies related to physical fitness in the 1970s.

3.2.2 Stage 2: Adaptability to Physical Environment and Physical Fitness Studies from the Late 1960s to the Late 1980s

Interest in physical fitness during this period focussed on studies of aerobic work capacity correlated with body physique, including body composition. Studies of differences in cardiovascular function in relation to maximal work capacity were investigated with respect to age, sex, and regional locality (Ikeda et al. 1977, 1978; Yamasaki et al. 1978). Fatigue studies in the 1960s greatly influenced research trends on human adaptability to various elements in the modern living environment. In this there were strong links with the International Biological Program (IBP). Research on human adaptability involving both laboratory and field studies continued to flourish until the 1980s.

This field of study developed further in the 1970s with the work of Professor Sato and his team on biotronics. Biotronics is the study of basic adaptability to the environment by observing the human physiological response to an artificial environment, where ambient temperature, humidity, air-flow, radiation, air pressure, illumination, and so on are artificially adjusted within a chamber or Biotron. Data on physiological responses, such as physical work capacity and body temperature regulation, under stressful conditions—including hyperbaric, hypobaric hyperthermic, or hypothermic conditions—have been accumulated over the years (Katsuura et al. 1977; Seki et al. 1978; Sato et al. 1974, 1979, 1980; Tochihara 1984; Yasukouchi et al. 1988) and have yielded important reference data in evaluating modern living environments (Yasukouchi et al. 1984, 2000; Katsuura 1985).

3.2.3   STAGE 3: MISSION-ORIENTED STUDIES FROM THE LATE 1980s TO THE EARLY 2000s

By the 1980s much basic research on physiological adaptability under various work and living environments had been undertaken. Since the establishment of the Conversazione of Physiology Anthropology in 1978, domestic meetings were held twice a year, and membership increased to about 1,000 people. With increasing divergent research fields, the organisation format was changed to incorporate the integrated sciences of anthropology. Studies in physiological anthropology up to this stage were on the verge of establishing truly evidence-based comfortable living environment specifications. However, concern about the rapid change in living environments due to technological developments had already heightened in society, and this trend forced industries to explore materials that took health and practicality into consideration, in addition to the manufacture of sales-oriented products. In short, both physiological anthropologists and society converged on the common ground of urgently addressing the issue of adaptable living environments for humans.

Thus, research in physiological anthropology came to take a mission-oriented strategy to satisfy the needs of society at this stage. Research activity was, and continues to be, not only geared towards laboratory and field findings but also towards interaction and information exchange with industries, organisations, and institutions involved in generating friendly living environments. In order to realise this strategy, seven research-oriented institutes (study groups on Clothes, Offices, Comfort, Measurements, Artificial Environments, Physical Fitness, and Design) were established. By 2008, this had expanded to eleven. The new additions were the study groups on (1) Offices, (2) Comfort, (3) *Kansei* (Sensitivity) Sciences, (4) Dwelling Environmental Evaluation, (5) Living Environments for the Elderly, (6) System Bioengineering, (7) Illumination and Lighting, (8) Physical Fitness, (9) Posture, (10) Evaluation and Control Problems of Human Medical/Welfare Facilities, and (11) Wood-Human Interaction.

In a further attempt to realise such goals, the Japan Society of Physiological Anthropology in 1999 established the PA (Physioanthropology) Design Awards in recognition of product designs that adequately incorporate physiological anthropology–derived originality, methodology, and evaluation, as well as being of good protocol design and product development. Through these efforts, developments of products which respect the biological significance of humans have taken a further step forward. An example of a product which has received this award is that of a washing machine that exploits the penetration force of water derived from centrifugation to yield a user-friendly washing method that attenuates wear and tear on laundry. Another award-winning product is an alarm clock which emits low illumination thirty minutes prior to wake-up time. Illumination gradually intensifies thereafter, increasing arousal level in the brain prior to the alarm call, which is soft electronic sound or music, to fully awaken its user. In this way the product exploits sensory awakening effects to make awakening from sleep more humane.

3.2.4 STAGE 4: A NEW APPROACH TO PHYSIOANTHROPOLOGICAL STUDIES FROM THE EARLY 2000S TO THE PRESENT

Since 2000, a new stage of development in physioanthropology has involved preparations for integrating the various schools of study with the aim of studying ways in which environments can be created that will enable humans to live a better life, now and into the future (Katsuura and Yasukouchi 2000). The Japan Society of Physiological Anthropology has identified areas of particular focus for research: physiological polytypism, functional potentiality, whole-body coordination, technological adaptability, and environmental adaptability. These areas are related but offer new ways to structure studies of adaptability, its diversity, and its mechanisms in modern society.

3.3 CHARACTERISTICS OF THE FIELD OF PHYSIOANTHROPOLOGY

Research in physiological anthropology in Japan has taken, in the main, a different path than that in the United States and Europe. A comparison of articles published in the *Journal of Physiological Anthropology* (JPA) of Japan and *Annals of Human Biology* (AHB), the journal of the Society for the Study of Human Biology of the United Kingdom, reveals some interesting differences (Yasukouchi 2004). Based on a survey of 230 articles from 1983 to 1991 in JPA and 385 articles from 1993 to 2002 in AHB, the major difference is in publications on stress (Figure 3.1).

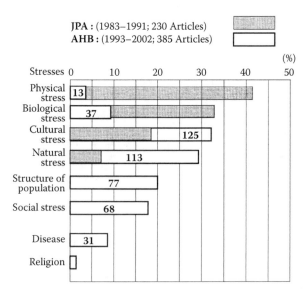

FIGURE 3.1 Type of stress investigated in the articles of *Journal of Physiological Anthropology* (JPA) and *Annals of Human Biology* (AHB).

While physical, environmental, and biological stresses are discussed in over 70% of documented studies in JPA, a mere 15% of articles in AHB consider the same issue. While stress induced by population structure (related to immigration, ethnicity, language, genes, and inbreeding, among other factors) and society-derived stress (such as parental occupation, education, and family size) accounted for 40% of all documented articles, very few of these were reported in JPA. In addition, natural environmental factors, such as geographical conditions, seasonal climatic changes, and nutritional aspects, accounted for 7% and 30% of documents in JPA and AHB, respectively. The main difference in special features between these two journals is that laboratory findings are more likely to have been investigated in studies documented in JPA, while field findings are more often reflected in studies published in AHB. Of the 385 articles published in AHB, 309 were reports of field studies conducted in about seventy different local and foreign sites.

In JPA, the majority of articles (52) investigated physiological and morphological adaptability to physical environment-related factors, followed by 39 articles on the special features of physiological and morphological differences with respect to age, sex, and region. In AHB, the highest number of articles (96) were concerned with demographic characteristics of population structure including genetic specificities, followed by 49 articles on differences in particular biological features with respect to age, sex, and region. This includes articles on genetic variation, a topic not published on in JPA. Studies describing the secular trend in growth (23 articles) and growth and development (48 articles) are particularly dominant in AHB.

The common themes shared between these two journals focus on adaptability to environments as well as attempts to further clarify mechanisms related to differences between individuals or groups. Articles in JPA concentrate on laboratory findings with physical environment-related factors as stressors and evaluate individual or small-group differences by intricately measuring physiological parameters in healthy subjects. In addition, in cases where a modern living environment is envisioned, studies using environmental factors as stressors shift to investigations in which subjects seek comfort. In contrast, articles in AHB focus on field studies, and investigations tend to evaluate adaptability to complex environmental factors. Numerous studies specifically deal with developing regions, concentrating on features related to demographic characteristics, diseases, and maladaptation under stressful environments.

In the United States, adaptability research underwent bifurcation around 1975 (Steegmann 2006). At that time, Albert Damon of Harvard University published his book on physiological anthropology (Damon 1975). This focussed on the major environmental factors such as ultraviolet light, heat, cold, desert environments, and altitude. At this time, Baker and his students proposed that differences in physiological responses between groups with respect to altitude and climate may not necessarily be observed in experimental findings even if studies are carried out in thoroughly controlled climatic chambers. Professor Baker also wrote of the possible influence of prenatal developmental stress on post-adulthood health. These concepts extended and expanded the use of biocultural models in field studies by many European and American physiological anthropologists since (Steegmann 2006).

3.4 POLYTYPISM IN ADAPTABILITY AND ITS
SIGNIFICANCE FOR THE LIVED ENVIRONMENT
IN TECHNOLOGICALLY ADVANCED SOCIETIES

More than 95% of mankind's history has been as hunter-gatherers. Human physiology has not undergone significant change across recent prehistory, and there is a major discrepancy between the environment to which we have adapted and the environment to which we are now struggling to adapt to. It is therefore important to create environments and lifestyle systems that take our recent past into consideration while solving problems for the present and future.

3.4.1 RELATIONSHIPS BETWEEN HUMAN BEHAVIOUR
AND ENVIRONMENT IN MODERN SOCIETY

With the discovery of intracranial self-stimulation, Olds and Milner (1954) and Olds (1956) were able to induce adaptation behaviour (access and avoidance) based on the reward and aversion system. In the case of human behaviour, reactions to stimuli are founded on emotion that encompasses pleasantness or unpleasantness. According to Eccles (1989), the size of the pleasantness centre in the human brain is the largest among primates. The emotional behaviour of *Homo sapiens* led to innovation of tools, which when incorporated in supporting life activity, assembled eventually into systems that subsequently created civilisations. Therefore, the potently cultural adaptation-dependent background enabled humans to seek ways to make everyday life more comfortable and pleasant.

While humans have adapted to living environments as prehistoric hunter-gatherers, living in the new environments of industrialised societies requires considerable adaptive readjustment. Humans have the most challenging standing posture, but this enabled them to walk long distances to hunt and gather food. However, office work in modern society is different from the hunter-gatherer era and demands a consistent sedentary posture that eventually weakens muscle and generates brittle bones of the lower limbs with additional risks of lower back problems. With respect to thermal stress, degeneration of bodily responses to heat and cold due to widespread use of air-conditioning and heating facilities in tropical and cold regions respectively has generated much concern. As for stimulation by light, exposure to artificial illumination sources from late night until the early hours of the morning has resulted in insufficient sleep and confusion of phase shifts of the hitherto well-maintained biological rhythm. Although the development of science and technology has yielded convenience and comfort in our lifestyle and living environments, there is a concern that these may promote degeneration of bodily system functions so that we may be unable to cope and adjust to sudden environmental changes as a result of living in comfort over long periods.

Physiological anthropology attempts first to understand extensively the relationships between various human phenotypes and environmental adaptability, in the hope of identifying factors that might shape the human-environment system as a whole. Figure 3.2 gives an overview of adaptive relations among individual humans, society, and environment. Human action is seen as being driven by emotion and

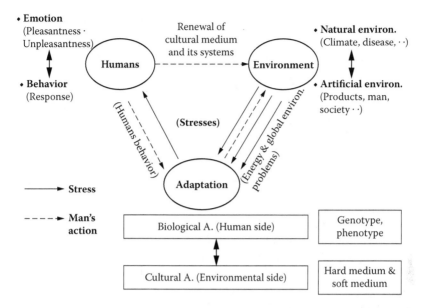

FIGURE 3.2 Relationship between human behaviour and stress from environment.

behaviour. At the broadest level, emotion may be categorised as either pleasantness or unpleasantness, while behaviour espouses visible action and invisible physiological responses. Adaptation, defined as the means employed to buffer stress from the environment, may be divided into biological (buffers on the human side) and cultural (buffers on the side of environment) media. Both of these buffers mutually influence human–environment relationships. Environment is posed as a stressor, which may be divided into artificial environment espousing the cultural medium and natural environment.

When the cultural medium is exploited positively for adaptation, the effects on the environment assume the form of energy and waste in global environmental issues (broken arrows). The greater the dependence on the cultural medium, the greater the environmental stress, resulting in various influences, either positive or negative, on humans. Long-term persistence of this process may exert novel effects on the genotype and phenotype moulding biological adaptation. Selection of human behaviours in moulding adaptation elicits potent effects on the balance between biological and cultural adaptation. Emotion intended to modify the cultural medium for higher comfort and satisfaction alters structural frameworks of artificial environments, thus initiating another form of stress. Persistent search for pleasantness and comfort in technological society may lead to: (1) excessive energy consumption and waste disposal, leading to environmental pollution; (2) degeneration of biological adaptability; and (3) excessive physiological tension derived from the discrepancy between the ancient environment to which we have adapted and the modern technological environment to which we struggle to adapt.

3.4.2 THREE TERMS OF PHYSIOLOGICAL ANTHROPOLOGY IN ENVIRONMENT ADAPTABILITY: PHYSIOLOGICAL POLYTYPISM, FUNCTIONAL POTENTIALITY, AND WHOLE-BODY COORDINATION

Homeostasis is maintained in all organisms for survival against various environmental stresses and the process of successful accommodation to various stresses results in adaptation. Homeostasis is functionally realised through accommodation of body systems or as a result of coordinated functions of many elemental organs in whole-body coordination. The body system accommodates, changes, and develops with a defined direction dependent on the behavioural history or on exposure to the type, magnitude, and frequency of stress, which as a whole moulds the characteristic phenotype (polytypism).

The genotype is affected by physical and cultural environments while the phenotype is derived from the genotype and is modified by behavioural history of exposure to certain stress from the living environment (Baker 1997). The living environment in turn affects the acquired characteristic phenotype to eventually develop physiological and/or morphological polytypism. Our interest focusses on the ability and necessity of the acquired characteristic phenotype to function as a phenotypic adjustment in technologically advanced environments of the present day.

Phenotypic adjustment is defined as the appropriate response elicited by proper maintenance of functional homeostasis. In brief, as human phenotype is modified by environmental factors (which eventually mould the acquired characteristic phenotype, incorporating morphological polytypism and physiological polytypism), a body system with well-maintained functional homeostasis elicits appropriate response to stress in the environment through a mechanism called phenotypic adjustment.

Figure 3.3 illustrates the relationships between acquired physiological polytypism and functional potentiality. The maximum tolerance to stress is basically determined by genetic composition. There are two different phenotypes of intrinsic tolerance: resistance adaptation (the ability to resist extreme stress) and capacity adaptation, which is the efficiency or effect of physiological response to trivial stress usually experienced in daily life (Precht 1958).

Tolerance levels change according to behavioural history; that is, tolerance is dependent on the incidence of severity and/or frequency of stress encountered (see Figure 3.4). Repeated heat exposures result in acclimatisation (Kuno 1956; Yoshimura 1960; Hori et al. 1976), physiological polytypism being based on differences in the relationships among climate and genetically-dependent maximum heat-tolerance capacity. After heat acclimatisation, these differences are accommodated, such that differences among groups are narrowed. The temperature maintenance ability of thermally temperate native Japanese is different from tropical natives such as Malaysians. However, when Malaysians are repeatedly subjected to physical exercise in a hot environment, their heat tolerance further improves (Saat et al. 2005), indicating that functional latency of temperature maintenance in tropical groups persists. Conversely, the sweat onset time of Malaysians long-resident in Japan becomes shorter when they are subjected to heat strain (Saat et al. 1999). In

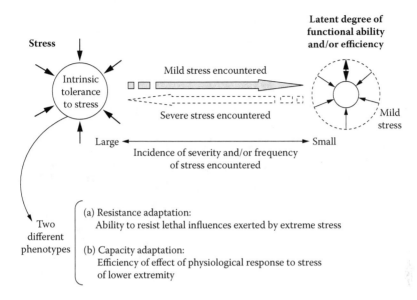

FIGURE 3.3   Relationship of a physiological polytypism to a functional potentiality.

this fashion, variations in functional latency may be moulded according to differences in frequency and intensity of stress encountered via individual behaviours, resulting in polytypism of apparent levels in actual situations.

Body temperature can be maintained through a range of output and coordination responses. For example, acclimatisation to heat elevates sweat ability in the

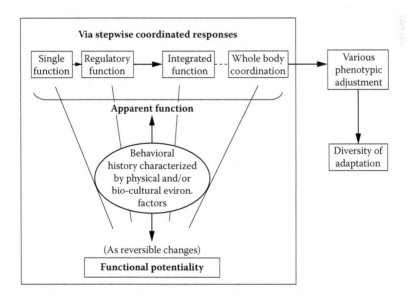

FIGURE 3.4   The subject of physiological polytypism and its relation to diversity in adaptation.

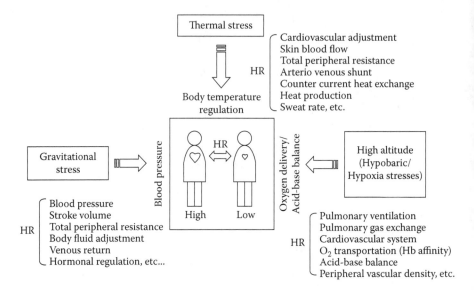

FIGURE 3.5 A parameter of coordinated response to a stress.

Japanese; however, this minimises water and salt loss by reducing the amount of sweat and salt concentration in sweat via elevation of dry heat loss. This is a coordinated response in reaction to hyperthermic stimulation (Allan and Wilson 1971; Kirby and Convertino 1986; Saat et al. 2005).

Physiological polytypism (Figure 3.4) may emerge from a single functionality to specific regulatory functions that mould whole-body coordination via stepwise coordinated responses. When the response maintains homeostasis, phenotypic adjustments ensue. Understanding characteristics of whole-body coordination may help understand the mechanisms of human variation and adaptability. In the case of thermal stress, heart rate (HR) responds via total peripheral resistance, stroke volume and cardiac output, heat production, and sweat rate, to effect favourable body temperature regulation (Figure 3.5). In the case of hypobaric hypoxia stress, HR responds via oxygen transportation in combination with the acid-base balance system. Thus, HR has to be understood as an elemental function of coordinated responses to a given stress, in the context of the characteristic pattern of coordinated response among individuals and/or populations.

3.4.3 Physiological Polytypism and Its Significance in Current Japanese Environments

Evaluating adaptability to modern environments is not a simple issue, as humans display multifaceted responses to different physical and cultural factors prevailing in the technologically comfortable environment of today. In the schematic figure

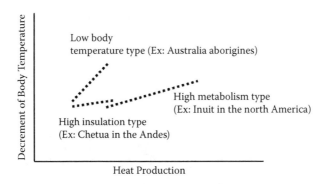

FIGURE 3.6 Schematic figure of physiological polytypism as seen in a coordinated response to cold stress.

of physiological polytypism as a coordinated response to cold stress (Figure 3.6), the x axis indicates heat production, while the y axis represents the decrement in body temperature. Generally, there are three well-known population types displaying different coordinated responses to a cold environment: one with declining body temperature such as the Australia aborigines, another with high insulation such as the Quechua living in the Andes, and a third with elevated metabolism such as the Inuit in the far north of Canada and Alaska. The different phenotypic adjustments are more likely to be moulded by the natural environment or genetic factors than by behavioural history.

Maeda et al. (2005) have found that basal metabolic rate (BMR) is reduced where snacking between meals is common and where daily physical activity is low. They have also shown BMR to be significantly lower among males who take night-time meals purchased at convenience stores, compared to those who feed on self-prepared meals (Maeda et al. 2005). The difference in BMR has been shown to be associated with coordinated responses to cold exposure (Maeda et al. 2007), males with a lower BMR having higher metabolic rate with a higher forearm blood flow.

A summary of the studies of Maeda et al. published between 2005 and 2007 (Figure 3.7) shows male low- and high-BMR types to have different coordinated responses to cold stress related to daily behavioural patterns. The low-BMR type tends to follow lifestyle practices such as eating snacks between meals, low physical activity, and feeding on food purchased at convenience stores. This may not be a bad response to mild cold stress but not to severe cold stress. The high-BMR type tends to not eat between meals, displays high physical activity, and feeds on self-prepared food.

Subjects who routinely display low physical activity have low maximum oxygen intake, which can be improved with aerobic training (Buskirk and Taylor 1957; Astrand and Rodahl 1970) via increase in the cardiorespiratory endurance (Taylor et al. 1955). According to Aoki (2008), not only an increase of maximum oxygen intake but in orthostatic tolerance (evaluated from head-up tilting with constant cardiac output accompanying by lowered HR and increased stroke volume) are also

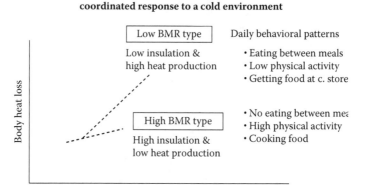

Difference in BMR associated with coordinated response to a cold environment

FIGURE 3.7   Difference in BMR and its relation to coordinated response to a cold environment.

established by twelve-week aerobic training in sedentary subjects. Furthermore, detraining for a twelve-month period followed by three-month aerobic training did not affect the characteristic orthostatic responses to postural change even when the aerobic ability is reduced. This shows the difference in persistence between functional abilities.

Melatonin release, which is promoted during night time without lighting, responds to change in intensity, duration, and time of the day in light exposure (Wehr 1991). In addition, nocturnal melatonin release is suppressed by light exposure; however, the extent of this nocturnal suppression depends on condition of diurnal light exposure as well (Park and Tokura 1999; Hebert et al. 2002). In a study by Torigoe et al. (2007), nocturnal suppression of melatonin is lower in subjects exposed to higher light intensity in the morning than in those exposed to lower light intensity. In another study by Higuchi et al. (2007) on the night-time melatonin release in relation to seasonal differences, the suppression rate of melatonin release by night-time lighting is greater in winter than summer, suggesting that light sensitivity in humans is elevated during winter, where light exposure is attenuated. Thus, people who spend more time indoors during daily daytime activity may have decreased nocturnal melatonin secretion and increased suppression of melatonin when exposed to night-time lighting.

Physiological polytypism is represented by the pattern of coordinated responses and functional potential of an individual to adapt (Figure 3.8). Both the apparent and latent components reflect physical and functional resources of humans. The apparent component is variable and related to behavioural history. It plays a minor part in maintaining homeostasis, and only in a comfortable environment can it be treated as a valid phenotypic adjustment. However, we do not know if this adjustment is desirable in relation to the ability to deal with unexpected stress in daily life.

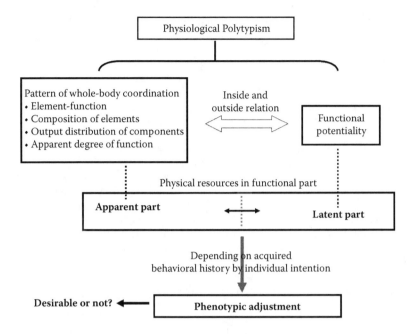

FIGURE 3.8 Summary of physiological polytypism and phenotypic adjustment.

3.5 FUTURE ISSUES

Mechanisms of human adaptability in respect of physiological polytypism, functional potentiality, and whole-body coordination have been described. Future issues to gain better insight into human adaptability include

1. The need to examine objectively the relationship between latent and apparent functionality
2. The need to confirm physiological polytypism at a population level, beyond current laboratory-level studies
3. Evaluation of physiological polytypism in relation to costs of dealing with unexpected stresses in technologically comfortable environments
4. Demonstration of the relationship between behavioural history and physiological responses
5. Identification of the genetic and environmental factors that contribute to apparent functions

REFERENCES

Allan, H.R. and C.G. Wilson. 1971. Influence of acclimatization on sweat sodium concentrations. *J. Appl. Physiol.* 30:708–12.
Aoki, K. 2008. Effects of physical training on cardiovascular responses to head-up tilt in sedentary men. PhD diss., Kyushu University, Fukuoka, Japan (in Japanese).

Astrand, P-O. and K. Rodahl. 1970. *Textbook of work physiology.* New York: McGraw-Hill Inc.

Baker, P.T. 1997. The Raymond Pearl memorial lecture. 1996: The eternal triangle-genes, phenotype and environment. *Am. J. Human Biol.* 9:93–101.

Buskirk, E. and H.L. Taylor. 1957. Maximal oxygen intake and its relation to body composition, with special reference to chronic physical activity and obesity. *J. Appl. Physiol.* 11:72–8.

Damon, A. 1975. *Physiological anthropology.* New York: Oxford University Press.

Eccles, J.C. 1989. *Evolution of the brain: creation of the self.* London and New York: Routledge.

Hayami, A., Ishii, M., Sato, H. et al. 1968. Activation level and cortical evoked responses. *Proc. 8th Intern. Cong. Anthropol. Ethnol. Sci.* 76–7.

Hebert, M., Martin, S.K., Lee, C. et al. 2002. The effect of prior light history on the suppression of melatonin by light in humans. *J. Pineal. Res.* 33:198–203.

Higuchi, S., Motohashi, Y., Maeda, T. et al. 2007. Less exposure to daily ambient light in winter increases sensitivity of melatonin to light suppression. *Chronobiol. Int.* 24:31–43.

Hori, S., Ishizuka, H. and M. Nakamura. 1976. Studies on physiological responses of residents in Okinawa to a hot environment. *Jpn. J. Physiol.* 26:235–44.

Ikeda, O., Takasaki, Y. and A. Yasukouchi. 1977. Factor of physical fitness test and relationship with physical characteristics in young adult females and males. *J. Anthrop. Soc. Nippon* 85:49–55 (in Japanese with English abstract).

Ikeda, O., Takasaki, Y. and A. Yasukouchi. 1978. The relations between body composition and aerobic work capacity in Fukuoka residents aged 12, 13 and 18, 19 years. *J. Anthrop. Soc. Nippon* 86:363–6 (in Japanese with English abstract).

Johnston, F.E. and M.A. Little. 2000. History of human biology in the United State of America. In *Human Biology—An Evolutionary and Biocultural Perspective*, ed. S. Stinson, B. Bogin, R. Huss-Ashmore, and D. O'Rourke, 27–46. New York: Wiley-Liss Inc.

Katsuura, T. 1985. Circulatory responses to work at simulated altitudes under different ambient temperatures. *Ann. Physiol. Anthropol.* 4:114–22.

Katsuura, T. and A. Yasukouchi. 2000. Meeting new challenges toward the 21st century. *J. Physiol. Anthropol. Appl. Human Sci.* 19:1–3.

Katsuura, K., Harada, H., Yasukouchi, A. and K. Yamazaki. 1977. Auditory and somatosensory evoked potential under hyperbaric environment. *J. Anthrop. Soc. Nippon* 86:119 (in Japanese).

Kirby, C.R. and V.A. Convertino. 1986. Plasma aldosterone and sweat sodium concentrations after exercise and heat acclimatization. *J. Appl. Physiol.* 61:967–70.

Kuno, Y. 1956. *Human Perspiration.* Springfield: Charles C. Thomas.

Maeda, M., Sugawara, A., Fukushima, T. et al. 2005. Effects of lifestyle, body composition, and physical fitness on cold tolerance in humans. *J. Physiol. Anthropol. Appl. Human Sci.*, 24:439–43.

Maeda, T., Fukushima, T., Ishibashi, K. et al. 2007. Involvement of basal metabolic rate in determination of type of cold tolerance. *J. Physiol. Anthropol.* 26:415–8.

Olds, J. 1956. Pleasure center in the brain. *Sci. Am.* 195:105–16.

Olds, J. and P. Milner. 1954. Positive reinforcement produced by electrical stimulation of septal area and other regions of rat brain. *J. Comp. Physiol. Psychol.* 47:419–27.

Park, S.J. and H. Tokura. 1999. Bright light exposure during the daytime affects circadian rhythms of urinary melatonin and salivary immunoglobulin A. *Chronobiol. Int.* 16:359–71.

Precht, H. 1958. Concepts of the temperature adaptation of unchanging reaction systems of cold-blooded animals. In *Physiological Adaptation*, ed. C.L. Prosser. Washington D.C.: American Physiological Society.

Saat, M.I., Lee, J.B., T. Matsumoto et al. 1999. Relationship between the duration of stay in Japan of Malaysian subjects and the suppression of sweat gland sensitivity by iontophoretically applied acetylcholine. *Acta Med. Nagasaki* 44:49–53.

Saat, M.I., Sirisinghe, R.G., Singh, R. and Y. Tochihara. 2005. Effects of short-term exercise in the heat on thermoregulation, blood parameters, sweat secretion and sweat composition of tropical-dwelling subjects. *J. Physiol. Anthropol. Appl. Human Sci.* 24:541–9.

Sato, M. 1959. Fatigue and exercise effects in the rapid repeating movement. *J. Jap. EMG Soc.* 1110–1.

Sato, M. 1963. An electromyographic study on skilled movements. *J. Faculty of Science, Univ. Tokyo, VII.* 4:323–69.

Sato, M. 1966. Muscle fatigue in the half rising posture. *J. Anthropol. Soc. Nippon,* 74:195–201.

Sato, M. 1995. The progress of physiological anthropology in Japan. *J. Physiol. Anthropol. Appl. Human Sci.* 14:1–4.

Sato, M. 2005. The development of conceptual framework in physiological anthropology. *J. Physiol. Anthropol.* 24:289–95.

Sato, M. and S. Tsuruma. 1967. Cardiac and muscular factors in the micro-vibration on the human body surface. *J. Anthropol. Soc. Nippon* 75:19–31.

Sato, M. and T. Sakate. 1974. Combined influences on cardiopulmonary functions of simulated high altitude and graded workloads. *J. Human Ergol.* 3:55–66.

Sato, M., Katsuura, T. and A. Yasukouchi. 1979. The lower and upper critical temperatures in male Japanese. *J. Human Ergol.* 8:145–53.

Sato, M., Yasukouchi, A. and H. Harada. 1980. Prediction of maximal oxygen intake over a wide range of air temperature conditions. *J. Human Ergol.* 9:81–9.

Sato, M., Takasaki, Y., Harada, H., Yamasaki, K. and S. Watanuki. 1983. Physiological anthropology in Japan between 1948 and 1982. *Ann. Physiol. Anthropol.* 2:3–8.

Seki, K., Yamasaki, M. and H. Nakayama. 1978. Studies on the sleep of man under the hyperbaric condition (21 ATA, Heliox). *J. Anthropol. Soc. Nippon* 87:193 (in Japanese).

Steegmann, T.A. Jr. 2006. Physiological anthropology: Past and future. *J. Physiol. Anthropol.* 25:67–73.

Taylor, H.L., Buskirk, E. and A. Henschel. 1955. Maximal oxygen uptake as an objective measure of cardiorespiratory performance. *J. Appl. Physiol.* 8:73–80.

Terada, K. 1981. *Anthropology in Japan.* Tokyo: Kadokawa Bunko (in Japanese).

Tochihara, Y. 1984. Physiological responses of men and women during prolonged 40% VO$_2$max exercise at different ambient temperatures. *J. Anthropol. Soc Nippon* 92:1–12.

Tokizane, T. and S. Kondo. 1954. Male voice and female voice. *Hum. Sci.* 6:18–27.

Tokizane, T. and H. Shimazu. 1964. *The Functional Differentiation of Human Skeletal Muscle.* Tokyo: Tokyo University Press.

Tomita, M. 1963. Activity patterns of four leg muscles during voluntary or involuntary body movements. *J. Anthropol. Soc. Nippon* 71:18–22.

Torigoe, H., Kozaki, T. and A. Yasukouchi. 2007. Effect of sunlight on light-induced suppression of nocturnal melatonin. *Jap. J. Physiol. Anthrop.* 12(Suppl 2):100–1.

Wehr, T.A. 1991. The durations of human melatonin secretion and sleep respond to changes in day length (photoperiod). *J. Cli. Endocrinol. Metabolism* 73:1276–80.

Yamasaki, M., Takasaki, Y., Sakate, T. et al. 1978. Physical fitness in Amamioshima residents aged 11–12 and 13–14 years. *J. Anthropol. Soc. Nippon* 86:131–132 (in Japanese).

Yasukouchi, A. 2004. Physiological anthropology and its possibility to be international. *Jap. J. Physiol. Anthropol.* 9 (Suppl 2):16–17 (in Japanese).

Yasukouchi, A., Inoue, K. and M. Sato. 1984. Effects of hypoxic and cool environment on pulmonary diffusing capacity for CO. *Ann. Physiol. Anthropol.* 3:11–18 (in Japanese with English abstract).

Yasukouchi, A., Inoue, K. and M. Sato. 1988. Changes in pulmonary diffusing capacity at simulated high altitudes under different ambient temperatures. In *Environmental Ergonomics*, ed. I.B. Mekjavic, E.W. Banister, and J.B. Morrison, 376–83. London: Taylor & Francis.

Yasukouchi, A., Yasukouchi, Y. and K. Ishibashi. 2000. Effects of color temperature of fluorescent lamps on body temperature regulation in a moderately cold environment. *J. Physiol. Anthropol.* 19:125–34.

Yoshimura, H. 1960. Acclimatization to heat and cold. In *Essential Problem in Climate Physiology*, ed. H. Yoshimura, K. Ogata, and S. Itoh, 61–106. Tokyo: Nankodo (in Japanese).

4 Tissue and Cell Adaptability to Physical and Chemical Factors

Susumu Kudo[*1] *and Kazuo Tanishita*[2]
[1]Department of Mechanical Engineering, Shibaura Institute of Technology, Tokyo, Japan
[2]Department of System Design Engineering, Keio University, Yokohama, Japan

CONTENTS

4.1 INTRODUCTION

The human body is organised hierarchically into organ systems which perform major physiological functions such as circulation, respiration, and digestion. Among these, the circulatory system delivers a continual flow of blood to each functional unit while

---

* Address all correspondence to kudous@sic.shibaura-it.ac.jp.

facilitating the filtration of metabolites and waste products. The circulatory system also serves as a conduit for communication between distant tissues through the transport of hormones. Oxygen and hormone concentrations in the blood, blood pressure, and blood flow are known to change when the external environment changes. The functions and morphology of the vascular system are regulated by haemodynamic stresses such as fluid frictional force (shear stress) induced by blood flow (Skalak and Price 1996). Wall shear stress affects vascular remodelling; Kamiya and Togawa (1980) report that wall shear stress due to blood flow induces adaptive changes in the blood vessel lumen in vivo, such that a constant wall shear stress is maintained within physiological bounds. Wall shear stress is thus a major factor influencing adaptive vessel regulation for physiological processes. To understand variations in human adaptability, the detailed mechanisms of human adaptability must be elucidated. Understanding adaptability of this system is constrained by the limited possibility of direct measurement. More generally, understanding what responses and adaptations to the external environment occur inside the body is not easily determined, as many factors interact with each other in complex ways. Simplification of systems is useful for understanding physiological mechanisms in detail. To this end, cell (tissue) studies are useful, since they can control experimental variables. With respect to the blood vessels of the circulatory system, endothelial cells (ECs) line the inner surface of blood vessels and are continuously exposed to physical factors such as fluid frictional force (shear stress) induced by blood flow and are also exposed to chemical factors such as oxygen and hormones in the blood.

Once thought to simply represent a passive barrier, ECs are now recognised as dynamic participants in the biology of blood vessels. In addition to their response to chemical ligands, ECs respond to physical factors such as fluid frictional force (shear stress), and the function and morphology of ECs have been shown to adapt to such factors.

This chapter describes EC adaptability to shear stress in the circulatory system focusing on morphological adaptability, mass transport adaptability, cell signalling (calcium ion signalling) adaptability, energy synthesis [adenosine triphosphate (ATP) synthesis], and adaptability to the magnitude of shear stress. In addition, the effect of shear stress and basic fibroblast growth factor (bFGF) on microvessel networks of ECs (angiogenesis) is discussed.

4.2   MATERIALS AND METHODS

4.2.1   Cell Culture

ECs must be cultured to allow study of the effects of shear stress on their function. Several types of ECs from human, bovine, and porcine sources are commercially available. In the studies described here, ECs were plated and cultured in a 25-cm^2 culturing flask in Dulbecco's Modified Eagle Medium (DMEM) containing 10% foetal bovine serum (FBS). After the ECs reached confluence, cells were detached using 0.05% trypsin-EDTA subcultured at a 1:4 ratio and passaged five to ten times. ECs were identified by their characteristic polygonal shape and their uptake of tetramethyl indocarbocyanine-percholorate-labelled acetylated low-density lipoprotein (DiI-Ac-LDL; Biomedical Technologies).

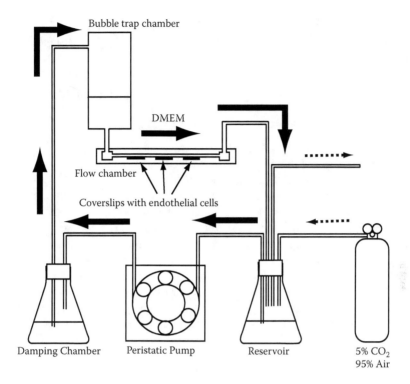

FIGURE 4.1 Schematic of flow circuit for producing shear stress on endothelial cells (ECs). ECs on the upper side of coverslips were placed in the flow chamber, then subjected to shear stress (i.e., flow) for 48 h. (Reprinted from Kudo et al. 2000, with permission from Elsevier.)

4.2.2 Flow Chamber Studies on EC Function

The effects of physical factors on EC functions, particularly the effects of shear stress [force per unit area, measured in units of dyne/cm^2 (1 dyne/cm^2 = 0.1 Pa)], were examined. Shear stress induced by blood flow acts parallel to the EC surface and is a product of fluid viscosity and the velocity gradient between adjacent layers of flowing fluid. ECs were cultured on clean coverslips and, after reaching confluence, were placed in a parallel-plate flow chamber for shear stress loading (Figure 4.1). The flow circuit was filled with Dulbecco's Modified Eagle's Medium (DMEM) containing foetal bovine serum (FBS), and steady laminar flow was generated using a peristaltic pump. A depulsator was used to eliminate pulsations from flow. To maintain a pH of 7.4 in the medium, a dilution gas mixture of 5% CO_2 and 95% air was introduced into a reservoir and then mixed with the DMEM flow as needed. With the exception of the peristaltic pump, temperature in the system was maintained at 37°C using a water bath. With this flow circuit, ECs were subjected to a given magnitude of shear stress, as follows:

$$\tau = 6\,\mu\,Q\,/\,b\,h^2 \tag{4.1}$$

where τ is shear stress; μ is the viscosity of DMEM containing FBS at 37°C; Q is volume flow (cm^3/s); b is cross-sectional width of the flow path; and a is distance between cells on the upper side of coverslips and the upper plate of the flow chamber.

4.3 EFFECTS OF PHYSICAL AND CHEMICAL FACTORS ON ECS

4.3.1 EFFECT OF SHEAR STRESS ON EC MORPHOLOGY

Numerous studies have examined the effects of shear stress on EC morphology (Levesque and Nerem 1985; Dewey et al. 1981; Eskin et al. 1984). At the initiation of shear stress exposure (0 h), ECs were relatively round and randomly aligned. After long-term shear stress exposure, ECs were elongated and aligned to the direction of flow. The effect of 24 h exposure to 2 Pa (20 dyne/cm^2) of shear stress on EC morphology and F-actin filament (cytoskeleton) was examined (Kudo et al. 2003). At the end of this period, F-actin was stained using BODIPY phalloidin (Molecular Probes) to obtain fluorescent images. F-actin fluorescent images and phase-contrast images of ECs were obtained using an MRC-1000 confocal laser scanning microscope (CLSM; Bio-Rad) mounted on an inverted microscope. Figure 4.2 shows the effect of shear stress on ECs morphology and F-actin filaments in ECs. At the initiation of shear stress exposure (0 h), F-actin filaments were mainly localised peripherally. However, after 24 h of shear stress exposure, F-actin filaments were mainly located

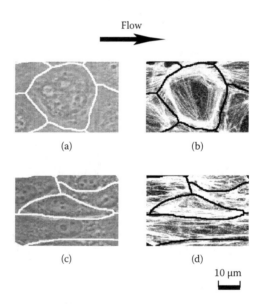

FIGURE 4.2 Fluorescent images of F-actin and phase-contrast images of BAECs exposed to shear stress. a) Phase-contrast image of BAECs at initiation of shear-stress exposure (0 h). b) Fluorescent image of F-actin at initiation of shear-stress exposure (0 h). c) Phase-contrast image of BAECs exposed to shear stress of 2 Pa for 24 h. d) Fluorescent image of F-actin exposed to shear stress of 2 Pa for 24 h. (Reprinted from Kudo et al. 2003, with permission from the Japan Society of Mechanical Engineers.)

in the central part of the elongated cells, and similar to the cells, filaments were oriented parallel to flow direction.

In vivo experiments (Yoshida et al. 1995; Sato and Ohshima 1994) have shown that ECs appear to align to flow direction and elongate due to flow-imposed shear stress, although some ECs have an almost round shape and seemed to be exposed to relatively low shear stress or complicated flow conditions. Stress fibres in ECs were oriented parallel to the direction of blood flow and were prominent in ECs from regions exposed to high-velocity flow (White et al. 1983; Wong et al. 1983). It has been suggested that stress fibre alignment with flow direction at regions exposed to high-velocity flow (high shear stress) may help ECs withstand haemodynamic stress (shear stress). Application of shear stress to ECs thus reorganises the cytoskeleton and morphology to adapt to the haemodynamic environment.

4.3.2 Effect of Shear Stress Magnitude on Macromolecule Uptake into ECs

Atherosclerotic lesions occur preferentially at low shear stress regions located at the walls of bifurcations and the inner curved surfaces of arteries (Caro et al. 1969). In these regions, macromolecules such as low-density lipoprotein, albumin, and horseradish peroxidase accumulate in the intima of the arterial wall (Caro et al. 1969; Bafakat et al. 1992; Packham et al. 1967; Schwanke and Carew 1989; Somer and Schwartz 1971; Yoshida et al. 1990).

ECs continuously exposed to shear stress that line the inner surfaces of blood vessels provide barriers to the transport of various substances to or from vessel walls, and thus regulate the exchange of molecules between the blood and vessel wall. This transport function has been implicated in the initiation and progression of atherosclerosis. The effect of the magnitude of shear stress on macromolecule (albumin) uptake into ECs was thus examined (Kudo et al. 1998). After shear stress loading for 48 h, coverslips were removed from the flow chamber, soaked in phosphate-buffered saline (PBS) including tetramethylrhodamine (a fluorescent label)-conjugated albumin (TRITC-albumin), then incubated for 60 min to take fluorescent images. Fluorescent and phase-contrast images of cells were obtained using a CLSM system as described above. Uptake for a no-stress (that is, no flow) sample from the same lot and the same generation of cells but without exposure to shear stress was measured as a control. Macromolecule uptake into ECs depended on imposed shear stress (Figure 4.3). At 10 dyne/cm^2, albumin uptake showed a 1.3-fold increase. Albumin uptake decreased with increasing shear stress, and the minimum uptake was one quarter of that under no-flow conditions at 60 dyn/cm^2.

4.3.3 Effect of Shear Stress on Albumin Permeability across an EC Monolayer

The effect of shear stress on macromolecule permeability across an EC monolayer was also analysed using a modified flow chamber (Figure 4.4) (Kudo et al. 2005). Concentrations of TRITC-albumin were maintained at 0.05 mg/ml in the DMEM

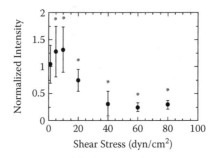

FIGURE 4.3 Effect of various shear stresses on albumin uptake into endothelial cells. Mean ±SD (*$p < 0.05$ vs. no flow). (Reprinted from Kudo et al. 1998, with permission from the Japan Society of Mechanical Engineers.)

solution circulating in the perfusion chamber circuit, where a 'blank' solution of PBS circulating through the circuit washed out the TRITC-albumin transported across the EC layer. The flow rate of blank solution in the diffusion chamber circuit was 15 ml/min and the disc in the diffusion chamber was installed 0.2 mm above the membrane. This disc protected the membrane from the direct impact of the circulating solution and had the effect of thinning the diffusion boundary layer. In this experimental system, length and size of the connecting tubes in the diffusion and perfusion circuits were regulated to keep the pressure in the diffusion chamber equal

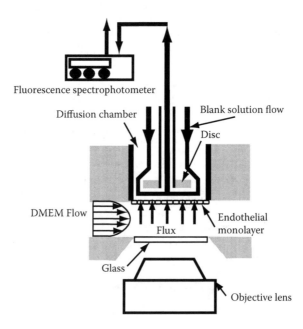

FIGURE 4.4 Scheme of the experimental system used to measure macromolecule permeability across PAECs (porcine aortic endothelial cells) subjected to flow. (Reprinted from Kudo et al. 2005, with permission from the Japan Society of Mechanical Engineers.)

to that in the perfusion chamber. A fluorescence spectrophotometer (M-450; Sequoia Turner) in the diffusion chamber circuit measured changes in fluorescence intensity over time, and this value was converted into the concentration of TRITC-albumin.

Total permeability to albumin (P_{tot}, cm/s) according to the method of Jo et al. (1991) was

$$P_{tot} = \Delta C_a \cdot V_a / (\Delta t \cdot A \cdot C_L) \qquad (4.2)$$

where C_a is the concentration of TRITC-albumin in the circuit of the diffusion chamber; V_a is the volume of the circuit of the diffusion chamber (12 cm^3); Δt is the sampling time (300 s); A is the area of the membrane (1 cm^2); and C_L is the concentration of TRITC-albumin in the circuit of the flow chamber (0.05 mg/cm^3).

The permeability resistance measured was the sum of the resistances of the EC layer, porous filter, and fluid boundary layer in the flow chamber. Transport resistance of the boundary layer was < 11.5% of the endothelial layer resistance under all experimental conditions (Jo et al. 1991), so permeability of ECs can be determined using the following formula:

$$1/P_{tot} = 1/P_e + 1/P_{filt} \qquad (4.3)$$

where P_{tot} is the permeability of the membrane and the EC monolayer; P_{filt} is the permeability of the porous membrane filter; and P_e is the permeability of the cell layer. P_{filt} was measured in a preliminary experiment, yielding a value of 9.5×10^{-5} cm/s as the P_{filt} of albumin.

Figure 4.5 shows mean ± standard error of the mean (SEM) values of albumin P_e as a function of shear intensity and time. At a shear stress of 1 Pa (10 dyn/cm^2), P_e increased gradually but did not show any significant change at any time up to 30 h.

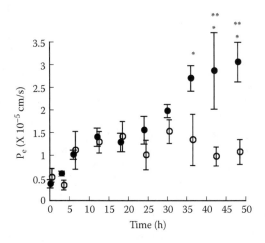

FIGURE 4.5   Effect of shear stress on endothelial permeability to albumin. Black dot (•) indicates shear stress of 1 Pa; white dot (○) indicates shear stress of 4 Pa. Values are mean ±SEM. *$p < 0.05$ compared with value at 0 h. **$p < 0.05$, 1 Pa vs. 4 Pa at the same time point. (Reprinted from Kudo et al. 2005, with permission from the Japan Society of Mechanical Engineers.)

After 36 h, a significant increase in P_e was observed relative to the value at 0 h. P_e was 2.7×10^{-5} cm s^{-1} ($p < 0.05$), 2.9×10^{-5} cm s^{-1} ($p < 0.05$) and 3.1×10^{-5} cm s^{-1} ($p < 0.05$) at 36, 42, and 48 h, respectively. Under a shear stress of 4 Pa (40 dyn/cm^2), albumin P_e did not show any significant change ($p > 0.05$) over the entire duration of the experiment. Both macromolecule uptake into ECs and permeability across ECs are higher at low-shear stress than at high-shear stress. The transport function of ECs is thus shear stress-dependent.

4.3.4 EFFECT OF SHEAR STRESS ON MITOCHONDRIAL MEMBRANE POTENTIAL

ECs that line the inner surface of blood vessels are continuously exposed to shear stress induced by blood flow in vivo, and shear stress affects many EC functions. However, the mechanisms underlying shear-dependent functions are not yet thoroughly understood. A key to understanding the functional changes under shear-stress loading is that most such functions are energy-dependent processes. ATP generation activity was therefore estimated in ECs affected by shear stress (Kudo et al. 2000). Since ATP synthesis activity is directly coupled with the membrane potential of mitochondria, this potential was visualised using a CLSM system and a fluorescent dye, 5,5',6,6'-tetrachloro-1,1',3,3'-tetraethyl-benzimidazolycarbocyanine iodide (JC-1; Molecular Probes).

After ECs were subjected to shear stress for 48 h, ECs on the coverslips were removed from the flow chamber and incubated in DMEM containing 10 mg/ml JC-1 and 20 mM KCl prewarmed for 15 min at 37°C. At low membrane potential, JC-1 exists in monomeric form and emits green fluorescence (emission peak, 527 nm) at 490-nm excitation. At high membrane potential, JC-1 exists as aggregates and emits red fluorescence (emission peak, 590 nm) at 490-nm excitation (Smiley et al. 1991; Reers et al. 1991). To subtract the membrane potential-independent fluorescence of JC-1, porcine aortic endothelial cells (PAECs) were treated with 1 μM p-trifluoromethoxy-phenylhydrozone (FCCP; Sigma) and 2 _M rotenone (Sigma) for 60 min and a fluorescent image was acquired. The excitation wavelength was 488 nm from an argon ion laser. Fluorescence through a 515-nm long-pass filter was divided by a 565-nm dichroic mirror. Green (through a 525- to 555-nm band-pass filter) and red (through a 600-nm long-pass filter) fluorescences were detected by two photomultipliers at each bandwidth.

As the low membrane potential area was subtracted from fluorescent images of mitochondria stained by JC-1, the images in Figure 4.6 represent high membrane potential mitochondria. From these images, mean mitochondrial membrane potential per cell was estimated. This was 30% higher for the low-stress sample than for the no-stress sample, whereas the mean for the high-stress sample was 20% lower. ATP synthesis function in ECs is thus considered shear stress-dependent.

4.3.5 EFFECT OF SHEAR STRESS ON EC NETWORKS

Angiogenesis involves the formation of new microvessels (capillaries) by EC migration and proliferation from a preexisting vessel. This process is essential for numerous physiological events (Gerwins et al. 2000). Angiogenesis is also beneficial for tissue recovery by reperfusion of ischemic tissue, but is maladaptive for arteriosclerosis, diabetes, and tumour growth (Carmeliet and Jain 2000). Shear stress also

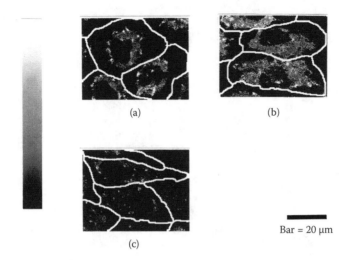

(a)

(b)

(c)

Bar = 20 μm

FIGURE 4.6 Effect of shear stress on mitochondrial membrane potential of endothelial cells. Bright region indicates high membrane potential mitochondria after: a) no-stress conditions (no flow); b) low-stress, 10 dyn/cm²; and c) high-stress, 60 dyn/cm². White lines represent cell contours. Bright regions show the ratio of red fluorescence to green fluorescence. This ratio is represented by a 256 grayscale (black, 0; white, 255). (Reprinted from Kudo et al. 2000, with permission from Elsevier.)

provides a significant stimulus for angiogenesis. In vivo studies by Ichioka et al. (1997) indicated that wound-healing angiogenesis is enhanced by adaptive responses of the microvasculature to shear stress, and Nasu et al. (1999) showed that increased blood flow causes tumour vascular enlargement.

The effect of shear stress on microvessel network formation of ECs was estimated using an in vitro three-dimensional (3D) model (Ueda et al. 2004). Collagen gels with three-dimensional (3D) networks were placed into a parallel-plate flow chamber made of polycarbonate (Figure 4.7), and ECs grown on these collagen gels were subjected to well-defined laminar fluid shear stress by the flow of DMEM. Flow of DMEM was provided by a sterile continuous-flow loop. Shear stress on ECs was calculated using Formula (4.1).

To investigate the effects of shear stress on network formation, cells were incubated with bFGF for 24 h and then subjected to laminar shear stress or left under static conditions for 48 h (Figure 4.8). Total length of the network was measured as an index of network formation. After about 10 h, applied shear stress started to enhance network formation compared with that under static conditions (Figure 4.9). Enhancement was significant after 24 h, with network length increased by a factor of 3.13 ± 0.46 under applied shear stress, compared with 2.17 ± 0.29 under static conditions ($p < 0.01$). Enhancement peaked after 48 h, increased by a factor of 6.17 ± 0.59 under applied shear stress, compared with 3.30 ± 0.41 for static conditions ($p < 0.01$), indicating that total length was 1.87 times longer under applied shear stress than under static conditions. These results clearly show that the 3D network formation is significantly enhanced by shear stress loading.

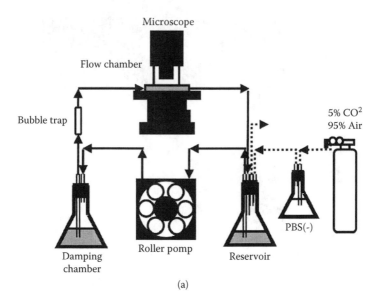

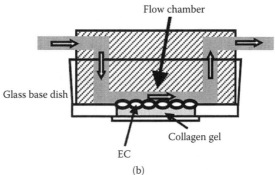

FIGURE 4.7   Apparatus to apply shear stress to endothelial cells (ECs). A) The flow chamber (dark shading) placed on the stage of a phase-contrast microscope was connected to a continuous flow circuit of culture medium (DMEM; arrows) and gas (5% CO_2–95% air; dashed line). B) ECs (ovals) on collagen gel (light shading) were subjected to shear stress by flow of DMEM (open arrows) for 48 h. Height and width of the flow chamber were 0.2 and 20 mm, respectively. (Reprinted from Ueda et al. 2004, with permission from The American Physiological Society.)

4.3.6   EFFECT OF SHEAR STRESS ON INTRACELLULAR CALCIUM IN ECs

In addition to physiological and biochemical processes described above, shear stress might also directly affect endothelial signal-transduction systems that initiate cellular responses. Intracellular calcium ($[Ca^{2+}]_i$) is an important second messenger that mediates other critical intracellular pathways after stimulation of ECs by various chemical and physical factors. The roles of second messengers, and particularly of $[Ca^{2+}]_i$, in shear stress-induced responses in ECs are not fully understood. The effect

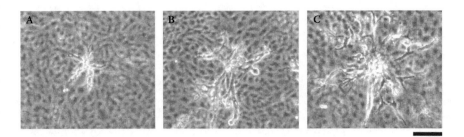

FIGURE 4.8 Typical networks under shear stress and static conditions. A) Network at the start of the experiment (0 h). B) and C) Network at the end of the experiment (48 h). B) Network under static condition. C) Network under shear-loaded condition. Direction of flow is left to right. The network under shear-loaded conditions expanded to wider than that under static condition. Bar scale, 100 μm. (Reprinted from Ueda et al. 2004, with permission from The American Physiological Society.)

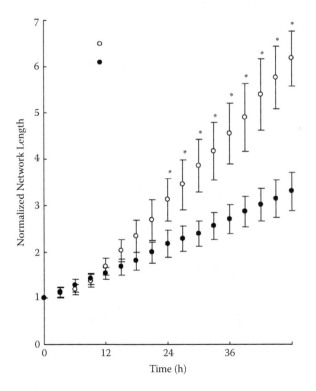

FIGURE 4.9 Effect of shear stress on network length. Total length of the network under shear stress and static conditions was calculated every 3 h. Total length was normalized to the initial total length (i.e., at 0 h). Data represent mean ±SD ($n = 6$). *$p < 0.01$ vs. static cultures. (Reprinted from Ueda et al. 2004, with permission from The American Physiological Society.)

of duration of exposure to frictional force (shear stress) on $[Ca^{2+}]_i$ response was therefore investigated (Kudo et al. 2003).

Cells were exposed to 2 Pa of shear stress for 0, 3, 6, 12, and 24 h using a flow chamber system. After exposure, cells were stained with Calcium Green-1/AM (Molecular Probes). Next, $[Ca^{2+}]_i$ of ECs under continued shear stress exposure (2 Pa) was measured across 5 min. To measure $[Ca^{2+}]_i$ in ECs, fluorescent and phase-contrast images of ECs were obtained using a CLSM system. The image sampling rate was one frame every 3 s. A total of 100 frames were acquired for each cell monolayer at each given time condition. All individual cells were manually outlined in phase-contrast images using Scion Image software (Scion). Average fluorescent intensity of the cell region was represented on a 256-gray scale. Fluorescent intensity of $[Ca^{2+}]_i$ in individual cells was then analysed as a normalised value calculated from the ratio of fluorescent intensity at the measured time to mean intensity over the total 5-min measurement time. An increase in fluorescent intensity of $[Ca^{2+}]_i$ was defined as a $[Ca^{2+}]_i$ response when fluorescent intensity was 30% larger than mean intensity.

Figure 4.10 shows representative single-cell $[Ca^{2+}]_i$ responses in cultured bovine aortic endothelial cell (BAEC) monolayers exposed to a shear stress of 2 Pa at the initiation of shear stress (0 h). Figure 4.10a shows fluorescent images of single-cell

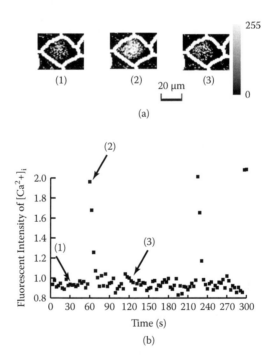

FIGURE 4.10 $[Ca^{2+}]_i$ responses of a single cell in a cultured BAEC monolayer at the initiation of shear stress (0 h). a) Fluorescent image of a single-cell, where $[Ca^{2+}]_i$ is visible as bright spots and cell boundaries as white lines. b) Fluorescent intensity of $[Ca^{2+}]_i$ in a single cell affected by shear stress for 5 min. (Reprinted from Kudo et al. 2003, with permission from the Japan Society of Mechanical Engineers.)

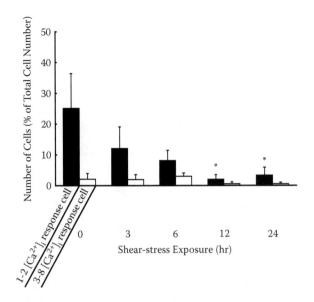

FIGURE 4.11 Histogram of single-cell $[Ca^{2+}]_i$ responses in cultured BAEC monolayers exposed to shear stress of 2 Pa for various exposure times. The histogram shows the relative percentage of cells exhibiting 1–2 or 3–8 $[Ca^{2+}]_i$ responses to the total number of cells during the 5 min after restarting shear stress. Black bar shows the number of cells that exhibited 1–2 responses, and white bar shows the number of cells that exhibited 3–8 responses. Values represent mean ±SD. *$p < 0.05$ versus 0 h. (Reprinted from Kudo et al. 2003, with permission from the Japan Society of Mechanical Engineers.)

$[Ca^{2+}]_i$ responses. In this case, $[Ca^{2+}]_i$ responses occurred at about 60 s, 240 s, and 300 s after initiation of shear stress (Figure 4.10a, Figure 4.10b). A total of three $[Ca^{2+}]_i$ responses were seen during the 5 min. Figure 4.11 shows a histogram and average frequencies computed to determine whether single-cell $[Ca^{2+}]_i$ responses are affected by duration of exposure to a shear stress of 2 Pa. The relative percentage of cells that exhibited $[Ca^{2+}]_i$ responses decreased with increasing shear-stress exposure time. At the initiation of shear-stress exposure (0 h), the total number of $[Ca^{2+}]_i$ responses was 27%, decreasing to about 4% after 24 h of shear-stress exposure.

4.3.7 EFFECT OF DISTURBED FLOW ON EC MORPHOLOGY AND MACROMOLECULE UPTAKE

Since ECs are exposed to disturbed flow at arterial curvatures and branches as well as steady flow, this might also affect their function. Focus was therefore placed on intracellular transport across ECs. Albumin uptake and morphology of ECs subjected to disturbed flow was therefore evaluated (Kudo et al. 1999), as macromolecule accumulation occurs at arterial bifurcations and curvatures where disturbed flow exists. First, ECs were subjected to 24 h in disturbed flow that models spatial variations in fluid shear stress found at arterial bifurcations, then incubated with albumin. Next, albumin uptake and morphological changes were evaluated from fluorescent and

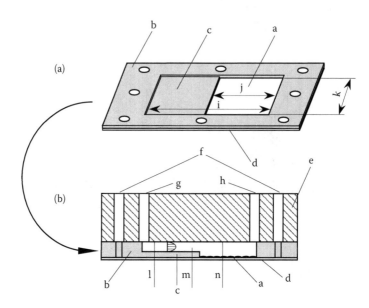

FIGURE 4.12 Apparatus used to expose endothelial cells to model disturbed flow. A) Glass coverslip with silicone gasket, and B) parallel-plate flow chamber for disturbed flow. a) endothelial monolayer; b) silicone gasket; c) step; d) coverslip; e) polycarbonate base plate; f) vacuum suction port; g) DMEM inlet port; h) DMEM outlet port; i) total channel length (40 mm); j) main channel length (20 mm); k) channel width (20 mm); l) entrance height (0.20 mm); m) step height (0.25 mm); n) main channel height (0.55 mm).

phase-contrast images, respectively, and compared for four different flow regions (shear stress and shear-stress gradient).

A flow chamber used to expose ECs to disturbed flow was fabricated as follows. Silicone liquid was poured into a mould to make a silicone gasket that had a step (Figure 4.12, see A). After solidification, the silicone gasket was adhered to a coverslip. ECs between passages 5 and 10 were seeded onto the coverslip (3.5 × 5.5 cm) with the silicone gasket to generate separated flow. Separated flow was created by combining two parallel flow channels of different channel heights (Figure 4.12, see B). The coverslip with the silicone gasket was fixed to a polycarbonate base plate using vacuum suction. In the test section, channel width (k) was 20 mm, entrance height (l) was 0.20 mm, and main channel height (n) was 0.55 mm. Total length of the channel (i) was 40 mm, and length of the main channel (j) was 20 mm.

Figure 4.13 shows phase-contrast images of ECs in each of the four flow regions. In the reversal flow and fully developed region where high shear stress and low shear stress gradients existed (Figure 4.13), ECs were aligned along the main flow direction and cell shape was elongated. In the stagnant flow and reattachment flow regions where low shear stress and high shear stress gradients existed (Figure 4.13), ECs were aligned randomly. Cell shape was round in the stagnant flow region and irregular in the reattachment flow region. Cell density was lowest in the reattachment flow region, among the four regions.

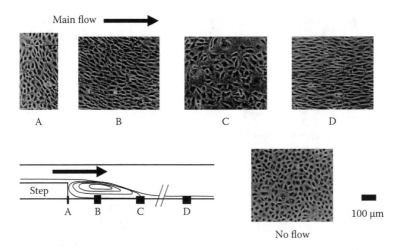

FIGURE 4.13 Phase-contrast images of ECs on noncoated coverslips in each region. A) Stagnant flow region; B) reversal flow region; C) reattachment flow region; and D) fully developed flow region. (Modified from Kudo et al. 1999, with permission from the Japan Society of Mechanical Engineers.)

Figure 4.14 shows fluorescent images of albumin uptake into ECs in each of the four flow regions. Albumin uptake into ECs in the reversal (Figure 4.14b) and fully developed flow regions (Figure 4.14d) where high shear stress and low shear stress gradients existed were smaller than in stagnant (Figure 4.14a) or reattachment (Figure 4.14c) flow regions where low shear stress and high shear stress gradients existed. In the recirculation eddy, the amount of albumin uptake in the

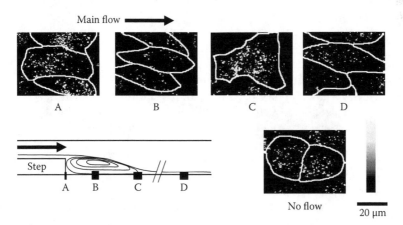

FIGURE 4.14 Fluorescent images of albumin uptake into ECs on noncoated coverslips in each region. A) Stagnant flow region; B) reversal flow region; C) reattachment flow region; and D) fully developed flow region. In these images, albumin is visible as bright spots, and cell boundaries as white lines. (Modified from Kudo et al. 1999, with permission from the Japan Society of Mechanical Engineers.)

stagnant (Figure 4.14a) and reattachment (Figure 4.14c) regions differed from that in the reversal flow region (Figure 4.14b). Disturbed flow affects EC function (Chiu et al. 1998; DePaola et al. 1992; DePaola et al. 1999; Tardy et al. 1997); these results also show that different flow patterns affect EC morphology and macromolecule uptake.

4.3.8 EFFECT OF CONCENTRATION GRADIENTS FOR GROWTH FACTORS ON EC NETWORK MORPHOLOGY

The concentration profile of extracellular growth factors is important in controlling 3D microvessel networks. Recent findings indicate that the morphology of such networks is influenced by extracellular growth factor distributions in vivo (Ruhrberg et al. 2002; Gerhardt et al. 2003). Focus was thus placed on the effects of concentration gradients of growth factors used in seeding ECs on network morphology in vitro (Ueda et al. 2006). First, ECs were seeded using two model environments: collagen gel containing bFGF and incubated without bFGF medium (gel-bFGF model), and collagen gel containing no bFGF and incubated with bFGF medium (medium-bFGF model). Networks were observed in 3D under confocal laser-scanning microscopy. Migration of ECs on collagen gel was analysed to study the effects of concentration gradients on the network formation process. ECs in the gel-bFGF model showed significantly greater migration distance and more sprouting points compared with those of the medium-bFGF model. Networks of the gel-bFGF model expanded mainly to a depth of 20–30 μm, and many reached a depth of 50–60 μm, whereas many networks in the medium-bFGF model expanded to a depth of only 10–20 μm. These results reveal that the initial growth factor distribution affects both EC migration of the network formation process and the number of sprouting points, and network morphology.

4.4 CONCLUSIONS

This article describes several EC functions that are affected by physical and chemical factors. In particular, various EC functions are changed by shear stress and depend on its magnitude, flow pattern, and exposure time. If blood flow changes, shear stress changes, and EC function will also change due to physical extracellular environments and may adapt to physical extracellular environments. Sessa et al. (1994) examined whether chronic exercise, which would presumably increase shear stress, could influence EC nitric oxide synthase (ECNOS) gene expression in vessels from dogs after chronic exercise. Nitric oxide is a mediator of vessel dilation. These authors identified a two- to threefold increase in ECNOS gene expression in chronically exercised dogs relative to controls. Chronic conditions change blood vessel function, and function is considered to adapt to the condition. Thus, because environments surrounding human beings differ widely, extracellular environments also differ. Physiological variation exists because cell functions are different. Linking EC function and adaptability to external environments shows how sensitive mechanisms of physiological and morphological adaptation can be.

REFERENCES

Bafakat, A.I., Uhthoff, P.A.F. and C.K. Colton. 1992. Topographical mapping of sites of enhanced HRP permeability in the normal rabbit aorta. *J. Biomech. Eng.* 114:283–92.

Carmeliet, P. and R.K. Jain. 2000. Angiogenesis in cancer and other diseases. *Nature* 407:249–57.

Caro, C.G., Fitz-Gerald, J.M. and R.C. Schroter. 1969. Arterial wall shear and distribution of early atheroma in man. *Nature* 223:1159–61.

Chiu, J.J., Wang, D.L., Chien, S., Skalak, R. and S. Usami. 1998. Effects of disturbed flow on endothelial cells. *J. Biomech. Eng.* 120:2–8.

DePaola, N., Gimbrone, M.A., Davies, P.F. Jr. and C.F. Dewey Jr. 1992. Vascular endothelium responds to fluid shear stress gradients. *Arterioscler. Throm.* 12:1254–7.

DePaola, N., Davies, P.F., Prichard, W.F., Florez, L. Jr., Harbeck, N., and D.C. Polacek. 1999. Spatial and temporal regulation of gap junction connexin43 in vascular endothelial cells exposed to controlled disturbed flows in vitro. *Proc. Natl. Acad. Sci. U.S.A.* 96:3154–9.

Dewey, C.F. Jr., Bussolari, S.R., Gimbrone. M.A. Jr. and P.F. Davies. 1981. The dynamic response of vascular endothelial cells to fluid shear stress. *J. Biomech. Eng.* 103:177–85.

Eskin, S.G. , Ives, C.L., McIntire, L.V. and L.T. Navarro. 1984. Response of cultured endothelial cells to steady flow. *Microvasc. Res.* 28:87–94.

Gerhardt, H., Golding, M., Fruttiger, M. et al. 2003. VEGF guides angiogenic sprouting utilizing endothelial tip cell filopodia. *J. Cell. Biol.* 23:1163–77.

Gerwins, P., Skoldenberg, E. and L. Claesson-Welsh. 2000. Function of fibroblast growth factors and vascular endothelial growth factors and their receptors in angiogenesis. *Crit. Rev. Oncol. Hematol.* 34:185–94.

Ichioka, S., Shibata, M., Kosaki, K., Sato, Y., Harii, K. and A. Kamiya. 1997. Effects of shear stress on wound-healing angiogenesis in the rabbit ear chamber. *J. Surg. Res.* 72:29–35.

Jo, H., Dull, R.O., Hollis, T.M. and J.M. Tarbell. 1991. Endothelial albumin permeability is shear dependent, time dependent, and reversible. *Am. J. Physiol.* 260:H1992–6.

Kamiya, A. and T. Togawa. 1980. Adaptive regulation of wall shear stress to flow change in the canine carotid artery. *Am. J. Physiol. Heart Circ. Physiol.* 239:H14–21.

Kudo, S., Ikezawa, K., Matsumura, S., Ikeda, M., Oka, K. and K. Tanishita. 1998. Effect of shear stress on macromolecule uptake into cultured endothelial cells. *Trans. Jpn. Soc. Mech. Eng. B* 64–618:367–74 (in Japanese).

Kudo, S., Sato, M., K. Machida et al. 1999. Macromolecule uptake into the cultured endothelial cells and the cell morphology in separated flow. *Trans. Jpn. Soc. Mech. Eng. B* 65–639:3705–12 (in Japanese).

Kudo, S., Morigaki, R., Saito, J., Ikeda, M., Oka, K. and K. Tanishita. 2000. Shear-stress effect on mitochondrial membrane potential and albumin uptake in cultured endothelial cells. *Biochem. Biophys. Res. Commun.* 270:616–21.

Kudo, S., Yamaguchi, R., Machida, K., Ikeda, M., Oka, K. and K. Tanishita. 2003. Effect of long-term shear stress exposure on calcium response and morphology of cultured endothelial cells, *JSME Int. C*, 46–2:1226–33.

Kudo, K., Tsuzaka, M., Ikeda, M. and K. Tanishita. 2005. Albumin permeability across endothelial monolayer's under long-term shear stress. *JSME Int. C* 48:419–24.

Levesque, M.J. and R.M. Nerem. 1985. The elongation and orientation of cultured endothelial cells in response to shear stress. *J. Biomech. Eng.* 107:341–7.

Nasu, R., Kimura, H., Akagi, K., Murata, T. and Y. Tanaka. 1999. Blood flow influences vascular growth during tumour angiogenesis. *Br. J. Cancer* 79:780–6.

Packham, M.A., Rowsell, H.C., Jorgensen, L. and J.F. Mustard. 1967. Localized protein accumulation in the wall of the aorta. *Exp. Mol. Pathol.* 7:214–32.

Reers, M., Smith, T.W. and L.B. Chen. 1991. J-aggregate formation of a carbocyanine as a quantitative fluorescent indicator of membrane potential. *Biochemistry* 30:4480–6.

Ruhrberg, C., Gerhardt, H., Golding, M. et al. 2002. Spatially restricted patterning cues provided by heparin-binding VEGF-A control blood vessel branching morphogenesis. *Genes Dev.* 16:2684–98.

Sato, M. and Ohshima, N. 1994. Flow-induced changes in shape and cytoskeletal structure of vascular endothelial cells. *Biorheology* 31:143–53.

Schwanke, D.C. and T.C. Carew. 1989. Initiation of atherosclerotic lesions in cholesterol-fed rabbits. *Arteriosclerosis* 9:895–907.

Sessa, W.C., Pritchard, K., Seyedi, N., Wang, J. and T.H. Hintze. 1994. Chronic exercise in dogs increases coronary vascular nitric oxide production and endothelial cell nitric oxide synthase gene expression. *Circ. Res.* 74:349–53.

Skalak, T.C. and R.J. Price. 1996. The role of mechanical stresses in microvascular remodeling. *Microcirculation* 3:143–65.

Smiley, S.T., Reers, M., Mottola-Hartshorn, C. et al. 1991. Intracellular heterogeneity in mitochondrial membrane potentials revealed by a J-aggregate-forming lipophilic cation JC-1. *Proc. Natl. Acad. Sci. U.S.A.* 88:3671–5.

Somer, J.B. and C.J. Schwartz. 1971. Focal ^{3}H-cholesterol uptake in the pig aorta. *Athelosclerosis* 13:293–304.

Tardy, Y., Rensnick, T.N., Gimbrone, M.A. Jr. and C.F. Dewey Jr. 1997. Shear stress gradients remodel endothelial monolayers in vitro via a cell proliferation-migration-loss cycle. *Arterioscler. Thromb. Vasc. Biol.* 17:3102–6.

Ueda, A., Koga, M., Ikeda, M., Kudo, S. and K. Tanishita. 2004. Effect of shear stress on micro-vessel network formation of endothelial cells with in vitoro three-dimensional model. *Am. J. Phisiol. Heart Circ. Physiol.* 287:H994–1002.

Ueda, A., Sudo, R., Ikeda, M., Kudo, S. and K. Tanishita. 2006. Initial bFGF distribution affects the depth of three-dimensional microvessel networks in vitro. *J. Biomech. Sci. Eng.* 1:136–146.

White, G.E., Gimbrone, M.A.Jr. and K. Fujiwara. 1983. Factors influencing the expression of stress fibers in vascular endothelial cells in situ. *J. Cell Biol.* 97:416–24.

Wong, A.J., Pollard, T.D. and I.M. Herman. 1983. Actin filament stress fibers in vascular endothelial cells in vivo. *Science* 219:867–9.

Yoshida, Y., Wang, S., Okano, M., Oyama, T., Yamane, T. and M. Mitsumata. 1990. The effect of augmented hemodynamic forces on the progression and topography of atherosclerotic plaque. *Ann. N.Y. Acad. Sci.* 598:256–73.

Yoshida, Y., Okano, M., S. Wang et al. 1995. Hemodynamic-force-induced difference of inter-endothelial junctional complexes. *Ann. N.Y. Acad. Sci.* 748:104–20.

5 Human Adaptation to Natural and Artificial Light
Variation in Circadian Photosensitivity

*Shigekazu Higuchi**
Kyushu University, Fukuoka, Japan

CONTENTS

5.1 INTRODUCTION

Humans adapt to the natural light–dark cycle according to the rotation of the Earth on its axis, and many physiological functions of humans have a circadian rhythm. However, natural light changes dynamically depending on latitude, season, and weather. Changes in natural light have a strong effect on mood and circadian rhythm in high-latitude regions in winter, when durations of sunshine are very short. In modern society, humans have succeeded in excluding night time from the life space and workplace by extensive use of artificial light. Despite the benefits of artificial light, recent studies have shown that artificial light at night has negative effects on human sleep, circadian rhythm, and health (Stevens et al. 2007).

In the field of human biology, many studies have shown human adaptation to ultraviolet light (Bell and Steegmann 2000; Frisancho 1993). However, little attention has been

---

* Address all corespondence to higu-s@design.kyushu-u.ac.jp.

paid to adaptation to visible natural light and artificial light from the viewpoint of human biological rhythm and sleep. The use of fluorescent lamps became popular after the 1940s, and only half a century has passed since humans spent significant parts of night time under bright light. It remains unclear how natural light and artificial light influence human physiology and health, and how humans adapt to artificial light. Biological and physioanthropological studies on human adaptation to light are therefore important.

There are individual differences and population differences in physiological responses to natural and artificial environments. In the field of physiological anthropology, these variations can be explained by five features, namely, environmental adaptability, functional potentiality, whole-body coordination, physiological polytypism, and technological adaptability (Iwanaga 2005; Katsuura 2005; Sato 2005; Yasukouchi 2005). In relation to individual and population variation, studies are needed 1) to elucidate physiological responses, 2) to clarify the factors that are responsible for them, 3) to determine their biological significance, and 4) to elucidate the physiological mechanisms underlying them. Here attention is paid to individual and population differences in suppression of melatonin induced by light exposure as a marker of circadian photosensitivity. Results of scientific studies on physiological adaptation to natural and artificial light are presented.

5.2 EFFECTS OF LIGHT ON MELATONIN AND CIRCADIAN RHYTHM

Many human physiological, endocrinal, and behavioural functions have a circadian rhythm. Human circadian rhythm is generated by an internal biological clock that exists in the suprachiasmatic nucleus (SCN). Melatonin secreted from the pineal gland during the night is controlled by the SCN and shows a clear circadian rhythm (Figure 5.1a). Melatonin is detected not only in serum but also in saliva and is a good marker of the phase of circadian rhythm because it is robust against any masking factor such as exercise, meals, and mental activity. However, melatonin secretion is easily suppressed by light exposure and is very sensitive to light (Figure 5.1b). There is a dose–response relationship between melatonin suppression and light intensity (Brainard et al. 1988; McIntyre et al. 1989), but there are large individual differences in the extent of melatonin suppression by light (Laakso et al. 1991; Lewy et al. 1985). Therefore, suppression of melatonin secretion is thought to be a good marker for evaluating the individual photosensitivity of the circadian system.

Light also has an impact on many physiological, endocrinal, and behavioural functions (Figure 5.2). Exposure to light induces suppression of melatonin secretion, increase in cortisol, phase shift of endogenous circadian rhythm, and enhancement of alertness and performance. These effects are thought to be different from visual effects (for example, image-forming effects) and they are called 'nonvisual effects' or 'non-image-forming effects'. The signal of light is transmitted to the SCN through the retinohypothalamic tract (RTH). The effect of light on the phase shift of circadian rhythm depends on the time of day. Light early at night delays the phase of circadian rhythm, whereas light late at night and early in the morning advances the phase of circadian rhythm. Light in daytime does not change the circadian phase. In

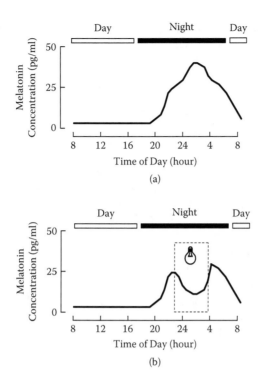

FIGURE 5.1 (a) Circadian profile of salivary melatonin concentration under a constant dim-light condition. (b) Melatonin is secreted during the night. Exposure to light at night acutely suppressed melatonin concentration.

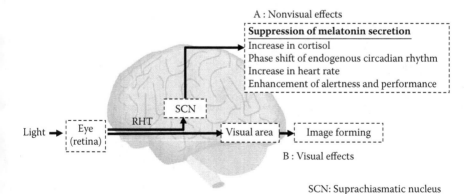

FIGURE 5.2 Visual and nonvisual effects of light. Light has an impact on many physiological, endocrinal, and behavioural functions through a nonvisual pathway.

daily life, exposure to ordinary artificial room light (approximately 100–200 lx) at night for a long time results in delay in circadian rhythm and delay in bedtime. On the other hand, advances in circadian rhythm by exposure to natural sunlight in the early morning adjust the internal clock to twenty-four hours.

5.3 ADAPTATION TO SHORT DURATION OF NATURAL SUNLIGHT

Natural sunlight changes dynamically depending on the weather, geographic conditions, and season. It is known that lack of exposure to sunlight is one of the risk factors of rickets and that white skin colour is a result of genetic adaptation to low exposure to ultraviolet rays. Other maladaptations to natural sunlight include seasonal affective disorder (SAD) and disturbance of circadian rhythm in winter (Rosenthal 1989). SAD is caused by short duration of natural sunlight in winter. It has been reported that there are ethnic or eye colour differences in prevalence of SAD (Goel et al. 2002; Suhail and Cochrane 1997). If short duration of sunshine acts as a stressor, there may be physiological adaptation to it. It is reasonable to assume that photosensitivity is increased in response to shortage of natural sunlight in winter and that there are seasonal and ethnic differences in suppression of melatonin by exposure to light as a maker of circadian photosensitivity (Figure 5.3).

5.3.1 SEASONAL DIFFERENCES IN PHOTOSENSITIVITY

SAD is caused by short duration of natural sunlight in winter and has symptoms of depression, excessive sleepiness in the morning, weight gain, and excessive consumption of carbohydrate substances (Rosenthal 1989). Although the prevalence of SAD is higher in high-latitude areas (Potkin et al. 1986), the prevalence depends not only on latitude but also on weather conditions. The prevalence is higher in areas of short duration of sunshine because of cloudy days in winter even in middle-latitude areas (Sakamoto et al. 1993).

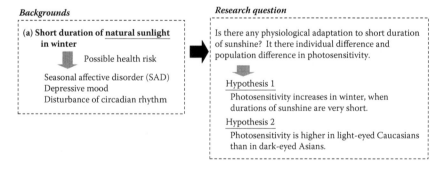

FIGURE 5.3 Physioanthropological study on adaptation to natural sunlight. Short duration of natural sunlight is a possible risk factor of mood- and circadian rhythm–related disorders.

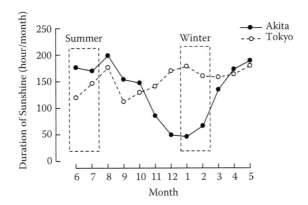

FIGURE 5.4 Annual variation in duration of sunshine in Akita (39° north). There is a large annual variation in Akita but not in Tokyo (35° north). The duration of sunshine in Akita in winter is approximately one third of that in summer. Data were obtained from databases of the Japan Meteorological Agency and National Astronomical Observatory of Japan.

It is possible that humans have physiological adaptability to short duration of natural sunlight in winter, if it acts as a stressor. An experiment, carried out to examine seasonal difference in the magnitude of suppression of melatonin secretion induced by exposure to light in the late evening, was carried out in Akita City (39° north, 140° east), in the northern part of Japan, where the duration of sunshine in winter is the shortest in Japan (Higuchi et al. 2007b). Ten healthy male university students volunteered to participate twice in the study, once in winter (from January to February) and once in summer (from June to July). According to Japanese meteorological data, the duration of sunshine in Akita in winter is approximately one third of that in summer (Figure 5.4). The subjects were exposed to light (1000 lx) at night for two hours using white fluorescent lamps. The starting time of exposure was set to two hours before the time of peak salivary melatonin concentration of each subject, which was determined in a preliminary experiment. Salivary melatonin concentration was measured before exposure to light and during exposure to light. The percentage of suppression of melatonin by light was calculated on the basis of melatonin concentration determined before the start of exposure to light. Salivary melatonin concentrations significantly increased before exposure to light and then decreased after exposure to light in both seasons. The interindividual difference in melatonin suppression was smaller in winter than in summer (Figure 5.5). The average percentage of suppression of melatonin two hours after the start of exposure to light was significantly larger in winter (66.6% ± 18.4%) than in summer (37.2% ± 33.2%). The level of daily ambient light of each subject was measured for twenty-four hours using a light sensor (Actiwatch-L, Mini-Mitter Co, Inc., USA) attached to the nondominant wrist of each subject. The level of daily ambient light to which the subjects were exposed from rising time to bedtime in summer was approximately two times greater than that in winter. Subjects who were exposed to a large amount of ambient light in summer showed a low percentage of melatonin suppression.

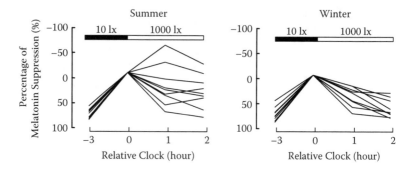

FIGURE 5.5 Interindividual variation in melatonin suppression by light in summer and winter. Interindividual variation was smaller in winter than in summer. The mean percentage of melatonin suppression was significantly larger in winter than in summer.

Studies have shown that shorter exposure to light in the recent past increases sensitivity of melatonin to light suppression even in the same season (Hebert et al. 2002; Smith et al. 2004). These results suggest that shorter exposure to daily ambient light in winter increases sensitivity of melatonin to light suppression. Increased light sensitivity in winter can be thought of as a human physiological adaptation, although its mechanism has not been elucidated. It has been reported that suppression of melatonin in SAD patients is higher than that in healthy controls (Thompson et al. 1990; McIntyre et al. 1990). Does increase in sensitivity to light in winter therefore help to prevent SAD? High sensitivity to light in SAD patients is thought to be caused by less exposure to natural sunlight. Further study is needed to clarify the causal relationships between individual differences in sensitivity to light and prevalence risk of SAD.

5.3.2 ETHNIC DIFFERENCES IN PHOTOSENSITIVITY

Variations in physiological and morphological traits of humans have been induced by human adaptation to given environments. It is known that white skin colour in Caucasians is the result of adaptation to short duration of sunshine (Bell and Steegmann 2000; Frisancho 1993). White skin facilitates absorption of ultraviolet rays, promotes vitamin D synthesis, and prevents rickets in high-latitude places with a short duration of sunshine in winter. However, it is not clear why Caucasians acquired blue eyes and whether there are some physiological advantages of having them.

There is ethnic or eye colour difference in the prevalence of SAD, with darker-eyed patients with SAD being significantly more depressed and fatigued than blue-eyed ones (Goel et al. 2002). Furthermore, the incidence of SAD was shown to be higher in an Asian female group than in a white English female group in England (Suhail and Cochrane 1997). These studies suggest that light-coloured eyes may have advantages in preventing SAD where there is a short duration of sunshine.

It is hypothesised that the magnitude of melatonin suppression is larger in Caucasians with blue and light-brown eyes than in Asians with dark-brown eyes

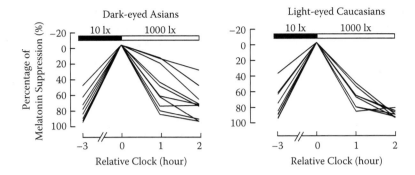

FIGURE 5.6 Interindividual variations in melatonin suppression by light in dark-eyed Asians and in light-eyed Caucasians. Interindividual variation was smaller in light-eyed Caucasians than in dark-eyed Asians. The mean percentage of melatonin suppression was significantly larger in light-eyed Caucasians than in dark-eyed Asians. (Reprinted from Higuchi et al. 2007a, with permission from the American Physiological Society.)

(Higuchi et al. 2007a). To this end a study was conducted in Japan in winter. The subjects were ten Caucasian males with blue/green irises or light-brown irises (light-eyed Caucasians) and eleven Asian males with dark-brown irises (dark-eyed Asians). The mean ages of the light-eyed Caucasians and dark-eyed Asians were 26.4 ± 3.2 and 25.3 ± 5.7 years, respectively. All subjects had been living in the same city (Akita City) for at least six months. Salivary melatonin concentrations significantly increased before exposure to light (1000 lx) and then decreased after exposure to light in both groups (Figure 5.6). There were large interindividual differences in the percentages of melatonin suppression in dark-eyed Asians in comparison with values for light-eyed Caucasians. The percentage of suppression of melatonin secretion two hours after the start of bright light exposure was significantly greater in light-eyed Caucasians (88.9 ± 4.2%) than in dark-eyed Asians (73.4 ± 20.0%) ($p < 0.01$).

What is the possible cause of the eye-colour-dependent difference found in this study? A previous study has shown that intraocular light scattering, which is called stray light, is dependent on pigmentation of the eye (IJspeert et al. 1990). It has been suggested that eye-colour-dependent difference in stray light is caused not only by light transmittance through the eye wall but also reflection from the fundus oculi (van den Berg et al. 1991). The role of the retinal pigment epithelium in the fundus oculi is to absorb light and to prevent light scattering. Therefore, the amount of dispersion of light in the fundus oculi is thought to be larger in light-eyed Caucasians with less pigmentation than in dark-eyed people. Recent studies have shown the existence of photosensitive retinal ganglion cells as circadian photoreceptors, in addition to rods and cones (Berson 2003; Hattar et al. 2002). In the present study, these photoreceptors might have received more light by dispersion of light in light-eyed Caucasians. Thus, greater suppression of melatonin secretion is thought to be induced in light-eyed Caucasians.

In the study just described, the percentages of melatonin suppression in some dark-eyed Asians were as large as those in light-eyed Caucasians, indicating that some ethnic factors other than eye colour may have contributed to the difference between melatonin suppression in dark-eyed Asians and that in light-eyed Caucasians. Further study is needed to examine the effects of eye colour and ethnic group separately. Although it is not clear whether Caucasians acquired blue eyes as a result of adaptation to a short duration of sunshine, blue eyes might have been advantageous for prevention of seasonal affective disorders in ancient times.

5.4   ADAPTATION TO ARTIFICIAL LIGHT

Humans in modern society benefit from artificial light in various ways. Artificial light enables twenty-four-hour working society. In the workplace, bright light increases alertness of night workers and rapidly adapts circadian rhythm for working at night (Campbell and Dawson 1990; Eastman and Martin 1999). However, a negative aspect of exposure to artificial light at night is suppression of melatonin by light, which has been suggested to increase cancer risk in night workers (Schernhammer et al. 2006; Stevens et al. 2007).

In 1980, it was first demonstrated that melatonin was suppressed during exposure to bright light at 2500 lx in humans (Lewy et al. 1980). More recent studies have shown that nocturnal melatonin secretion can be suppressed by exposure to a light of several hundred luces (Aoki et al. 1998; Zeitzer et al. 2000). It has been reported that there is a dose–response relationship between light intensity and human alertness during the early biological night, exposure to a light of ~100 lx increasing human alertness (Cajochen et al. 2000). This corresponds to the light intensity in the home. In Japan, sleep time decreases every year because of delayed bedtime, and about 20% of adults have a sleep problem (Kim et al. 2000). Furthermore, according to a Japanese white paper on information technology (Japanese Ministry of Posts and Telecommunications 1998), 53.7% of Internet users in Japan had delayed bedtimes and 45.4% of them had shortened sleeping hours. Recent research has shown that using the Internet, watching TV, and playing computer games delays bedtime and reduces sleep time (Alexandru et al. 2006; Paavonen et al. 2006; Van den Bulck 2004).

Artificial light at night and the use of electric media at night may negatively influence human sleep and circadian rhythm (Figure 5.7). A study examined the effects of playing a computer game with a bright display during night time on presleep physiology and sleep. It is known that there is a large interindividual variation in melatonin suppression by light, but the relationship between individual differences in photosensitivity to light and habitual bedtime is unclear. If there is a relationship between them, high sensitivity to light may be a risk factor for delayed sleep–wake cycle.

5.4.1   EFFECTS OF USING A COMPUTER AT NIGHT ON HUMAN PHYSIOLOGY AND SLEEP

What aspects of computer usage are physiologically related to sleep and circadian physiology? It is thought that gazing at a bright display while using a computer is one of the factors that affect the human biological clock. Another factor affecting circadian physiology and sleep is thought to be increased mental activity caused by

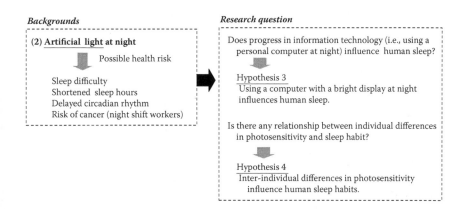

Backgrounds

(2) **Artificial light at night**

⬇ Possible health risk

Sleep difficulty
Shortened sleep hours
Delayed circadian rhythm
Risk of cancer (night shift workers)

Research question

Does progress in information technology (i.e., using a personal computer at night) influence human sleep?

⬇

Hypothesis 3
Using a computer with a bright display at night influences human sleep.

Is there any relationship between individual differences in photosensitivity and sleep habit?

⬇

Hypothesis 4
Inter-individual differences in photosensitivity influence human sleep habits.

FIGURE 5.7 Physioanthropological study on adaptation to natural sunlight. Exposure to bright light at night is a possible risk factor of disturbance of circadian rhythm and sleep.

using a computer. The effects of a bright display and increased mental activity were tested by examining the effects of playing a computer game on presleep physiology and sleep.

Seven young adult males played exciting computer games with a bright display (game-BD) and a dark display (game-DD) and performed simple tasks with low mental load as a control condition in front of a BD (control-BD) and DD (control-DD) between 23:00 p.m. to 1:45 a.m. in randomised order and then went to bed at 2:00 and slept until 8:00 (Figure 5.8). The 17-inch colour-display was placed at eye level 45 cm in front of the subject. The screen of the BD was white with 120 cd/m^2, and

FIGURE 5.8 A subject playing a computer game with a bright display in the experimental room. Physiological indices (EEG, heart rate, rectal temperature, and saliva melatonin) were measured.

the screen of the DD was black with 0.5 cd/m^2. The vertical light intensities of the BD and DD were 45 lx and 15 lx at each subject's eye level, respectively. Rectal temperature, electroencephalogram (EEG), heart rate, salivary melatonin concentration, and subjective sleepiness were recorded while the subject played the computer game and a polysomnogram was recorded during sleep.

Playing the computer game with both BD and DD significantly suppressed the nocturnal decreases in rectal temperature and heart rate and the nocturnal increase in sleepiness. The BD group had significantly lower nocturnal decreases in rectal temperature and heart rate, while nocturnal salivary melatonin concentration was significantly suppressed by the combination of the computer game and BD. Sleep latency was also significantly longer and REM sleep significantly shorter after playing games than after the control conditions. No significant effects of either computer games or BD were found on slow-wave sleep (SWS).

These results suggest that playing a computer game and using a bright display disturbs nocturnal changes in physiological variables and sleep. It has been reported that not only using a computer but also using electric media before bedtime might have negative effects on sleep and sleep habits (Punamaki et al. 2007; Sunagawa et al. 2007; Van den Bulck 2007). In the next decade, information technology will progress further. Both epidemiological and experimental studies are needed to determine the effects of using electric media (such as watching TV, using the Internet and computer, using a mobile phone) on human sleep and health.

5.4.2 Light Sensitivity and Habitual Bedtime

It has been reported that the magnitude of melatonin suppression by light in patients with delayed sleep phase syndrome (DSPS) is greater than in healthy controls (Aoki et al. 2001). It is thought that hypersensitivity to light is related to delayed bedtime because exposure to light at night has the potential to delay the circadian rhythm. However, there have been few studies in which the relationship between individual differences in melatonin sensitivity and habitual bedtime in normal healthy people has been examined. A study investigated individual variation in melatonin suppression in healthy subjects and examined the relationship between melatonin suppression by light and habitual bedtime in healthy subjects.

Twenty-one healthy male university students volunteered to participate in the study. The average habitual bedtime and average rising time were 01:22 a.m. ± 1.0 hours and 07:48 a.m. ± 0.7 hours, respectively. The subjects were exposed to light (1000 lx) at night for two hours using white fluorescent lamps placed on the ceiling. During the period of exposure to the light, melatonin suppression occurred in fifteen subjects, but not in two subjects. The relationship between percentage of melatonin suppression and habitual bedtime is shown in Figure 5.9. No significant correlation was found between rate of melatonin suppression and habitual bedtime in the fifteen subjects in whom melatonin suppression occurred ($r = 0.07$) (Figure 5.9, solid circles). However, the bedtime of the two subjects in whom melatonin suppression did not occur (Figure 9, open circles) was earlier than that of the other subjects (Figure 5.9, solid circles).

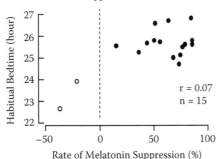

FIGURE 5.9 No significant correlation was found between rate of melatonin suppression and habitual bedtime in fifteen subjects with suppression of melatonin ($n = 15$) (solid circles). The two subjects (open circles) in whom melatonin was not suppressed had earlier habitual bedtimes than those of the subjects with suppression of melatonin (solid circles). (Reprinted from Higuchi et al. 2005, with permission from the International Academic Printing Co., Ltd.)

Contrary to expectation, there was no significant relationship between individual difference in rates of melatonin suppression and habitual bedtime. This result suggests that individual differences in percentage of melatonin suppression by light do not affect habitual bedtime among people in whom melatonin secretion is suppressed by light. There were two subjects in whom melatonin suppression did not occur. This suggests that there are some people who have very low melatonin sensitivity to suppression by light. Furthermore, the two subjects in whom melatonin suppression did not occur had earlier habitual bedtimes than those of the subjects with suppression of melatonin.

The fact that there were two subjects in whom melatonin was not suppressed by light may be an example of physiological polytypism. However, it is not clear what mechanisms (genetic factors, environmental factors, and cultural factors) are related to the low sensitivity of melatonin suppression by light. Further examination of more subjects in whom melatonin is not suppressed by light is needed to determine the causal relationships between nonsuppression of melatonin and early bedtime and its mechanisms.

5.4.3 Possible Cultural Adaptation to Risk of Artificial Light

Since artificial bright light at night has a negative impact on human sleep and health, what should be done to avoid unfavourable effects of light? There is a need to accumulate scientific evidence concerning the influence of artificial light on human sleep and health in terms of individual and population differences. Reliable scientific knowledge will enable cultural adaptation, including behavioural and instrumental adaptation, to avoid unfavourable effects of artificial light.

There have been noteworthy findings concerning human spectral sensitivity to nonvisual effects of light, such as suppression of melatonin secretion, phase shift of endogenous circadian rhythm, and enhancement of alertness. In 2001, an action spectrum for melatonin suppression by light in humans was reported by two groups (Brainard et al. 2001; Thapan et al. 2001). According to these studies, short-wavelength blue light (approximately 480 nm) has a greater impact on melatonin suppression, while the classical visual system is most sensitive to green light (555 nm). It was later shown that exposure to blue light also causes a phase shift of circadian rhythm and enhancement of alertness (Cajochen et al. 2005; Lockley et al. 2003; Lockley et al. 2006; Revell et al. 2005). This spectral sensitivity to nonvisual effects of light has been found using not only monochromatic light but also polychromatic light such as that emitted by fluorescent lamps. Exposure to high colour temperature florescent lamp (5000 K to 7000 K) has a greater impact on human physiology than does exposure to a low colour temperature lamp (approximately 3000 K) (Deguchi and Sato 1992; Morita and Tokura 1996; Mukae and Sato 1992; Noguchi and Sakaguchi 1999).

Exposure to not only bright light but also high colour temperature light should be avoided at night. Low colour temperature light can reduce suppression of melatonin secretion and prevent the delay of circadian rhythm and initiation of the natural sleep process. Experimental studies have been carried out to determine the effectiveness of the use of low colour temperature light or the use of goggles that block blue light for reducing the risk of cancer in shift workers (Kayumov et al. 2007). Exposure to sunlight during the daytime is another way to avoid unfavourable effects of nocturnal light. Some studies have shown that light history affects photosensitivity at night. Much exposure to sunlight during the daytime reduces suppression of melatonin by light at night (Hebert et al. 2002; Higuchi et al. 2007b; Rufiange et al. 2007).

On the other hand, artificial bright light is useful for avoiding unfavourable effects of short duration of natural sunlight. For example, light therapy is used for patients with SAD (Rosenthal et al. 1984; Wirz-Justice et al. 1993) and is effective for treating nonseasonal depression (Kripke 1998; Terman and Terman 2005) and sleep and circadian rhythm disorder (Lack and Wright 1993; Mishima et al. 1994; Rosenthal et al. 1990). Light therapy using white fluorescent lamps utilises bright light of more than 2500 lx, but greater effectiveness of blue light has been reported (Glickman et al. 2006).

5.5 CONCLUSIONS

There are both positive and negative effects of exposure of humans to artificial light. This chapter has argued that as an adaptation to natural sunlight, seasonal and ethnic variations in suppression of melatonin by exposure to light occur as an index of circadian photosensitivity. Even artificial light at night through a computer display impacts on human physiology and sleep. In addition, there is large interindividual difference in the physiological responses to light. These variations should be considered from the perspective of human adaptation to natural and artificial light in the modern society.

REFERENCES

Alexandru, G., Michikazu, S., Shimako, H. et al. 2006. Epidemiological aspects of self-reported sleep onset latency in Japanese junior high school children. *J. Sleep Res.* 15:266–75.

Aoki, H., Yamada, N., Ozeki, Y., Yamane, H. and N. Kato. 1998. Minimum light intensity required to suppress nocturnal melatonin concentration in human saliva. *Neurosci. Lett.* 252:91–4.

Aoki, H., Ozeki, Y. and N. Yamada. 2001. Hypersensitivity of melatonin suppression in response to light in patients with delayed sleep phase syndrome. *Chronobiol. Int.* 18:263–71.

Bell, C. and A. Steegmann. 2000. Human adaptation to climate: Temperature, ultraviolet radiation, and altitude. In *Human Biology: An Evolutionary and Biocultural Perspective*, ed. S. Stinson, B. Bogin, R. Huss-Ashmore and D. O'Rourke, 163–224. New York, NY: Wiley-Liss.

Berson, D.M. 2003. Strange vision: Ganglion cells as circadian photoreceptors. *Trends. Neurosci.* 26:314–20.

Brainard, G.C., Lewy, A.J., Menaker, M. et al. 1988. Dose-response relationship between light irradiance and the suppression of plasma melatonin in human volunteers. *Brain Res.* 454:212–8.

Brainard, G.C., Hanifin, J.P., Greeson, J.M. et al. 2001. Action spectrum for melatonin regulation in humans: Evidence for a novel circadian photoreceptor. *J. Neurosci.* 21:6405–12.

Cajochen, C., Brunner, D.P., Krauchi, K., Graw, P. and A. Wirz-Justice. 2000. EEG and subjective sleepiness during extended wakefulness in seasonal affective disorder: Circadian and homeostatic influences. *Biol. Psychiatry* 47:610–7.

Cajochen, C., Munch, M., Kobialka, S. et al. 2005. High sensitivity of human melatonin, alertness, thermoregulation, and heart rate to short wavelength light. *J. Clin. Endocrinol. Metab.* 90:1311–6.

Campbell, S.S. and D. Dawson. 1990. Enhancement of nighttime alertness and performance with bright ambient light. *Physiol. Behav.* 48:317–20.

Deguchi, T. and M. Sato. 1992. The effect of color temperature of lighting sources on mental activity level. *Ann. Physiol. Anthropol.* 11:37–43.

Eastman, C.I. and S.K. Martin. 1999. How to use light and dark to produce circadian adaptation to night shift work. *Ann. Med.* 31:87–98.

Frisancho, A. 1993. Skin color and adaptation to solar radiation. In *Human Adaptation and Accommodation*, 147–74. MI: University of Michigan Press.

Glickman, G., Byrne, B., Pineda, C., Hauck, W.W. and G.C. Brainard. 2006. Light therapy for seasonal affective disorder with blue narrow-band light-emitting diodes (LEDs). *Biol. Psychiatry* 59:502–7.

Goel, N., Terman, M. and J.S. Terman. 2002. Depressive symptomatology differentiates subgroups of patients with seasonal affective disorder. *Depress. Anxiety* 15:34–41.

Hattar, S., Liao, H.W., Takao, M., Berson, D.M. and K.W. Yau. 2002. Melanopsin-containing retinal ganglion cells: Architecture, projections, and intrinsic photosensitivity. *Science* 295:1065–70.

Hebert, M., Martin, S.K., Lee, C. and C.I. Eastman. 2002. The effects of prior light history on the suppression of melatonin by light in humans. *J. Pineal. Res.* 33:198–203.

Higuchi, S., Motohashi, Y., Maeda, T. and K. Ishibashi. 2005. Relationship between individual difference in melatonin suppression by light and habitual bedtime. *J. Physiol. Anthropol. Appl. Human Sci.* 24(4): 419–423.

Higuchi, S., Motohashi, Y., Ishibashi, K. and T. Maeda. 2007a. Influence of eye colors of Caucasians and Asians on suppression of melatonin secretion by light. *Am. J. Physiol. Regul. Integr. Comp. Physiol.* 292:R2352–6.

Higuchi, S., Motohashi, Y., Ishibashi, K. and T. Maeda. 2007b. Less exposure to daily ambient light in winter increases sensitivity of melatonin to light suppression. *Chronobiol. Int.* 24:31–43.

IJspeert, J.K., de Waard, P.W., van den Berg, T.J. and P.T. de Jong. 1990. The intraocular straylight function in 129 healthy volunteers; dependence on angle, age and pigmentation. *Vision Res.* 30:699–707.

Iwanaga, K. 2005. The biological aspects of physiological anthropology with reference to its five keywords. *J. Physiol. Anthropol. Appl. Hum. Sci* 24:231–5.

Katsuura, T. 2005. Looking at physiological anthropology from a historical standpoint. *J. Physiol. Anthropol. Appl. Hum. Sci.* 24:227–9.

Kayumov, L., Lowe, A., Rahman, S.A., Casper, R.F. and C.M. Shapiro. 2007. Prevention of melatonin suppression by nocturnal lighting: Relevance to cancer. *Eur. J. Cancer Prev.* 16:357–62.

Kim, K., Uchiyama, M., Okawa, M., Liu, X. and R. Ogihara. 2000. An epidemiological study of insomnia among the Japanese general population. *Sleep* 23:41–7.

Kripke, D.F. 1998. Light treatment for nonseasonal depression: Speed, efficacy, and combined treatment. *J. Affect. Disord.* 49:109–17.

Laakso, M.L., Porkka-Heiskanen, T., Stenberg, D. and A. Alila. 1991. Interindividual differences in the responses of serum and salivary melatonin to light. In *Role of Melatonin and Pineal Peptides in Neuroimmunomodulation*, ed. F. Fraschini and R. J. Reiter, 307–11. New York: Plenum Press.

Lack, L. and H. Wright. 1993. The effect of evening bright light in delaying the circadian rhythms and lengthening the sleep of early morning awakening insomniacs. *Sleep* 16:436–43.

Lewy, A.J., Wehr, T.A., Goodwin, F.K., Newsome, D.A. and S.P. Markey. 1980. Light suppresses melatonin secretion in humans. *Science* 210:1267–9.

Lewy, A.J., Nurnberger, J.I. Jr., Wehr, T.A. et al. 1985. Supersensitivity to light: Possible trait marker for manic-depressive illness. *Am. J. Psychiatry* 142:725–7.

Lockley, S.W., Brainard, G.C. and C.A. Czeisler. 2003. High sensitivity of the human circadian melatonin rhythm to resetting by short wavelength light. *J. Clin. Endocrinol. Metab.* 88:4502–5.

Lockley, S.W., Evans, E.E., Scheer, F.A., Brainard, G.C., Czeisler, C.A. and D. Aeschbach. 2006. Short-wavelength sensitivity for the direct effects of light on alertness, vigilance, and the waking electroencephalogram in humans. *Sleep* 29:161–8.

McIntyre, I.M., Norman, T.R., Burrows, G.D. and S.M. Armstrong. 1989. Human melatonin suppression by light is intensity dependent. *J. Pineal. Res.* 6:149–56.

McIntyre, I.M., Norman, T.R., Burrows, G.D. and S.M. Armstrong. 1990. Melatonin supersensitivity to dim light in seasonal affective disorder. *Lancet* 335:488.

Ministry of Posts and Telecommunications. 1988. (in Japanese).

Mishima, K., Okawa, M., Hishikawa, Y., Hozumi, S., Hori, H. and K. Takahashi. 1994. Morning bright light therapy for sleep and behavior disorders in elderly patients with dementia. *Acta Psychiatr. Scand.* 89:1–7.

Morita, T. and H. Tokura. 1996. Effects of lights of different color temperature on the nocturnal changes in core temperature and melatonin in humans. *Appl. Hum. Sci.* 15:243–6.

Mukae, H. and M. Sato. 1992 The effect of color temperature of lighting sources on the autonomic nervous functions, *Ann. Physiol. Anthropol.* 11:533–8.

Noguchi, H. and T. Sakaguchi. 1999. Effect of illuminance and color temperature on lowering of physiological activity. *Appl. Hum. Sci.* 18:117–23.

Paavonen, E.J., Pennonen, M., Roine, M., Valkonen, S. and A.R. Lahikainen. 2006. TV exposure associated with sleep disturbances in 5- to 6-year-old children. *J. Sleep Res.* 15:154–61.

Potkin, S.G., Zetin, M., Stamenkovic, V., Kripke, D. and W.E. Bunney Jr. 1986. Seasonal affective disorder: Prevalence varies with latitude and climate. *Clin. Neuropharmacol.* 9(Suppl 4):181–3.

Punamaki, R.L., Wallenius, M., Nygard, C.H., Saarni, L. and A. Rimpela. 2007. Use of information and communication technology (ICT) and perceived health in adolescence: The role of sleeping habits and waking-time tiredness. *J. Adolesc.* 30:569–85.

Revell, V.L., Arendt, J., Terman, M. and D.J. Skene. 2005. Short-wavelength sensitivity of the human circadian system to phase-advancing light. *J. Biol. Rhythms.* 20:270–2.

Rosenthal, N.E. 1989. *Seasons of the Mind.* 3rd edition. New York: Bantam Dell Pub Group.

Rosenthal, N.E., Sack, D.A., Gillin, J.C. et al. 1984. Seasonal affective disorder: A description of the syndrome and preliminary findings with light therapy. *Arch. Gen. Psychiatry* 41:72–80.

Rosenthal, N.E., Joseph-Vanderpool, J.R., Levendosky, A.A. et al. 1990. Phase-shifting effects of bright morning light as treatment for delayed sleep phase syndrome. *Sleep* 13:354–61.

Rufiange, M., Beaulieu, C., Lachapelle, P. and M. Dumont. 2007. Circadian light sensitivity and rate of retinal dark adaptation in indoor and outdoor workers. *J. Biol. Rhythms* 22:454–7.

Sakamoto, K., Kamo, T., Nakadaira, S., Tamura, A. and K. Takahashi. 1993. A nationwide survey of seasonal affective disorder at 53 outpatient university clinics in Japan. *Acta Psychiatr. Scand.* 87:258–65.

Sato, M. 2005. The development of conceptual framework in physiological anthropology. *J. Physiol. Anthropol. Appl. Hum. Sci.* 24:289–95.

Schernhammer, E.S., Kroenke, C.H., Laden, F. and S.E. Hankinson. 2006. Night work and risk of breast cancer. *Epidemiology* 17:108–11.

Smith, K.A., Schoen, M.W. and C.A. Czeisler. 2004. Adaptation of human pineal melatonin suppression by recent photic history. *J. Clin. Endocrinol. Metab.* 89:3610–4.

Stevens, R.G., Blask, D.E., Brainard, G.C. et al. 2007. Meeting report: The role of environmental lighting and circadian disruption in cancer and other diseases. *Environ. Health Perspect.* 115:1357–62.

Suhail, K. and R. Cochrane. 1997. Seasonal changes in affective state in samples of Asian and white women. *Soc. Psychiatry Psychiatr. Epidemiol.* 32:149–57.

Sunagawa, M., Kikuchi, T., Yanagi, K. et al. 2007. Using electronic media before sleep can curtail sleep time and result in self-perceived insufficient sleep. *Sleep Biol. Rhythms* 5:204–14.

Terman, M. and J.S. Terman. 2005. Light therapy for seasonal and nonseasonal depression: Efficacy, protocol, safety, and side effects. *CNS Spectr.* 10:647–63; quiz 672.

Thapan, K., Arendt, J. and D.J. Skene. 2001. An action spectrum for melatonin suppression: Evidence for a novel non-rod, non-cone photoreceptor system in humans. *J. Physiol.* 535:261–7.

Thompson, C., Stinson, D. and A. Smith. 1990. Seasonal affective disorder and season-dependent abnormalities of melatonin suppression by light. *Lancet* 336:703–6.

van den Berg, T.J., IJspeert, J.K. and P.W. de Waard. 1991. Dependence of intraocular straylight on pigmentation and light transmission through the ocular wall. *Vision Res.* 31:1361–7.

Van den Bulck, J. 2004. Television viewing, computer game playing, and Internet use and self-reported time to bed and time out of bed in secondary-school children. *Sleep* 27:101–4.

Van den Bulck, J. 2007. Adolescent use of mobile phones for calling and for sending text messages after lights out: Results from a prospective cohort study with a one-year follow-up. *Sleep* 30:1220–3.

Wirz-Justice, A., Graw, P., Krauchi, K. et al. 1993. Light therapy in seasonal affective disorder is independent of time of day or circadian phase. *Arch. Gen. Psychiatry* 50:929–37.

Yasukouchi, A. 2005. A perspective on the diversity of human adaptability. *J. Physiol. Anthropol. Appl. Hum. Sci.* 24:243–7.

Zeitzer, J.M., Dijk, D.J., Kronauer, R., Brown, E. and C. Czeisler. 2000. Sensitivity of the human circadian pacemaker to nocturnal light: Melatonin phase resetting and suppression. *J. Physiol.* 526 Pt 3:695–702.

6 Effect of Environmental Light on Human Gastrointestinal Activity
From Laboratory Study to Fieldwork

*Yoshiaki Sone**
Graduate School of Human Life Science,
Osaka City University, Japan

CONTENTS

6.1 INTRODUCTION

It has been shown that daytime dim-light exposure has a negative effect on the efficiency of dietary carbohydrate absorption in the evening, while evening time dim-light exposure has a beneficial effect on it. Following from this, it is possible that

---

* Address all correspondence to sone@life.osaka-cu.ac.jp.

seasonal changes in environmental light may affect gastrointestinal activity, and that there might be seasonal variation in the efficiency of dietary carbohydrate absorption. To examine this, the amount of dietary carbohydrate unabsorbed from the intestine after a breakfast in young and elderly Japanese and Polish subjects who were living their everyday life during four seasons of the year in 2004, 2005, and 2006 was measured, using a combination of laboratory and fieldwork methods.

6.2 GENERAL INTRODUCTION

An advantage of laboratory studies is that controlled experiments can be set up that are difficult to perform in the field (Mai et al. 2005). However, laboratory and fieldwork studies are complementary in studies of physiological anthropology. In this chapter we describe results of fieldwork carried out to examine seasonality in the efficiency of dietary carbohydrate absorption in two populations (Japan and Poland).

Carbohydrate is usually the least expensive and most available source of dietary energy, the most important source of glucose, and the only energy source for brain tissue. The efficiency of dietary carbohydrate absorption in the intestine was measured by means of the breath hydrogen test. This method has been widely used as a noninvasive and simple method of detecting carbohydrate malabsorption in humans (Levitt and Donaldson 1970) as well as determining small intestinal and orocecal transit time (Bond and Levitt 1975; Hirakawa et al. 1988). It is based on the ability of the anaerobic microflora of the colon to ferment carbohydrate that has travelled unabsorbed through the small intestine and to produce hydrogen as a metabolic by-product.

Figure 6.1 shows a typical curve of hydrogen concentration vs. breath sampling time, where the variable 'hydrogen concentration' (ppm) is the sum of the

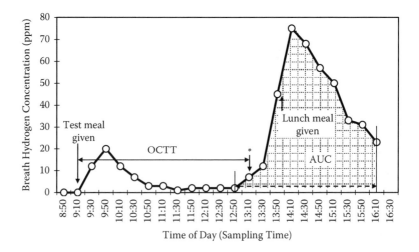

FIGURE 6.1 Typical change in concentration of hydrogen in the breath vs. time of day. The time marked by an asterisk (*) is when the head of the chime is assumed to enter the caecum.

hydrogen and methane concentrations. It has been shown that there is a roughly linear correlation between the excretion of hydrogen in the breath and the quantity of unabsorbed carbohydrate (Fritz et al. 1985); the area under the curve (AUC) when hydrogen concentration in the breath was plotted against time was therefore used to represent the amount of hydrogen due to fermentation of unabsorbed carbohydrate by the microflora of the large bowel. The AUC was calculated according to the trapezoidal rule (Rumessen et al. 1989) and expressed in units of parts per million hour (ppm·h). The AUC values corresponding to the breakfast meal or lactosucrose solution were defined as the area under the curve for 3 h and 20 min, commencing with the point 20 min before the rise in breath hydrogen level above the individual baseline value. This time of rise was defined as one of more than 5 ppm and one which was followed by at least two more rises (Hirakawa et al. 1988); this point is marked by an asterisk in Figure 6.1. At that point, the head of chyme was considered to have first entered the cecum. The length of time between 09:10 h and the time of rise (see above) was defined as the orocecal transit time (OCTT, expressed in min).

6.3   LABORATORY WORK

6.3.1   EFFECTS OF DIM- OR BRIGHT-LIGHT EXPOSURE DURING THE DAYTIME ON HUMAN GASTROINTESTINAL ACTIVITY (SONE ET AL. 2003)

On the basis of previous findings that bright-light exposure during daytime has profound influence on physiological parameters such as melatonin secretion and tympanic temperature in humans, Tokura et al. (Lee 2001) proposed the hypothesis that bright- or dim-light exposure during the daytime might have different influences on the activity of the digestive system via endocrine factors and/or the autonomic nervous system. In order to examine this hypothesis, we conducted a series of counterbalanced experiments in which subjects stayed in daytime conditions from 7:00 a.m. to 15:00 p.m. under either dim (80 lx) or bright (5,000 lx) light. This laboratory work was designed and conducted with consideration of the following issues. Careful control of the composition and time of meals was undertaken so as to elaborate the malabsorption of dietary carbohydrate in the evening meal. A counterbalanced experimental design accounted for a variety of time-related variables such as subjects' simple adaptation to the laboratory setting. Inclusion of a control phase scheduled between the bright-light phase (day 1) and dim-light phase (day 3) was incorporated to avoid the relatively long-lasting influence of bright light on human physiology.

Gastrointestinal activity was measured by means of a breath hydrogen test (indicator of carbohydrate malabsorption) and electrogastrography (EGG, an indicator of gastric myoelectric activity). Figure 6.2 compares the two AUCs of each subject. In eight out of eleven subjects, the breath hydrogen under the dim phase was more than that under the bright one. The figure also shows that the order of light exposure in this experiment had no effect on the ratio of the two AUCs because four subjects were exposed to the bright light first and the other four, the dimmer light first. Postprandial breath hydrogen excretion during the following night-time period after experiencing daytime under the dim-light condition was significantly

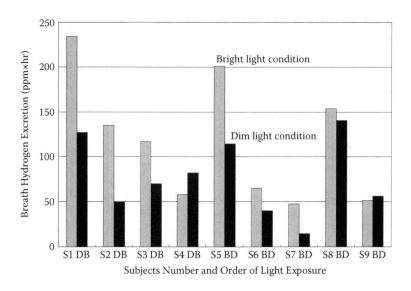

FIGURE 6.2 Comparison of AUC of each subject; Black bar and gray bar show AUC under dim-light condition and bright-light condition, respectively. S1–S9, subjects' numbers; DB, dim-light condition first, then bright light condition; BD, bright first, then dim.

higher than that experienced under the bright-light condition ($p < 0.05$). In addition, the spectrum total power of EGG recorded after taking an evening meal was significantly lower in the dim phase than in the bright phase ($p < 0.05$). These results support the hypothesis that the dim-light exposure during the daytime suppresses the digestion of evening meal, resulting in malabsorption of dietary carbohydrates in the evening meal.

6.3.2 EFFECT OF EVENING EXPOSURE TO DIM OR BRIGHT LIGHT ON THE DIGESTION OF CARBOHYDRATE IN THE EVENING MEAL (HIROTA ET AL. 2003)

A previous study (Sone et al. 2003) found that daytime exposure to bright as compared with dim light exerted a beneficial effect on the digestion of the evening meal. This finding prompted us to examine whether the digestion of the evening meal is also affected by evening light intensity. Subjects lived in light of 200 lx during the daytime (08:00 a.m. to 17:00 p.m.) and took their evening meal at 17:00 p.m. under 20 lx (evening dim-light condition: 17:00 p.m. to 02:00 a.m.) or 2000 lx (evening bright-light condition: 17:00 p.m. to 02:00 a.m.) until retiring at 02:00 a.m. Assessment of carbohydrate digestion of the evening meal was determined by the breath hydrogen test. Figure 6.3 shows the AUCs under the two lighting conditions for each subject. In seven of the nine subjects, hydrogen excretion in the breath under bright light was greater than under dim light. A repeated-measures ANOVA with one within-subjects factor (light condition) and one between-subjects factor (order of treatment) was not statistically significant.

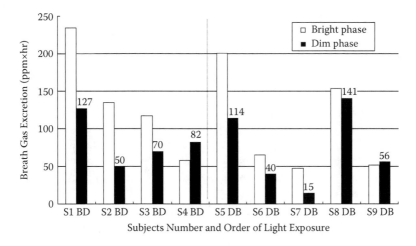

FIGURE 6.3 Comparison of breath hydrogen excretion (AUC, ppm × h) of each subject. S1–S9, subjects' numbers; DB, dim-light condition first, then bright light condition; BD, bright first, then dim.

Thus the order of light exposure did not influence the result, but hydrogen excretion in the evening under the dim-light condition was significantly lower than under the bright-light condition ($p < 0.05$). This finding is the opposite to that obtained in previous experiments (Sone et al. 2003) in which subjects were exposed to the different intensities of light during the daytime, and indicates that the exposure to dim in the evening exerts a greater effect on carbohydrate digestion in evening meal than does exposure to bright light.

These two laboratory studies considered the effect of dim and bright light on the absorption of dietary carbohydrate mainly from the view point of the effect of light on the autonomic nerve system. However, previous studies have showed that exposure to bright light relative to dim light during daytime enhanced the nocturnal decline of core temperature, an increase in leg skin temperature (Kim and Tokura 1998), a rise of salivary melatonin secretion (Park and Tokura 1999), and a decrease in urinary noradrenaline concentration (Kanikowska et al. 2001). Those findings strongly suggest parasympathetic nervous activity is enhanced by daytime bright-light exposure. It is possible that exposure to bright light during the daytime can activate the parasympathetic nervous system the following evening and night-time period (Nishimura et al. 2002), although the suppression of nocturnal rise of melatonin under the influence of evening and night-time bright-light exposure is accompanied with an attenuated nocturnal decline in core temperature and reduced increment of lower extremity skin-temperature elevation (Kim and Tokura, unpublished data). This is suggestive of a concomitant enhanced activity of the sympathetic nervous system and/or reduced activity of parasympathetic nervous system. Suppression of the cholinergic tone of the parasympathetic nervous system under the influence of evening bright-light exposure decreases the excretion of digestive juices and activity of gastrointestinal muscle (Ganong 1999), which is not beneficial for the digestion and absorption of a meal taken in the evening.

6.4   FIELDWORK

Extending observations made in the laboratory to the field, a plausible hypothesis important for everyday life might be as follows. Seasonal changes in sunshine duration (daytime light exposure) may affect gastrointestinal activity, and there may be seasonal variation in the efficiency of dietary carbohydrate absorption from the intestine. In order to examine this hypothesis, the amount of dietary carbohydrate unabsorbed from the intestine after breakfast in healthy Japanese subjects was measured over the four seasons of the year.

6.4.1   SEASONAL VARIATION IN AMOUNT OF DIETARY CARBOHYDRATE UNABSORBED FROM THE INTESTINE AFTER BREAKFAST IN JAPANESE YOUNG SUBJECTS (TSUMURA ET AL. 2005)

Figure 6.4 presents the experimental protocol. Subjects were requested to choose an easily-digestible, low-fat evening meal for consumption at 20:00 p.m. the day before (day 0) and the first day (day 1) of participation. Fasting periods of at least 12 h duration prior to the first breath sample were required on days 1 and 2. Subjects were also required to retire earlier than 23:30 p.m. on the nights of days 0 and 1.

Subjects wore a small waist accelerometer (Lifecorder, Suzuken, Nagoya, Japan) for one week before their participation, to record their normal daily physical activity and to be sure they adhered as close as possible to their normal lifestyle. Analysis of the five-day physical activity records of each subject prior to their participation showed that there was no significant seasonal variation in their daily physical activity [$F (3, 97) = 1.315$, $p = 0.274$; one-factor independent measure analysis of variance]. Subjects were also asked to keep a dietary record for the week prior to participation using a standard

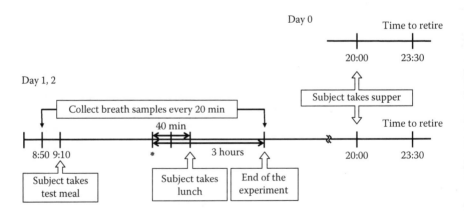

FIGURE 6.4   Experimental protocol (the same in the four seasons) *OCTT (Orocecal transit time).

TABLE 6.1
Nutrient Composition of the Breakfast

Total weight	540 g
Total energy	367 kcal
Protein	12.4 g
Fat	2.5 g
Carbohydrate*	72 g
* dietary fibre	7 g

dietary questionnaire (Tsumura et al. 2003). Subjects (usually in pairs) entered the air-conditioned room at 08:30 a.m. in the morning of day 1 and day 2 and remained sedentary under normal fluorescent lamps (300–400 lx) until the end of the breath hydrogen test, at 17:30 p.m. Approximate average room temperatures in Osaka and Nagano were 24°C, 25°C, 27°C, and 24°C in the winter, spring, summer, and autumn, respectively. Subjects were instructed in how to collect their end-alveolar breath sample at 08:40 a.m., before eating a normal breakfast at 09:10 a.m. on day 1. This meal was prepared from commercially available ready-to-eat minestrone, boiled potato, and macaroni (nutrient composition shown in Table 6.1). On day 2, subjects ingested an indigestible trisaccharide solution (200 ml of water containing 6 g of commercially available lactosucrose; Ensuiko Co., Ltd. Shizuoka, Japan) at the same time as breakfast on day 1. Every subject took the two meals in the same order in every season. Because lactosucrose is a growth factor for colonic bacteria (Ohkusa et al. 1995) and causes diarrhoea in some people, it was planned for subjects to ingest the lactosucrose solution on day 2 in order to avoid altering the normal microflora and physiological condition of the gastrointestinal tract during the study. End-alveolar breath samples were collected every 20 min, beginning at 08:50 a.m., into special airtight bags (TERAMECS, Kyoto, Japan), and hydrogen concentrations (and also methane) were measured by gas chromatography (Breath Gas Analyzer model TGA-2000, TERAMECS, Kyoto, Japan).

Subjects ate lunch 40 min after rising from bed. The rise in the hydrogen concentration in the breath just after breakfast, as observed in Figure 6.1, may be due to hydrogen accumulation in the colon during sleep (Solomonsand Viteri 1978) and/or fermentation of food that had not been absorbed during the previous day and had been expelled by gastro-ileal reflex after ingestion of breakfast (Levitt et al. 1987). The amount of unabsorbed dietary carbohydrate (UDC) from the breakfast meal was evaluated as the 'lactosucrose equivalent (g)', calculated as following:

$$6 \times (\text{AUC for the breakfast meal}/\text{AUC for 6 g of lactosucrose})$$

The ratio of the UDC of the breakfast meal to its total carbohydrate content was calculated assuming that 7 g of dietary fibre in the breakfast meal are also fermented in the colon by its microflora (Eastwood 1999). Therefore, the percentage inhibition of intestinal digestion of the breakfast meal was calculated by the following equation:

$$\text{UDC (lactosucrose equivalent, g} / 72 \times 100$$

TABLE 6.2
Comparison between the Two Subgroups and at the Four Seasons of OCTT Values (Min) for Both Test Meals and the Amount of UDC for the Breakfast

Subgroup	Season	OCTT (Min) Breakfast	OCTT (Min) Lactosucrose	UDC (g)	N
Osaka	winter	231 ± 63	157 ± 35	8.1 ± 2.2	7
	spring	229 ± 66	134 ± 50	6.9 ± 2.5	7
	summer	226 ± 73	146 ± 71	7.7 ± 5.0	7
	autumn	197 ± 61	130 ± 70	4.6 ± 2.3	6
Nagano	winter	248 ± 32	178 ± 89	7.1 ± 3.2	8
	spring	244 ± 49	185 ± 76	6.3 ± 2.8	11
	summer	223 ± 23	153 ± 70	5.9 ± 1.5	8
	autumn	228 ± 40	143 ± 63	4.5 ± 1.9	8

Note: Values are mean ± SD.

In the breath hydrogen test, the OCTT value for the breakfast meal indicates the length of time taken for the chyme to pass through the stomach and the small intestine, while the OCTT for the lactosucrose solution indicates the length of time taken for the oligosaccharide to pass through the small intestine due to lack of retention of oligosaccharides in the stomach (Kondo et al. 1994). Table 6.2 compares the OCTT for both test meals and the UDC for the breakfast in the two subgroups. The mean OCTT values for the breakfast as well as the lactosucrose solution were not significantly different between the two subgroups in any season [F (3, 54) = 0.258, p = 0.855 and F (3, 54) = 0.339, p = 0.797, respectively]. Analysis of seasonal variation in OCTT for the two test meals (using pooled data) showed that there was no significant difference in OCTT for either test meal or between the four seasons [F (3, 58) = 0.839, p = 0.478 for the breakfast and F (3, 58) = 0.685, p = 0.565 for the lactosucrose solution]. One-factor (season) independent measures ANOVA showed no significant difference between the four seasons [F (3, 58) = 0.018, p = 0.997], indicating that the length of time during which the chyme from the breakfast was retained in the stomach was almost constant throughout the year.

The main purpose of this study was to examine the seasonal variation in UDC. This was evaluated as a lactosucrose equivalent value (g) because the ecological population of colonic microflora may vary from season to season (Conway 1995), and, thus, the capability of colonic microflora to ferment the undigested dietary carbohydrate may change from season to season. Therefore, hydrogen production due to fermentation of a fixed amount of nondigestible carbohydrate (6 g of lactosucrose) by colonic microflora in each season was estimated, and then the amount of UDC in the breakfast was evaluated by comparing the AUC with that for lactosucrose solution. Figure 6.5 depicts UDC values as percentages for each subject and also the means (± SD) for each season. Table 6.2 shows the UDC values (g lactosucrose equivalent) of the two subgroups over the four seasons. Two-factor (season and subgroup)

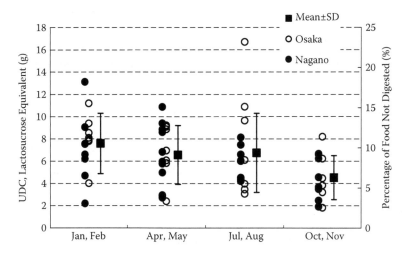

FIGURE 6.5   Amount of UDC (gram, lactosucrose equivalent) and percentage of nondigestion of each subject after ingestion of the breakfast (○ : Osaka and ● : Nagano), means of UDC (with SD bars) shown for the four seasons.

independent measures ANOVA showed there was no significant difference in the amount of UDC between the two subgroups in any season [F (3, 54) = 0.213, p = 0.887]. Accordingly, the data were pooled and the UDC was analysed for each season (Table 6.3). This table shows that there was a significant seasonal change in the amount of UDC [F (3, 58) = 3.134, p = 0.032], the post hoc multiple comparison by Tukey's Honestly Significant Difference (HSD) test indicated that there was a significant difference between winter and autumn (p = 0.022).

6.4.2   SEASONAL VARIATION IN GASTRIC MYOELECTRICAL ACTIVITY IN YOUNG JAPANESE FEMALES (TSUMURA ET AL. 2007)

Figure 6.6 shows the general protocol as well as the times for performing electrogastrography. Subjects were requested to choose an easily-digestible, low-fat evening meal at 20:00 p.m. on the day before participation. Because the breath hydrogen test,

TABLE 6.3
ANOVA Summary Table for One-Factor (Season)
Independent Measures Analysis of Variance
for the Amount of UDC in Four Seasons

	Sum of Squares	DF	Mean Square	F	Sig.
Between groups	72.92	3	24.305	3.134	p = 0.032
Within groups	449.81	58	7.755		
Total	522.72	61			

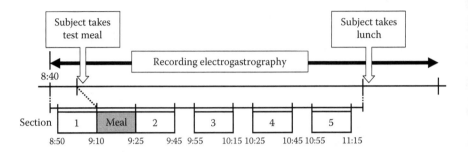

FIGURE 6.6 Experimental protocol for electrogastrography. (This was the same for all four seasons.)

which was conducted on the same day as the electrogastrography, required fasting periods of at least 12 h, subjects were required to retire earlier than 23:30 p.m. on the night before participation.

Subjects (usually in pairs) entered an air-conditioned room at 08:30 a.m. of the experimental day and remained sedentary in the room under normal fluorescent lamps (300–400 lx) until the end of experiment. Average room temperatures in Osaka and Nagano were about 24°C, 25°C, 27°C, and 24°C in winter, spring, summer, and autumn, respectively. Subjects took a normal breakfast at 09:10 a.m. This meal was prepared from commercially available ready-to-eat minestrone, boiled potato, and macaroni (its nutrient composition is shown in Table 6.1).

The cutaneous electrogastrogram (EGG) was recorded from 08:40 a.m. (30 min prior to taking breakfast) until the end of experiment. The skin of the anterior abdominal wall was rubbed with gel to reduce impedance prior to the attachment of the electrodes. Five electrodes (standard disposable Ag/AgCl adhesive electrocardiogram electrodes) were placed over the abdominal wall according to the EGG instruction manual. The reference electrode was placed over a point midway between the xiphoid process and the umbilicus. The four active electrodes were placed at the corners of a horizontal rectangle of ~12 cm × ~6 cm, centred on the reference electrode. The five electrodes were connected to a portable recording device (Nipro EG, Nipro, Osaka, Japan) and four-channel monopolar recordings were carried out. The EGG signals were isolated using a band-pass filter of between 1.6 and 6.0 cpm, transferred to a DOS/V computer. Spectral analysis was performed by Nipro EG data-processing software (EGS2, Gram Co., Ltd., Urawa, Saitama), to give the dominant frequency of the EGG (DF) and percentage of the 2–4 cpm gastric slow wave (Normal %). These two parameters of gastric myoelectrical activity are defined as follows (Chang 2005): The dominant frequency (DF) is that which shows peak power in the spectral analysis. It is very similar to the natural slow-wave rhythm, which is approximately 3 cpm. The slow wave is not always stable in its absolute frequency and amplitude, particularly in motor disorders. Hence, a normal range of computed frequencies is used to diagnose stomach motor dysfunction. Currently, bradygastria is defined as the values below normal range, whereas tachygastria is that above the normal range. The Normal %, the normal range of slow-wave frequency, was defined as 2–4 cpm.

This was computed as the number of frequencies within this range expressed as a percentage of the total number of frequencies present in the spectral analysis.

Most of the signals that were analysed were obtained from channels 1 and 3, which were located on the right side of the abdomen. The EGG was recorded from 08:40 a.m. (30 min prior to taking breakfast) until the end of the breath hydrogen test (see Figure 6.1), but the spectral analyses were performed upon five sections of this data span, as shown in Figure 6.6: section 1, 08:50–09:10 a.m.; section 2, 09:25–9:45 a.m.; section 3, 09:55–10:15 a.m.; section 4, 10:25–10:45 a.m.; and section 5, 10:55–11:15 a.m.

Among the EGGs obtained from day 1, the number of high-quality recordings (with low artefact signals) that could be subjected to spectral analysis was 8, 7, 9, and 10 in the spring, summer, autumn, and winter, respectively (Osaka subgroup), and 10, 5, 9, and 7 in the spring, summer, autumn, and winter, respectively (Nagano subgroup). Analysis of variance showed there was no significant difference between the two subgroups nor was there significant interaction between season and subgroup (p values being above 0.10). Therefore, the data from the two subgroups were pooled for DF and Normal %.

The pooled results are shown in Figure 6.7 and Figure 6.8, which show that dominant slow-wave frequency (DF, Figure 6.7) and percentage of normal wave (Normal %, Figure 6.8) both increased just after the meal and then gradually deceased to pre-ingestion levels. Table 6.4 and Table 6.5 show associated statistics. For DF (Table 6.4), there was a significant main effect for time, but no significant main effect for seasons and no interaction between time and season. The post hoc multiple comparison indicates that the DFs of sections 2–5 (postprandial) were significantly higher than section 1 (pre-prandial) ($p < 0.01$) and sections 3 and 4 were significantly higher than section 2 ($p < 0.05$).

Table 6.5 (Normal %) shows that there was a significant main effect for time, but no significant main effect for seasons and no interaction between time and season. The post hoc multiple comparisons indicate that sections 3–5 (postprandial) were significantly

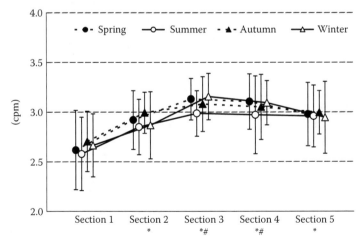

FIGURE 6.7 Means (with SD bars) of DF, dominant slow-wave frequency at each section of the four seasons. Section1: preprandial, sections 2–5: postprandial. * and # show the significant differences ($p < 0.05$) from section 1 and section 2, respectively, by post hoc multiple comparisons.

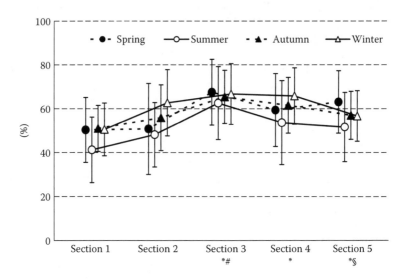

FIGURE 6.8 Means of normal % (with SD bars) at each section of the four seasons. Section 1: preprandial, sections 2–5: postprandial. *, #, and § show the significant differences ($p < 0.05$) from section 1, section 2, and section 3, respectively, by post hoc multiple comparisons.

TABLE 6.4
Two-Factor (Time and Season) ANOVA of DF

	Sum of Squares	DF	Mean Square	F	Sig.
Between groups					
Season	0.368	3	0.123	0.667	$p = 0.576$
Within groups					
Time	7.966	4	1.991	30.548	$p < 0.001$
Time x season	0.373	12	0.031	0.477	$p = 0.927$

TABLE 6.5
Two-Factor (Time and Season) ANOVA of Normal %

	Sum of Squares	DF	Mean Square	F	Sig.
Between groups					
Season	3036.638	3	1012.213	1.857	$p = 0.146$
Within groups					
Time	10442.617	4	2610.654	20.589	$p < 0.001$
Time x season	1988.439	12	165.703	1.307	$p = 0.215$

higher than section 1 (preprandial) ($p < 0.01$), section 3 was significantly higher than section 2 ($p < 0.01$), and section 5 was significantly lower than section 3 ($p < 0.05$).

There was no seasonal variation in the young Japanese female subjects in the two parameters of gastric myoelectric activity measured, dominant slow-wave frequency (DF), and percentage of the 2–4 cpm gastric slow wave (Normal %). It has previously been shown that no seasonality in orocecal transit time (OCTT) of the normal breakfast (solid meal) as well as of soluble indigestive oligosaccharide (lactosucrose) (Tsumura et al. 2005) occurs. The present result supports this previous demonstration of a lack of seasonal change in gastric myoelectric activity. Therefore, it is concluded that seasonal variation in digestion and absorption of dietary carbohydrate is caused by factors other than gastrointestinal motility and gastric myoelectric activity.

6.5   GENERAL CONCLUSION

Many scientists have investigated the seasonality (circannual rhythm) of human physiological behaviours. Over ten years ago, Reinberg (1997) reviewed the circannual rhythms of human nutrition. For example, Haus et al. (1988) found a seasonal change in the insulin response to normalised diets in normal subjects, the effect being strongest and fastest in autumn, and weakest and slowest at the end of winter. Griffith found a circannual rhythm of the human Respiration Quotient (RQ), it being highest from the middle of summer to the middle of autumn, indicating that human beings metabolise carbohydrate most efficiently in these periods of the year. These results obtained from both laboratory and fieldwork described here are consistent with the results of previous studies and support the hypothesis that the seasonality of human behaviour could be an important physiological adaptation to annual environmental change, and could be related to the behavioural rhythm of animals that hibernate (Reinberg 1997).

ACKNOWLEDGEMENTS

This fieldwork was supported by Grants-in Aid for Scientific Research (No. 15207024) from the Japan Society for the Promotion of Science.

REFERENCES

Bond, J.H. and M.D. Levitt. 1975. Investigation of small bowel transit time in man utilizing pulmonary hydrogen (H_2) measurements. *J. Lab. Clin. Med.* 85:546–55.
Chang, F-Y. 2005. Electrogastrography: Basic knowledge, recording, processing and its clinical applications. *J. Gastroenterol. Hepatol.* 20:502–16.
Conway, P.L. 1995. Microbial ecology of the human large intestine. In *Human Colonic Bacteria: Role in Nutrition, Physiology, and Pathology*, ed. G.R. Gibson and G.T. Macfarlane, 1–24. Boca Raton: CRC Press Inc.
Eastwood, M.A. 1999. Structure and function of the colon, In *Encyclopedia of Human Nutrition*. Volume 2, ed. M.J. Sadler, J.J. Strain and B. Caballero, 945–53. San Diego: Academic Press.
Fritz, M., Siebert, G. and H. Kasper. 1985. Dose dependence of breath hydrogen and methane in healthy volunteers after ingestion of a commercial disaccharide mixture, Palatinit®. *Br. J. Nutr.* 54:389–400.

Ganong, W.F. 1999. The autonomic nervous system. In *Review of Medical Physiology*, 19th edition, 214–20. Stanford, Connecticut: Appleton & Lange, a Simon & Schuster Company, Prentice-Hall International, Inc.

Haus, E., Nicolau, G.Y., Lakatua, D. and L. Sackett-Lundeen. 1988. Reference values for chronopharmacology. *Annu. Rev. Chronopharmacol.* 4:333–425.

Hirakawa, M., Iida, M., Kohrangi, N. and M. Fujishima. 1988. Hydrogen breath test assessment of orocecal transit time; comparison with barium meal study. *Am. J. Gastroenterol.* 83:1361–3.

Hirota, N., Sone, Y. and H. Tokura. 2003. Effect of evening exposure to dim or bright light on the digestion of carbohydrate in the supper meal. *Chronobiol. Int.* 20:853–61.

Kanikowska, D., Hirata, Y., Hyun, K.J. and H. Tokura. 2001. Acute phase proteins, body temperature and urinary melatonin under the influence of bright and dim light intensities during the daytime. *J. Physiol. Anthropol.* 20:333–8.

Kim, H.E. and H. Tokura. 1998. Visual allieshesia-cloth color preference in the evening under the influence of different light intensities during the daytime, *Physiol. Behav.* 65:367–70.

Kondo, T., Liu, F. and Y. Toda. 1994. Milk is a useful test meal for measurement of small bowel transit time. *J. Gastroenterol.* 29:715–20.

Lee, Y-A., Hyun K-J., Sone Y. and H. Tokura. 2001. Effects of bright or dim light during the daytime on digestive activity in humans. *Naturwissenschften.* 88:126–8.

Levitt, M.D. and Donaldson, R.M. 1970. Use of respiratory hydrogen (H_2) excretion to detect carbohydrate malabsorption. *J. Lab. Clin. Med.* 75:937–45.

Levitt, M.D., Hirsh, P., Fetzer, C.A., Sheahan, M. and A.S. Levine. 1987. H_2 excretion after ingestion of complex carbohydrates. *Gastroenterology* 92:383–9.

Mai, L.L., Owl, M.Y. and M.P. Kersting. 2005. *The Cambridge Dictionary of Human Biology and Evolution.* 291. Cambridge: Cambridge University Press.

Nishimura, S., Hyun, K.J., Lee, Y.A. and H. Tokura. 2002. Increase in parasympathetic nerve activity during the nighttime following bright light exposure during the daytime. *Biol. Rhythm Res.* 34:233–40.

Ozaki, Y., Sato, C., Mikuni, K., and H. Ikeda. 1995. Long-term ingestion of lactosucrose increases Bifidobacterium sp. in human fecal flora, *Digestion.* 56:415–20.

Park, S.-J. and H. Tokura. 1999. Bright light exposure during the daytime affects circadian rhythms of urinary melatonin and salivary immunoglobulin A. *Chronobiol. Int.* 16:359–71.

Reinberg, A. 1997. Chronobiologie et nutrition. In *Les Rythmes Biologiques, QUE SAIS-JE?* No. 734, 106–16. Paris: Presses Universitaires de France.

Rumessen, J.J., Hamberg, O. and E. Gudmad-Hoyer. 1989. Influence of orocecal transit time in hydrogen excretion after carbohydrate malabsorption. *Gut* 30:811–4.

Solomons, N.W. and F. Viteri. 1987. Development of an interval sampling hydrogen (H_2) breath test for carbohydrate malabsorption in children: Evidence for a circadian pattern of breath H_2 concentration. *Pediatr. Res.* 12:816–23.

Sone, Y., Hyun, K.J., Nishimura, S., Lee, Y.A. and H. Tokura. 2003. Effects of dim or bright-light exposure during the daytime on human gastrointestinal activity. *Chronobiol. Int.* 20:85–95.

Tsumura, Y., Hirota, N., Tokura, H., Rutkowska, D. and Y. Sone. 2005. Seasonal variation in amount of unabsorbed dietary carbohydrate from the intestine after breakfast in Japanese subjects. *Chronobiol. Int.* 22:1107–19.

Tsumura, Y., Hirota, N., Tokura, H., Rutkowska, D. and Y. Sone. 2007. Seasonal variation in gastric myoelectrical activity in young Japanese females. *Biol. Rhythm Res.* 38:383–90.

Tsumura, Y., Ogino, T., Hirota, N. and Y. Sone. 2003. Study on dietary life of the elderly from the viewpoint of food consumption pattern, *J. Hum. Life Sci.* 8:53–7.

7 Cold Tolerance and Lifestyle in Modern Society

Takafumi Maeda[*]
Laboratory of Environmental Ergonomics,
Hokkaido University, Sapporo, Japan

CONTENTS

7.1 Introduction 99
7.2 Air Conditioning... 100
7.3 Dietary Habits... 101
7.4 Exercise Habits 103
7.5 Cold Tolerance and Lifestyle 103
7.6 Effects of Basal Metabolic Rate on Cold Tolerance.................................... 104
7.7 Effects of Maximum Oxygen Uptake on Cold Tolerance 105
7.8 Conclusion 106
References... 108

7.1 INTRODUCTION

Physiological reactions to cold temperatures, as with other stimuli, vary greatly from individual to individual. Investigating factors influencing human variation is useful in assessing the environmental adaptability of human beings. Systemic adaptability to cold temperatures (that is, systemic cold tolerance) can be assessed by the ability of the body to suppress heat loss and to produce heat. These abilities are not independent of each other; however, when assessing human variation with respect to cold tolerance, these two abilities need to be investigated separately.

Lifestyle contributes to risk factors for various diseases (Azegami et al. 2006; Forman and Bulwer 2006; Irigaray et al. 2007; Iseki 2008; Slattery et al. 1999), with certain aspects of lifestyle, particularly exercise and dietary habits, affecting various physiological functions (Dawson et al. 2008; Jennings et al. 1997; Kaufman et al. 2008; Okamoto et al. 2007). Regular exercise improves not only muscular strength and the ability for instantaneous force but also respiratory and cardiovascular functions, including improvement of the autonomic nervous system, which

---

[*] Address all correspondence to maeda@eng.hokudai.ac.jp.

controls respiratory and cardiovascular functions. On the other hand, overeating and a fatty and unbalanced diet can induce hyperlipidemia and arteriosclerosis (Akiyama et al. 1996; Ruixing et al. 2008). These physiological functions are closely related to thermoregulation. Thus, the thermoregulatory ability, which can produce variation in various physiological functions, is affected by a person's lifestyle.

Humans can change the surrounding thermal environment by using technology. For example, air conditioning and super-insulated houses help to maintain a comfortable temperature at all times. Therefore, people are spending more time in artificial, 'comfortable' environments throughout the year. Is living in this 'comfortable' environment good for human beings, and how does an artificial environment influence cold tolerance? The effects of living in an artificial environment, exercise, and dietary habits on the ability of thermogenesis and an ability to suppress heat loss are introduced and discussed here.

7.2 AIR CONDITIONING

One of the factors influencing not only cold tolerance but also thermoregulation, including heat tolerance, is the ambient temperature that a person is exposed to on a daily basis. In many modern societies, including Japan, indoor temperatures are controlled to remain within comfortable levels by air conditioning throughout the year, and people spend much of their time in artificial indoor environments. It would seem that the amount of heat stress placed on the body is low, and the amount of stress on physiological function is also low. How does spending the entire year at a comfortable temperature affect cold tolerance?

Living in an environment which is always at a comfortable temperature will affect the ability to produce heat. As being in this controlled environment means that the core body temperature can be controlled without increasing heat production by shivering, the body may lose some of its ability to produce heat. In addition, the basal metabolic rate, which is the basis of heat production in the body, is also affected by living at constantly comfortable temperatures. Basal metabolic rate changes with seasonal variations (Plasqui and Westerterp 2004; Umemiya 2006). As a general rule, basal metabolic rate is higher in winter than in summer. This is due to acclimatisation to cold stimuli in winter. However, living in heated rooms during the day in winter lowers basal metabolic rate (Maeda et al. 2005) (Figure 7.1).

Evidence of this can be found in data from previous studies about the seasonal variation of basal metabolic rate, which were conducted over the past fifty years in Japan. Yurugi et al. (1972) reported that the seasonal variation of basal metabolic rate was less than 10% in 1967, although the variation had been about 20% in 1950 (Yurugi et al. 1972). The variation in 1987 continued to decline and was about 11% (Shimaoka et al. 1987). Additionally, Shinya et al. (2006) reported that no variation was found in 2006 (Shinya et al. 2006). The decrease in seasonal variation of basal metabolic rate from 1950 to 1967 might be caused mainly by the change of energy intake and dietary habits.

However, one may speculate that the change of indoor thermal conditions was responsible for the disappearance in the seasonal variation of basal metabolic rate

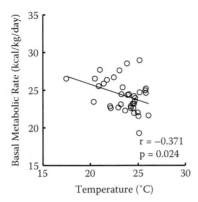

FIGURE 7.1   Relationship between daytime temperature around subject in winter and basal metabolic rate.

because energy status did not change in the duration. Thus, the adverse effects of a constant comfortable indoor temperature (produced by using air conditioning) on basal metabolic rate may be great, according to the results of Maeda et al. (2005), which demonstrated that basal metabolic rate is decreased by living in heated rooms during the day in winter. Living in an environment with a comfortable range of temperatures may thus weaken not only the ability to generate heat in a cold environment but also basal metabolic rate—which is the basis for thermogenesis.

7.3   DIETARY HABITS

In Japan, the number of people who do not eat breakfast is increasing. For example, 28.3% of young people, in general, and 49.4% of young people who live alone do not eat breakfast (Ministry of Health, Labour and Welfare 2007). Additionally, many young people in Japan buy ready-to-eat meals at convenience stores. Most convenience stores are open twenty-four hours a day, and a wide variety of foods, snacks, and beverages can be purchased anywhere at any time. People who buy foods at convenience stores tend to purchase their favourite foods, resulting in a poor diet with marked dietary deficiencies.

   When considering cold tolerance, it is evident that diet affects both heat production and heat loss. Eating temporarily increases heat production (Garrow 1986), while nutrient intake is known to affect basal metabolic rate (Garrow 1986; Kouda et al. 2006; Maeda et al. 2005; Yamashita and Hayashi 1989). Total energy intake, protein, carbohydrate, and fat intake have been shown to be positively correlated with basal metabolic rate (Figure 7.2). Excessive eating increases basal metabolic rate, while insufficient eating decreases basal metabolic rate. In addition, according to a study of thirty-seven Japanese students, methods of meal preparation affect basal

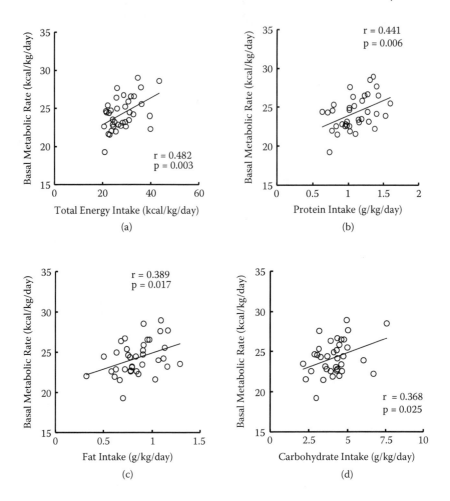

FIGURE 7.2 Relationships between basal metabolic rate and total energy intake (a), protein intake (b), fat intake (c), and carbohydrate intake (d).

metabolic rates. The basal metabolic rate of male students who bought prepared foods at stores, including convenience stores, was lower than for those who cooked at home (Figure 7.3a). Furthermore, basal metabolic rate of students who ate snacks daily was lower than that for students who rarely ate them (Figure 7.3b).

Dietary habits also affect the ability to suppress heat loss. First, dietary habits affect vessel hardening and flexibility. High-fat diets induce arteriosclerosis, affecting the suppressive function of vasoconstriction on heat loss. Second, dietary habits influence subcutaneous fat development. Excessive nutrient intake on a regular basis causes body fat to accumulate. Body fat, particularly subcutaneous fat, affects heat loss by acting as an insulator.

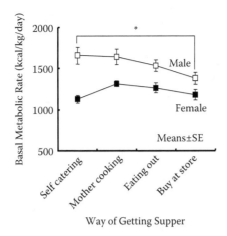

Way of Getting Supper

FIGURE 7.3 Eating behaviours [way of getting supper (a) and snacking (b)] and basal metabolic rate.

7.4 EXERCISE HABITS

According to the 2005 National Health and Nutrition Examination Survey in Japan, 18.5% of Japanese men in their twenties and 14.6% of Japanese women in their twenties exercise, indicating that relatively few of today's young people exercise. Young Japanese people also spend a lot of time watching TV, playing videogames, and using computers, and these activities may result in reduced levels of exercise (Koezuka et al. 2006). Inactive lifestyles and less exercise lower aerobic exercise capacity. However, in terms of the suppression of heat loss and the promotion of heat production, previous studies have not necessarily agreed about the relationship between aerobic capacity and cold tolerance.

Several studies have shown that exercise training enhances the increments of metabolic heat production in cold environments (Andersen 1966; Bittel 1992; Bittel et al. 1988; Hirata and Nagasaka 1981; Kashimura 1988; Yoshida et al. 1998) and diminishes the suppression of the heat loss (Bittel et al. 1988; Falk et al. 1994), while other investigations have found that exercise training suppresses the increment of heat production (Armstrong 1998) and enhances the suppression of heat loss (Yoshida et al. 1998).

7.5 COLD TOLERANCE AND LIFESTYLE

Lifestyle factors such as living in a thermal environment, exercise, and dietary habits influence various physical functions, which then affect the ability of the body to tolerate cold. Cold tolerance was generally estimated by the abilities of cold-induced metabolic heat production and vasoconstriction during cold air exposure and also by basal metabolic rate. As discussed previously, basal metabolic rate is affected by air conditioning, dietary habits, and body composition (which in turn is affected by dietary and exercise habits); vasoconstriction is affected by exercise habits; and cold-induced metabolic heat production is affected by air conditioning, dietary habits, and exercise habits. The relationships between physical functions affected by lifestyle factors, especially basal metabolic rate and maximum oxygen uptake, and the toler-ance for a cold environment are discussed in Section 7.6.

7.6 EFFECTS OF BASAL METABOLIC RATE ON COLD TOLERANCE

Basal metabolic rate (obligatory nonshivering heat production) served as one of the indicators for cold tolerance in the present study and also affects cold-induced metabolic heat production and vasoconstriction. According to a study of thirty-seven Japanese university students in winter (Maeda et al. 2005), basal metabolic rate was influenced by muscle mass, fat intake, and daytime temperature around them. Muscle mass and fat intake were positively correlated with basal metabolic rate, while daytime temperature was negatively correlated with it (Figure 7.1 and Figure 7.2c). These factors are related to air conditioning, diet, and exercise habits. How does basal metabolic rate influence cold tolerance and cold-induced metabolic heat production?

In one study, ten healthy male university students were exposed to a temper-ature of 10°C for ninety minutes, and the relationship between basal metabolic rate and cold-induced metabolic heat production was investigated (Maeda et al. 2007). There was a negative correlation between basal metabolic rate and cold-induced metabolic heat production (Figure 7.4a). When the students were exposed

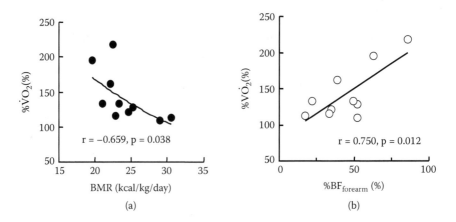

FIGURE 7.4 Relationships between basal metabolic rate (a), skin blood flow at forearm (b), and oxygen uptake at the end of cold exposure (10°C) for 90 min. BMR means basal meta-bolic rate. $\%BF_{forearm}$ and $\%VO_2$ mean changing rate of skin blood flow at forearm and oxygen uptake from baseline data before cold exposure, respectively. (From Maeda et al. 2007, with permission from the Japan Society of Physiological Anthropology.)

to 10°C for ninety minutes, those with a basal metabolic rate of 30 kcal/kg/day showed a small increase in cold-induced heat production and those with a basal metabolic rate of 20 kcal/kg/day had a large increase in heat production of over 70%. Compared with students with low basal metabolic rates, those with high basal metabolic rates possessed higher levels of heat production and were able to regulate body temperatures without markedly elevating metabolic heat production even when exposed to cold temperatures. These findings indicate that cold tolerance was more developed in individuals with a high basal metabolic rate than in those with a lower one. Maeda et al. (2007) also reported that the degree of change in blood flow did not correlate with basal metabolic rate but exhibited a positive correlation to the degree of increase in heat production (Figure 7.4b). Thus the degree of increase in heat production was low in students with marked decreases in skin blood flow. The above findings suggest that basal metabolic rate influences the degree of increase in cold-induced metabolic heat production. Furthermore, since the degree of increase in cold-induced metabolic heat production is influenced by changes in skin blood flow, basal metabolic rate appears to influence indirectly skin blood flow regulation.

7.7   EFFECTS OF MAXIMUM OXYGEN UPTAKE ON COLD TOLERANCE

Endurance training increases maximum oxygen uptake and cold tolerance. However, previous studies do not agree on the relationship between aerobic performance and the tolerance of cold temperature in terms of suppression of heat loss and promotion of heat production. According to Bittel et al. (1988), skin heat conductance (an indicator of heat loss) and metabolism following exposure to cold was positively correlated to maximum oxygen uptake. However, Yoshida et al. (1998) compared individuals with comparable degrees of subcutaneous fat and reported that heat loss was suppressed by exercise training. Furthermore, when students were exposed to a temperature of 10°C for ninety minutes, the degree of decrease in toe skin temperature was negatively correlated to maximum oxygen uptake (Figure 7.5). Those results indicate that heat loss for individuals with high maximum oxygen uptake is lower than that for individuals with low maximum oxygen uptake, contradicting the results obtained by Bittel et al. (1988).

In these studies, the ability to suppress heat loss was estimated by skin temperature or by the skin heat conductance calculated from skin, core, and ambient temperatures. This is likely to be the reason for the disagreement in results of these studies, because skin temperature is influenced by not only vasoconstriction but also subcutaneous fat thickness. Yoshida et al. (1998) indicated that heat loss during cold exposure in a trained person was less than that in an untrained person with the same ratio of body fat from the relationship between body fat content and skin heat conductance. Furthermore, blood vessels of the fingers in individuals with high maximum oxygen uptake were more constricted after cold exposure, which was 10°C for ninety

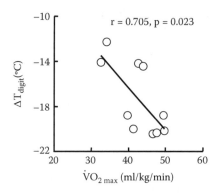

FIGURE 7.5   Relationship between maximum oxygen uptake and decreasing value of skin temperature of digit (ΔT_{digit}) at the end of cold exposure (10°C) for 90 min. (From Maeda et al. 2007, with permission from the Japan Society of Physiological Anthropology.)

minutes, than in those with low maximum oxygen uptake (unpublished data). These data suggest that the ability to suppress heat loss by vasoconstriction is improved by sustained physical activity. It is possible that cold-induced vasoconstriction is influenced by vascular function, including flexibility. Active lifestyle also improves vascular and autonomic nervous function, while sedentary lifestyle impaired both (Hagg et al. 2005; Hirayanagi et al. 2004; Ng et al. 1994; Watanabe et al. 1997). Furthermore, thermoregulatory responses to hot and cold are changed by exercise training and bed rest (Andersen 1966; Falk et al. 1994; Smorawinski et al. 2005; Watanabe et al. 1997).

7.8   CONCLUSION

Previous findings suggested that the abilities of heat production and suppression of heat loss in the cold are affected by lifestyle factors. Figure 7.6 summarises the relationship between the adaptability to cold temperature and three lifestyle factors: air conditioning, dietary habits, and exercise habits. Air conditioning decreases basal metabolic rate by formation of a comfortable indoor thermal condition, resulting in increase of metabolic heat production in the cold. Exercise habits improve vascular function as well as aerobic physical capacity, resulting in the improvement of vascular contractility. Exercise also develops skeletal muscles which influence basal metabolic rate and shivering thermogenesis. Dietary habits change basal metabolic rate, body fat, and vascular function. Appropriate dietary habits and good nutritional status change those for the better, but poor eating habits change those for the worse. Thus, the abilities of thermogenesis and to suppress heat loss are easily changed by some lifestyle factors, showing this aspect of human variation to be extremely plastic.

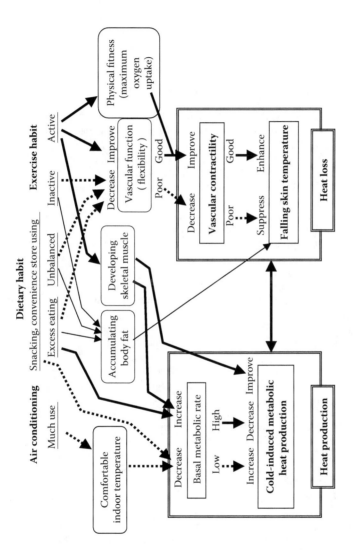

FIGURE 7.6 The scheme of the relation between lifestyle and cold tolerance.

REFERENCES

Akiyama, T., Tachibana, I., Shirohara, H., Watanabe, N. and M. Otsuki. 1996. High-fat hyper-caloric diet induces obesity, glucose intolerance and hyperlipidemia in normal adult male Wistar rat. *Diabetes Res. Clin. Pract.* 31:27–35.

Andersen, K.L. 1966. Metabolic and circulatory aspects of tolerance to cold as affected by physical training. *Fed. Proc.* 25:1351–6.

Armstrong, D.W. III. 1998. Metabolic and endocrine responses to cold air in women differing in aerobic capacity. *Med. Sci. Sports Exerc.* 30:880–4.

Azegami, M., Hongo, M., Yanagisawa, S. et al. 2006. Characteristics of metabolic and lifestyle risk factors in young Japanese patients with coronary heart disease: A comparison with older patients. *Int. Heart J.* 47:343–0.

Bittel, J. 1992. The different types of general cold adaptation in man. *Int. J. Sports Med.* 13(Suppl 1):S172–6.

Bittel, J.H., Nonotte-Varly, C., Livecchi-Gonnot, G.H., Savourey, G.L. and A.M. Hanniquet. 1988. Physical fitness and thermoregulatory reactions in a cold environment in men. *J. Appl. Physiol.* 65:1984–9.

Dawson, E.A., Whyte, G.P., Black, M.A. et al. 2008. Changes in vascular and cardiac function after prolonged strenuous exercise in humans. *J. Appl. Physiol.* 105(5):1562–8.

Falk, B., Bar-Or, O., Smolander, J. and G. Frost. 1994. Response to rest and exercise in the cold: Effects of age and aerobic fitness. *J. Appl. Physiol.* 76:72–8.

Forman, D. and B.E. Bulwer. 2006. Cardiovascular disease: Optimal approaches to risk factor modification of diet and lifestyle. *Curr. Treatment Options Cardiovasc. Med.* 8:47–57.

Garrow, J.S. 1986. Chronic effects of over- and under-nutrition on thermogenesis. *Int. J. Vitam. Nutr. Res. (Internationale Zeitschrift fur Vitamin- und Ernahrungsforschung)* 56:201–4.

Hagg, U., Wandt, B., Bergstrom, G., Volkmann, R. and L.M. Gan. 2005. Physical exercise capacity is associated with coronary and peripheral vascular function in healthy young adults. *Am. J. Physiol.* 289:H1627–34.

Hirata, K. and T. Nagasaka. 1981. Enhancement of calorigenic response to cold and to norepi-nephrine in physically trained rats. *Jpn. J. Physiol.* 31:657–65.

Hirayanagi, K., Iwase, S., Kamiya, A., Sasaki, T., Mano, T. and K. Yajima. 2004. Functional changes in autonomic nervous system and baroreceptor reflex induced by 14 days of 6 degrees head-down bed rest. *Eur. J. Appl. Physiol.* 92:160–7.

Irigaray, P., Newby, J.A., Clapp, R. et al. 2007. Lifestyle-related factors and environmental agents causing cancer: An overview. *Biomed. Pharmacother.* 61:640–58.

Iseki, K. 2008. Metabolic syndrome and chronic kidney disease: A Japanese perspective on a worldwide problem. *J. Nephrol.* 21:305–12.

Jennings, G.L., Chin-Dusting, J.P., Kingwell, B.A. et al. 1997. Modulation of vascular func-tion by diet and exercise. *Clin. Exp. Hypertens.* 19:727–37.

Kashimura, O. 1988. Positive cross-adaptation between endurance physical training and gen-eral cold tolerance to acute cold exposure in rats. *Nippon Seirigaku Zasshi* 50:753–60 (in Japanese).

Kaufman, C.L., Kaiser, D.R., Kelly, A.S., Dengel, J.L., Steinberger, J. and D.R. Dengel. 2008. Diet revision in overweight children: Effect on autonomic and vascular function. *Clin. Auton. Res.* 18:105–8.

Koezuka, N., Koo, M., Allison, K.R. et al. 2006. The relationship between sedentary activi-ties and physical inactivity among adolescents: Results from the Canadian Community Health Survey. *J. Adolesc. Health* 39:515–22.

Kouda, K., Nakamura, H., Kohno, H. et al. 2006. Metabolic response to short-term 4-day energy restriction in a controlled study. *Environ. Health Prev. Med.* 11:89–92.

Maeda, T., Sugawara, A., Fukushima, T., Higuchi, S. and K. Ishibashi. 2005. Effects of life-style, body composition, and physical fitness on cold tolerance in humans. *J. Physiol. Anthropol. Appl. Human Sci.* 24:439–43.

Maeda, T., Fukushima, T., Ishibashi, K. and S. Higuchi. 2007. Involvement of basal metabolic rate in determination of type of cold tolerance. *J. Physiol. Anthropol.* 26:415–8.

Ministry of Health, Labour and Welfare. 2007. Report of the National Health and Nutrition Survey Japan, 2005.

Ng, A.V., Callister, R., Johnson, D.G. and D.R. Seals. 1994. Endurance exercise training is associated with elevated basal sympathetic nerve activity in healthy older humans. *J. Appl. Physiol.* 77:1366–74.

Okamoto, T., Masuhara, M. and K. Ikuta. 2007. Combined aerobic and resistance training and vascular function: Effect of aerobic exercise before and after resistance training. *J. Appl. Physiol.* 103:1655–61.

Plasqui, G. and K.R. Westerterp. 2004. Seasonal variation in total energy expenditure and physical activity in Dutch young adults. *Obes. Res.* 12:688–94.

Ruixing, Y., Jinzhen, W., Yaoheng, H. et al. 2008. Associations of diet and lifestyle with hyper-lipidemia for middle-aged and elderly persons among the Guangxi Bai Ku Yao and Han populations. *J. Am. Diet. Assoc.* 108:970–6.

Shimaoka, A., Machida, K., Kumae, T. et al. 1987. Seasonal variation of basal metabolism. *Jpn. J. Biometeor.* 24:3–8.

Shinya, H., Yoshida, T., Yorimoto, A. and S. Nakai. 2006. Seasonal variation of resting meta-bolic rate in adult females. *Jpn. J. Biometeor.* 43:S60.

Slattery, M.L., Edwards S.L., Boucher, K.M., Anderson, K. and B.J. Caan. 1999. Lifestyle and colon cancer: An assessment of factors associated with risk. *Am. J. Epidemiol.* 150:869–77.

Smorawinski, J., Mlynarczyk, C., Ziemba, A.W. et al. 2005. Exercise training and 3-day head down bed rest deconditioning: Exercise thermoregulation. *J. Physiol. Pharmacol.* 56:101–10.

Umemiya, N. 2006. Seasonal variations of physiological characteristics and thermal sensation under identical thermal conditions. *J. Physiol. Anthropol.* 25:29–39.

Watanabe, F., Takenaka, K., Suzuki, Y. et al. 1997. Prolonged bed rest impairs response of the small arteries to cold pressor test. *J. Gravit. Physiol.* 4:S72–4.

Yamashita, J. and S. Hayashi. 1989. Changes in the basal metabolic rate of a normal woman induced by short-term and long-term alterations of energy intake. *J. Nutr. Sci. Vitaminol.* 35:371–81.

Yoshida, T., Nagashima, K., Nakai, S., Yorimoto, A., Kawabata, T. and T. Morimoto. 1998. Nonshivering thermoregulatory responses in trained athletes: Effects of physical fitness and body fat. *Jpn. J. Physiol.* 48:143–8.

Yurugi, R., Sasaki, T. and M. Yoshimura. 1972. Seasonal variation of basal metabolism in Japanese. In *Advances in Climatic Physiology*, ed. S. Itoh, K. Ogata and H. Yoshimura, 395–10. Tokyo: Igaku Shoin Ltd.

8 Human Adaptability to Emotional and Intellectual Mental Stresses

*Koichi Iwanaga**
Department of Design Science, Graduate School
of Engineering, Chiba University, Chiba, Japan

CONTENTS

8.1 PHYSIOLOGICAL ANTHROPOLOGY AND THE KEYWORDS

In industrialised countries, the daily life of human beings is supported by numerous scientific technologies and artificial assistance techniques. Individuals living in such environments are interesting subjects of biological anthropology study, although this has not been common except in Japan where physiological anthropology has been developing independently, focussing on humans living in the scientific technological civilisation.

Simulation of various conditions of modern living environments in laboratories has allowed researchers to study human adaptability to modern life. Such adaptability is called techno-adaptability by physiological anthropologists. Advanced technologies have allowed stable continuity of lives, finance, and production, through use of video display terminals; automation of manufacturing plants, offices, and homes; maintenance of rooms at constant temperature and light intensity; and ensuring that

---

* Address all correspondence to iwanaga@faculty.chiba-u.jp

the information network systems are unaffected by geographical location. However, such technologies may also cause new stresses and health problems to humans.

Techno-adaptability is predicted by functional potentiality, whole-body coordination, and physiological polytypism, three interrelated characteristics that cannot be defined individually (Iwanaga et al. 2005; Sato 2005).

8.1.1　Techno-Adaptability

Techno-adaptability indicates the characteristics of human adaptability from the viewpoint of cultural adaptation. The earliest examples of cultural adaptation in the process of human evolution can be observed in houses and clothes and material products of human mind, which contributed to protecting our ancestors from their enemies and harsh natural environments. Techno-adaptability can be defined in two ways. The first is as an ability with which humans can, with technological support, maintain the efficiency of their lives, compensate for their limited or declined physical abilities, and develop living conditions useful in sustaining life. For example, the invention and development of houses, clothes, and air conditioning systems have enabled humans to protect themselves from potential life-threatening natural phenomena, including heat or cold. Transportation devices, such as cars and aircraft, have enabled humans to travel long distances over short time periods, enabling humans to travel far beyond their physical capacities. These devices can be considered as artificial organs in addition to natural human ones, which extend biological function. In this context, such devices are occasionally called exosomatic organs, while biological internal organs are called endosomatic organs. Considering its ability to extend the information processing capacity of the human brain, the computer may be viewed as the ultimate exosomatic organ.

An alternative definition of techno-adaptability is that it is an ability with which humans can maintain or improve their biological function against a variety of stresses resulting from the intervention of scientific technology in daily life. In approximately 6 million years of human evolution, our ancestors spent most of their time living in environments at very low levels of technology. The human body therefore evolved in natural environments. The living environment in present-day industrialised societies is different from the natural environment in many ways. Room temperature can be maintained at a constant level throughout the day, and sufficient light intensity for work and social interaction is available anytime during the day or night. Human behaviour is mediated by technological transmission of information. Such environments subject human beings to several kinds of stress, which can cause disorders in biological rhythm, continuous optical and mental tensions resulting from the use of computers, and spatial and temporal oppression of dwellers of densely-populated cities. In the 1980s, the new term *techno-stress* was initially employed in Japan for a variety of physical disorders among computer operators. This definition has been expanded to incorporate stress resulting not only from the incompatibility of humans and machines (mainly with computers) but also from an increasingly complicated daily life supported by highly advanced scientific technology (Iwanaga et al. 2005). Studies of human adaptability in such stressful environments contribute to the knowledge base needed for the construction of desirable living environments for humans.

8.1.2   WHOLE-BODY COORDINATION, FUNCTIONAL POTENTIALITY, AND PHYSIOLOGICAL POLYTYPISM

Biologically, animals, including humans, possess various physiological functions to maintain the healthy continuation of individuals, populations, and species. To understand physiological mechanisms acting as biological functions, it is important to consider them at the systems level. For instance, the mechanism of erect bipedalism of humans is associated not only with the mechanism of muscle contraction at a peripheral level but also with many other physical functions, such as neurological adjustments including that of the central nervous system, feedbacks from peripheral sensory receptors, and the provision of oxygen and energy to skeletal muscles. These numerous physiological functions not only act according to the commands sent by individual organs and tissues but also maintain their cooperativeness with one another while functioning at an individual level (whole-body coordination).

Adaptation and evolution involve changes in biological function such that the living organism is able to sustain life in altered environmental conditions. Continuous physical training contributes to the hypertrophy of muscle fibres, an increase in level of muscle contraction force and muscular endurance. The haemoglobin concentration of whole blood of high-altitude dwellers in the Andes is greater than that of sea-level dwellers; this suggests that these highlanders have physiological features that are suited to low-oxygen environments. Thus, humans have the potential to alter physiological function to adapt according to environmental requirements. In physiological anthropology, such human potential for adaptive alterations in physiological functions is referred to as functional potentiality.

When two different populations that exist under different environmental conditions are compared, it is natural that their physiological function phenotypes, acquired as the result of their adaptations, differ from each other. For instance, compared with individuals who are born in cold regions, those born in hot regions are known to produce less sweat and demonstrate lower electrolytic concentration of sweat when exposed to warm environments. The characteristics of the sweating function of people born in warm environments come as a result of acquisition of more effective perspiration function, via reduction in the amount of ineffective perspiration for heat radiation and via the reduction in electrolyte loss by perspiration. In physiological anthropology, such difference in physiological function phenotype is referred to as physiological polytypism.

In the realm of biological anthropology, the methodology of physiological anthropology is unique in that it focusses on measuring various human physiological reactions under well-controlled laboratory conditions, where physiological mechanisms as reactions to environmental stresses are observed at the level of the individual. Sato (2005) has described this as follows:

> Physiological anthropologists have come to realize the importance of individual thinking and the inadequacy of essentialistic concept such as the ideal man, and now infer that all populations are polytypic. Physiological anthropologists have refined the conceptual framework of their science and composed a set of keywords characterizing it. These are technological adaptability, environmental adaptability, functional potentiality, whole body coordination, and physiological polytypism. These keywords are mutually interdependent and do not form any orthogonal relations. (p. 289, with permission from The Japan Society of Physiological Anthropology)

One unique aspect of humans is that they possess a highly-developed cerebral neo-cortex, which enables them to perform extremely cognition-intensive activities. By adapting themselves skilfully to environments via such activity, humans in industria-lised societies have achieved extraordinary levels of prosperity. Ironically, they are forced to adapt themselves to new stresses caused by the consequent rapid change in environment. In this chapter, stresses in modern society and human adaptability are explored with a focus on mental stress.

8.2 HUMAN ACQUISITION OF INTELLIGENCE AND MENTAL STRESSES

In the process of evolution, humans have been exposed to many stresses. Baker (1984) has described that 'an interlocked process' created new stresses which evoked further adaptive responses in the history of human evolution. The advanced intel-ligence of humans, which contributed to cultural adaptation, is likely to have been acquired through an interlocking process as suggested by Baker.

Compared with other primates, the brain of humans has become significantly enlarged. Its size tripled over a period of approximately 2 million years and acquired dimensions of approximately 1,400 cm^3, while that of the chimpanzee brain is approximately 370 cm^3 (Holden 2006). Compared with other primates, the fron-tal polar cortex (Brodmann's area 10) of humans has become significantly larger; currently, it is approximately 6.5 times larger in volume than that of chimpanzee (Allman et al. 2002). Arguments put forward for this increase include increasingly complicated societal conditions (Lindenfors 2005; Sawaguchi 1992; Dunbar 1992; Kudo and Dunbar 2001; Wyles et al. 1983; Wilson 1985; Lefebvre et al. 1997; Jolly 1966; Humphrey 1976; Byrne and Whiten 1988; Whiten and Byrne 1997; Barrett and Henzi 2005), the production and use of tools (Ambrose 2001), and genetic features (Gagneux and Varki 2001; Tang 2006; Bradley 2008). However, most of such studies are comparative ones that employed animals other than primates. The human brain is not considered to have enlarged further over the past 2 million years (Holden 2006). Moreover, there is a low correlation (0.4) between intelligence quo-tient (IQ) and brain size (Holden 2006). It is, therefore, not easy to explain human intelligence, including its definition and content, in relation to the complexity of human society.

Increased efficiency of life via the acquisition of tools and languages is the essence of human techno-adaptability. Qualitative changes in living environments are likely to have increased demand for intellectual ability. Such demand is likely to have worked as a selective pressure, which may have contributed to the acceleration of human intelligence. However, this is not due to change in brain size (Balter 2002). Since human brain-size expansion is probably limited by narrow maternal pelvic size, subsequent development of intelligence is more likely to have been achieved via increased brain surface area (Balter 2005) and of the development of neural networks (Allman et al. 2002).

While the development of scientific technology, which underlies techno-adapt-ability, has increased convenience and efficiency of human life and has contributed to the prosperity of industrialised human society, it has produced certain types of stresses; these stresses had not been experienced by humans until now. Although sociopsychological hypotheses put forward the view that human intelligence has advanced in cycles of cultural adaptation and responses to new stresses generated by such adaptation (Baker 1984), its physiological mechanism has not been clarified. In Section 8.3, I put forward one possible mechanism.

8.3   MENTAL STRESS MODEL EMPLOYED IN THE LABORATORY

Human mental stress can be broadly classified into two categories: emotional and intellectual. Emotional stress is a state of mental strain caused by instinctive/intuitive sensations. This is commonly produced in humans and other animals mainly by the action of the limbic system. Intellectual stress is a state of mental strain caused by the recognition and processing of advanced information. This stress is considered to be mainly generated by the actions of the neocortex and its neural connection, namely, the limbic system. Intellectual stress is likely to have served as a mainspring of human brain development and as a selective pressure.

Many studies on mental stress in the realms of psychology and social science make no specific distinction between emotional and intellectual stress. It is important, however, to distinguish between these two types of stress in order to discuss human stress in an evolutionary framework, despite the fact that it is difficult to separate them, given the complex neural network connecting the neocortex to the limbic system. Furthermore, these two stress types are frequently related to each other when they occur in daily life. In order to study their physiological features, therefore, it is best to adopt simplified stress models in controlled laboratory environments.

A variety of experimental models have been proposed and employed to determine the psychological and physiological characteristics of mental stress. Mental arithmetic tasks (addition and subtraction) and colour–word tests are frequently employed as intellectual mental stress models. Anger recall and cold pressure tests, the public speech task, and the international affective picture system (IAPS) are employed as emotional mental stress models. There is an apparent difference in brain information handling between these two types of stress. In intellectual stress models, brain information handling is linguistic and logical, while in emotional stress models, it is nonlinguistic and intuitive.

Some stress tests employed in the laboratory consist of simple stressors, as seen in the mental arithmetic task and cold pressure test. Some other stress tests use multiple combinations of stressors that serve as models of commonly observed complex social stresses. For example, the Trier Social Stress Test (TSST), which is frequently employed for study on social psychological stress, requires examinees to deliver a speech before an audience and to perform serial mental arithmetic (Chong et al. 2008).

A good measure of physiological response to these two types of stress is that of blood pressure elevation. Blood pressure elevation associated with intellectual stress reflects a cardiac response characterised by an increase in cardiac output. Blood pressure response to emotional stress is induced by vascular response characterised by an increase in total peripheral resistance. These mechanisms of blood pressure elevation in response to different stresses are common phenomena. A difference can also be observed between the responses of the heart and the blood vessels to elevated blood pressure due to stress. At the individual level, blood pressure elevation as a response to intellectual mental stress cannot be explained only by an increase in cardiac output.

8.4   EXPERIMENTAL TRIAL IN LABORATORY

Mental arithmetic tasks and white-noise exposure as the laboratory models for intellectual and emotional stresses, respectively, have been used to examine physiological responses. In these experiments, cardiovascular reactions, particularly alteration in blood pressure, have been examined. More recently brain function, via electroencephalogram and brain haemodynamics, has been studied. A number of studies on physiological responses to mental stress have also been conducted by other academic disciplines, such as psychology and medical science. The respective cardiovascular reactions to white-noise exposure as an emotional stress model and to mental arithmetic task as an intellectual mental stress model have been compared (Liu et al. 2007a). Figure 8.1 shows changes in mean arterial pressure at the time of mental arithmetic task and white-noise exposure. The data represent mean values from ten subjects. The subjects were required to either perform a five-minute mental arithmetic task or to be exposed to white noise for five minutes, for four times each, with a three-minute break following each task or exposure, respectively. As shown in Figure 8.1, mean arterial pressure was elevated at the start of the arithmetic task or white-noise exposure, respectively. As a reaction to the mental arithmetic task, mean arterial pressure declined at the start of the third three-minute break period. There was no decline in mean arterial pressure in response to white-noise exposure at the start; rather, it remained elevated. At the start of the second white-noise exposure, a cumulative increase in mean arterial pressure was observed. The degree of

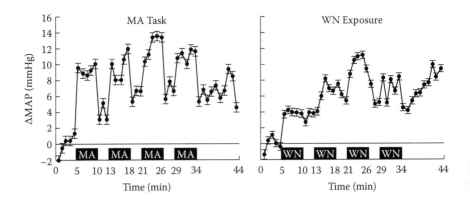

FIGURE 8.1   Changes in mean arterial pressure (MAP) during experimental period with intermittent mental arithmetic task (MA task) and white-noise exposure (WN exposure). Data are the mean and SE of ten subjects. (From Liu et al. 2007a, with permission from the Japan Society of Physiological Anthropology.)

elevation in mean arterial pressure, whose baseline was set at the pressure value measured prior to the first exposure, increased in stages up to the time of the third exposure; it subsequently declined at the termination of the third exposure and at the third break. The phased increase in degree of elevation was no longer observed at the time of the fourth exposure. One-way analysis of variance revealed that the degree of elevation of mean arterial pressure was significantly affected by the task period and that the degree of elevation at the time of the third exposure was significantly higher than that at the first exposure.

Concerning the performance of the mental arithmetic tasks, a change in mean arterial pressure was observed in clear response to the commencement and termination of the respective tasks, though the pressure value did not decline to its baseline level during the break periods. One-way analysis of variance showed no significant effects of task period on the degree of elevation in mean arterial pressure during each task period.

Figure 8.2 and Figure 8.3 show changes in cardiac output and total peripheral resistance during the mental arithmetic task and white-noise exposure. In response to mental arithmetic task, an increase in cardiac output and total peripheral resistance was observed. With respect to a change in cardiac output, an elevated level was maintained throughout the trial period except for the first break period and the first half of the third task period, when cardiac output declined. Regarding a change in total peripheral resistance, highly responsive patterns of increase and decrease were observed at the start and end of each task, respectively; however, the value declined almost to the baseline during the break periods.

With white-noise exposure, the changes in elevation of cardiac output tended to be random, being high at the first exposure but showing no further increase with following exposures. With respect to the changes in total peripheral resistance, the degree of elevation at second exposure, and all the exposures that followed, was higher than that at the first exposure. Total peripheral resistance remained higher than the baseline even during the break periods.

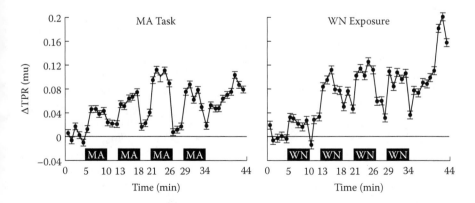

FIGURE 8.2   Changes in total peripheral resistance (TPR) during experimental period with intermittent mental arithmetic task (MA task) and white-noise exposure (WN exposure). Data are the mean and SE of ten subjects.

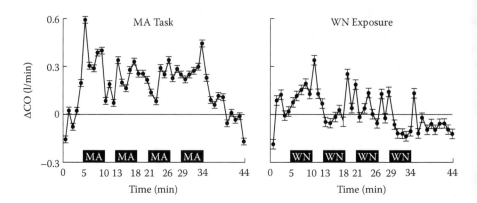

FIGURE 8.3   Changes in cardiac output (CO) during experimental period with intermittent mental arithmetic task (MA task) and white-noise exposure (WN exposure). Data are the mean and SE of ten subjects.

In terms of the degree of change at each task and exposure period, correlations with mean arterial pressure, cardiac output, and total peripheral resistance were examined. The degree of change was significantly and positively correlated with cardiac output in response to the mental arithmetic task ($r = 0.66$, $p < 0.001$) and with total peripheral resistance in response to white noise exposure ($r = 0.43$, $p < 0.001$).

An increase in mean arterial pressure at the time of the mental arithmetic task was mainly due to increased cardiac output. With the swift recovery of elevated total peripheral resistance to the baseline level with the completion of the task and commencement of the break period, the pattern of change in mean arterial pressure responded to both the task and the break. With respect to white-noise exposure, increased arterial pressure was mainly due to increased total peripheral resistance. The degree of elevation in total peripheral resistance peaked at the second exposure and all the exposures that followed it. Elevated total peripheral resistance did not recover to the baseline level with the completion of the task and may contribute to an accumulated increase in mean arterial pressure.

Sympathetic nervous system responses to stress can be classified into beta-adrenergic and alpha-adrenergic ones (Gregg et al. 1999). It has been reported that the stress responses to mental arithmetic tasks are generally beta-adrenergic and that they increase cardiac output (Sherwood et al. 1990; Kasprowicz et al. 1990). The stress responses to the cold pressure test and anger recall interview are generally alpha-adrenergic ones and are reported to cause an increase in total peripheral resistance (Bongard et al. 2002; Lawler et al. 2002). Intellectual mental stress possesses beta-adrenergic characteristics while emotional mental stress possesses alpha-adrenergic characteristics.

The different physical responses to the two types of mental stresses are observed at the mean value level but are not reflected among individuals. At the individual level, there is considerable diversity in response to both qualitatively different stresses, as well as to the same stressor.

8.5   PHYSIOLOGICAL POLYTYPISM OF STRESS RESPONSES

Figure 8.4 shows change in mean arterial pressure, cardiac output, and total peripheral resistance measured in two subjects while taking a five-minute mental arithmetic task. A change in blood pressure resulting from sudden exposure to stress is reflected in a change in cardiac output and total peripheral resistance. The mean arterial pressures and cardiac output of both subjects were elevated during the test. However, while the total peripheral resistance of one subject increased, that of the other decreased, and the possibility of different mechanisms at the individual level should not be discounted. The simultaneous measurement of multiple physiological responses and the observation of their inter-relativity as whole-body coordination is important, if such interindividual variation is to be understood.

The responses of cardiac output and total peripheral resistance to mental arithmetic tasks of forty subjects (Figure 8.4) were examined. Increases in cardiac output and total peripheral resistance were observed in twenty-four subjects, while an increase in cardiac output and a decrease in total peripheral resistance were observed in fourteen of them. A decrease in cardiac output was observed in two subjects, although they experienced an increase in total peripheral resistance. Such individual-based differences in response have been handled in much past research by averaging to a group mean value, without considering variation. In my view, however, such individual differences in physiological response reflect the physiological characteristics of the population as much as the average does. From the viewpoint of physiological

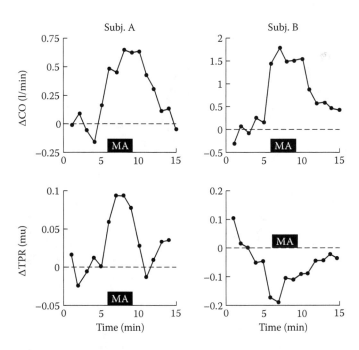

FIGURE 8.4   Changes in cardiac output (CO) and total peripheral resistance (TPR) during experimental period with single mental arithmetic task for five minutes of two subjects.

polytypism, such differences may mirror individual differences in adaptive strategy and degree of adaptation to environmental stress.

Approaches to physiological polytypism have been actively discussed by many physiological anthropologists in Japan. For example, a number of studies on individual differences in blood pressure response have been conducted in terms of cardiovascular responses to mental stress. If blood pressure response were the only variable considered, individual differences in stimulus–response reaction would be explained by a single distribution. As it is likely that individual differences in response to such a single parameter reflect differences in reactivity within a certain normal distribution, they do not show characteristics of physiological polytypism.

Organisms have numerous physiological functions which exhibit mutual cooperation or antagonism, and a physiological phenotype is determined by the manner of integration of these functions. From these, it is reasonable to consider the diversity of physiological phenotype (physiological polytypism) as differences in multiple physiological mechanisms in their responses to stimuli. To illustrate diversity of physiological phenotypes from the viewpoint of adaptation, however, it is necessary to demonstrate biological relative advantage based on the phenotype differences (Mazess 1975). An example of this is given below.

Thirty-seven subjects were classified into two groups according to their differences in reactivity (increase or decrease) of total peripheral resistance to the mental arithmetic task, and their cardiovascular reactions compared (Figure 8.5). Although it is expected that a significant difference would exist in total peripheral resistance

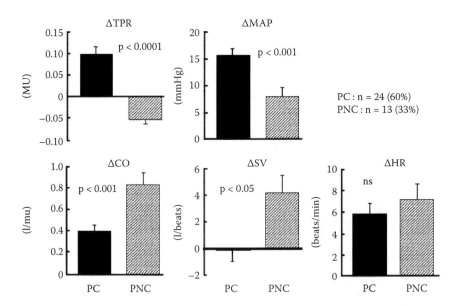

FIGURE 8.5 Comparison of cardiovascular responses between peripheral contributor [PC; total peripheral resistance (TPR) showed increase] and nonperipheral contributor (NPC; TPR showed decrease). MAP, CO, SV, and HR mean arterial pressure, cardiac output, stroke volume, and heart rate, respectively.

variation, the other parameters between the two groups showed statistically significant differences too. Compared to the group with a decrease in total peripheral resistance (the NPC group), the group with an increase in total peripheral resistance (the PC group) had a significantly lower cardiac reactivity, including cardiac output and stroke volume, and a significantly higher level of increase in mean arterial pressure. (There was no significant difference in heart rate, although the heart rate of the PC group tended to be lower than that of the NPC group.)

Blood pressure elevation, sympathetic nerve responses, and the risk of development of cardiovascular diseases have been much researched (Matthews et al. 1998). Stewart and France (2001) have reported that the high reactivity of heart rate to mental arithmetic tasks and the recovery of systolic blood pressure from tourniquet ischemia caused by cold pressure test and the use of a cuff were correlated with the systolic blood pressure elevation that would occur three years later. Using a videogame as a laboratory stress test, Markovitz et al. (1998) reported that the high reactivity of systolic blood pressure was significantly correlated with the systolic blood pressure elevation in the subsequent five years.

Recently, Flaa et al. (2008) has reported that concentrations of plasma norepinephrine and epinephrine as sympathetic nerve responses to laboratory stress tests (mental arithmetic tasks and cold pressure tests) predicted the level of systolic blood pressure eighteen years after the tests were taken. In addition, borderline hypertensive patients have been known to demonstrate a tendency to show a higher increase in blood pressure in response to laboratory stress test (mental arithmetic task) (Nestel 1969; Baumann et al. 1973; Light and Obrist 1980; Jern 1982; Eliasson et al. 1983). Furthermore, subjects with hypertensive family members demonstrate a higher increase in blood pressure in response to laboratory stress tests (colour–word tasks) than those without such family members (Widgren et al. 1992).

It can be presumed, therefore, that a group demonstrating an increase in total peripheral resistance as a response to mental stress is biologically disadvantaged and has a higher risk of developing cardiovascular diseases in the future than a group demonstrating no increase in total peripheral resistance under the same circumstances.

Differences in blood pressure reactivity to laboratory stress tests have been studied from numerous different viewpoints, including socioeconomic status (Steptoe et al. 2002, 2003), immigrant acculturation (Cruz-Coke 1987; Hackenberg et al. 1983; Hull 1979; Bongard et al. 2002), medical history of family (in terms of hypertension) (Lawler et al. 1998), sex difference (Cheng et al. 1997; Sabban 2007; Hermans et al. 2007), heredity (Cheng et al. 1997; Ijzerman et al. 2000), and personality (Habra et al. 2003). The contributing factors for our present results have not yet been clarified. With respect to various aspects of human intelligence and the diversity of intellectual demands in current life, further studies should be conducted to clarify whether stress reactions to mental arithmetic tasks reflect adaptability (techno-adaptability) to scientific technology.

8.6 WHOLE-BODY COORDINATION IN INTELLECTUAL STRESS

In recent years, central nervous system responses to intellectual and emotional stresses have been measured using P300 (the event-related potential of an electro-encephalogram) and brain haemodynamics (change in oxygenated haemoglobin concentration) measured by near-infrared spectroscopy. Simultaneously, cardiovascular reactions, with specific focus on blood pressure, were measured, and attempts made to elucidate the relationship between the central nervous system and whole-body circulatory dynamics in association with mental stress from the viewpoint of whole-body coordination.

In one experiment, a five-minute mental arithmetic task and exposure to white noise for five minutes were respectively conducted four times with a three-minute break following each test. An oddball task with the use of light for P300 derivation was conducted before and after the experiment. Cardiovascular reaction and brain haemodynamics were consistently measured throughout the experimental period.

Figure 8.6 shows a subject with electrodes for the derivation of electroencephalogram and probes for the detection of a change in haemoglobin concentration of the brain cortex. In the experiment, a pair of probes, which was designed specifically for the detection of haemoglobin concentration, was covered with black rubber to provide protection against ambient light. The electroencephalogram was detected at F7, F8, Fz, Cz, and Pz. A pair of probes to transmit and receive near-infrared light for haemoglobin concentration detection was placed by clipping the electrodes for F7 and F8. A past study showed that F7 and F8 are regions of the cerebral cortex which are activated by short-term memory and executive attention (Kane and Engle 2002).

Figure 8.7 shows changes in mean arterial pressure and oxygenated haemoglobin concentration of two subjects. The left column presents the respective results of mental arithmetic tasks, and the right column presents results of white-noise exposure. Changes in blood pressure as a response to the mental arithmetic task corresponded to those in haemoglobin concentration of the brain cortex. With white-noise exposure, blood pressure and brain haemodynamics did not correspond to each other.

Figure 8.8 shows the change in P300 of one subject. The thin line indicates the P300 measured prior to mental arithmetic task and white-noise exposure, while the thick line indicates the P300 measured following the task and exposure. P300 amplitude increased after the mental arithmetic task. This tendency was most noticeable at F7 and F8, located in the dorsolateral prefrontal cortex. Interestingly, P300

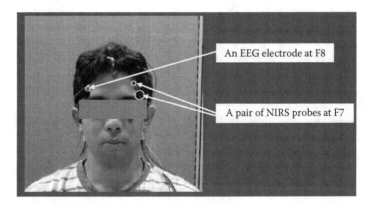

An EEG electrode at F8

A pair of NIRS probes at F7

FIGURE 8.6   A subject attached to EEG electrodes and near-infrared spectroscopy (NIRS) probes.

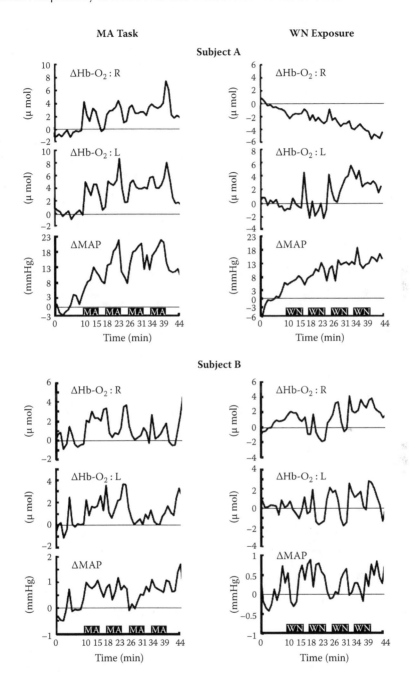

FIGURE 8.7 Changes in oxygenated haemoglobin concentration (Hb-O$_2$) and mean arterial pressure (MAP) during experimental period with intermittent mental arithmetic task (MA task) and white-noise exposure (WN exposure) of two subjects. Hb-O$_2$:R and Hb-O$_2$:L are Hb-O$_2$ at right and left hemisphere, respectively.

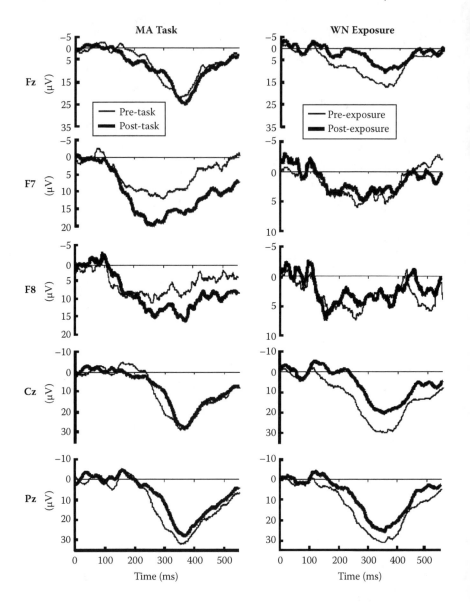

FIGURE 8.8 P300 obtained pre- and postexperimental period with intermittent mental arithmetic task (MA task) and white-noise exposure (WN exposure) of one subject.

amplitude tended to decrease after the white-noise exposure. The P300 amplitude at Cz and Pz tended to decrease after both the mental arithmetic task and the white-noise exposure, though the level of decrease caused by the white-noise exposure was higher than that by the mental arithmetic task.

As shown in Figure 8.7, the prefrontal cortex haemodynamics during the mental arithmetic task exhibited a pattern of change which corresponded to the change in mean arterial pressure. Such a tendency was not observed in white-noise exposure. Immediately after the mental arithmetic task, cognitive function demonstrated in the form of P300 amplitude, clearly peaked. These results indicate that the blood pressure elevation caused by intellectual mental stress ensures a sufficient amount of blood supply corresponding to the activation of the brain cortex. It is also suggested here that intellectual stress has been a selective pressure in human evolution and that this has contributed to the advancement of intellectual ability. In contrast, it is suggested here that emotional mental stress without intellectual information processing has had little effect on human intellectual advancement.

8.7   HUMAN ADAPTABILITY TO MENTAL STRESS: A HYPOTHESIS

As shown in Section 8.6, intellectual and emotional mental stresses cause cardiovascular and central nervous reactions, human adaptability to emotional mental and intellectual stresses being essentially different. With the repetition of white-noise exposure, the level of blood pressure elevation was cumulatively heightened across the first three exposures, but declined during the fourth. In this manner, the physiological strain resulting from continual emotional stress increased cumulatively, while the physiological response gradually declined with time regardless of the existence of the concerned stressor.

This phenomenon can be understood as the occurrence of an adaptive reaction to emotional stress. The cumulative response to emotional stress apparently renders the continual exposure to such stresses harder and induces active adaptation to avoid stress. If an individual fails to avoid stress actively and continues to be exposed to it, acclimation occurs and adaptation is established in a manner that eases physical strain. This hypothesis is supported by the experimental finding that the P300 amplitude at Cz and Pz significantly declined with white-noise exposure. Such a decrease in sensitivity is considered to not be caused only by auditory stimulation but also by olfactory and visual stimulation. Experimental findings also included elevation of blood pressure in response to repetitious mental arithmetic tasks and lowering of blood pressure during breaks in these tasks. This blood pressure reaction is useful for humans because it enables them to conduct the activity of cognition or information processing continually and for long durations, as long as short break periods are taken. In real life, it is possible to control the continuation of intellectual mental activities, by varying the number and duration of breaks; this is of great adaptive significance.

Considering the possibility that the change in blood pressure could be a response to the intermittent mental arithmetic task corresponding to the change in brain haemodynamics, discussion on human adaptability to intellectual mental activity can be developed further. In human evolution, whole-body coordination in relation to the haemodynamics of the circulatory system and central nervous systems has been developed in such a way that continual logical and mental activity is enabled. Figure 8.9 illustrates a hypothesis concerning reaction to mental arithmetic tasks, based on current insights into the coordination of the central nervous, sympathetic nervous, and cardiovascular systems. Performance of intellectual mental activity increased blood supply in response to an increase in demand for energy of the prefrontal cortex, which plays a role in decision making as a consequence of information processing. Performance of

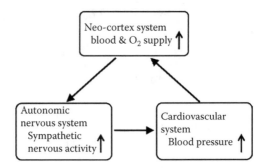

FIGURE 8.9   Schematic presentation of hypothesis of whole-body coordination to intellectual mental stress.

intellectual mental activity also provokes on the sympathetic nervous system, which results in blood pressure elevation. Such blood pressure elevation is coordinated in a way that corresponds to energy supply to the prefrontal cortex. Across evolutionary time, increased blood supply to the prefrontal cortex, as a response to intellectual mental activity, not only corresponds to an increase in demand accompanying information processing but would also contribute to physiological development, including the enlargement of the neocortex and the formation of a nervous network via the supply of various nutritional substrates to the prefrontal cortex.

The data presented here suggest two patterns of increased mean arterial pressure response to mental arithmetic tasks. With the first type, increased total peripheral resistance contributes to increased mean arterial pressure. With the second type, it does not. There was a significant difference between the two types in the level of increase in mean arterial pressure as a response to mental arithmetic tasks: That in the contributing type was higher than that in the noncontributing type. Many studies have shown that reactivity related to blood pressure elevation, such as cardiovascular reaction, to laboratory stress tests reflects the future risk for cardiovascular diseases. These two types of reactivity may reflect an important difference in reactivity to intellectual mental activity more generally. Reasons for these differences can be speculated upon and may involve genetics and environmental factors, such as differences in the level of technical intervention in the lived environment, and differences in personal history of the engagement in intellectual mental activity.

8.8   CONCLUSION

While the data shown in this chapter is limited, it does provide a satisfactory base for the hypothesis of techno-adaptability proposed. In the experiment described in this chapter, intellectual mental stress induced the activation of cortex function that corresponded with the demand for cognition and handling of information, indicating that intellectual demand across human evolution was a selective pressure for the enlargement of the neocortex and the advancement of intelligence.

The emotional stress investigated here was a physical one generated by environmental factors. Emotional mental stress in real social life, however, may be stronger and harsher than the stress model used in our study. It is therefore important to investigate the extent to which the experimental stress models employed in the laboratory reflect stress in daily life.

Many past studies have supported the relativity between data obtained in the laboratory and results produced in real life. It is necessary to explore this relationship between the stress models employed in various laboratories and the characteristics of adaptability in daily life. If the level of blood pressure elevation is related to the risk of developing cardiovascular disease, the vascular response type with higher level of blood pressure elevation would have a higher risk of developing hypertension or other cardiovascular diseases. Since most of the responses to emotional stress are vascular, adaptability to emotional mental stress may be inferior to the adaptability to intellectual mental stress.

Emotional stress frequently occurs in both humans and other animals, and exposure to emotional stress is very likely to precede intellectual stress in the course of human evolution. If the hypothesis that the adaptability of modern humans to emotional stress is inferior to that to intellectual stress is accurate, it may be that emotional stress in modern society has been changing in terms of quality and that humankind needs to find new adaptive powers to these new types of stresses.

The laboratory stress tests presented in this study are valid in examining the physiological mechanisms of response to constant stress stimuli, as loaded in a controlled environment. In order to examine these results from a biological viewpoint, which includes adaptation and evolution, however, it is necessary to view them in collaboration with the achievements of field work with a large volume of samples, long-term follow-up, and data on health in daily life. In this sense, what has been discussed in this study should serve as a bridge between specific laboratory research and fieldwork.

ACKNOWLEDGEMENT

This study was supported in part by a Grant-in-Aid for Scientific Research (No. 20370097, No. 15207021, and No. 15107005) from the Japan Society for the Promotion of Science.

REFERENCES

Allman, J., Hakeem, A. and K. Watson. 2002. Two phylogenetic specializations in the human brain. *Neuroscientist* 8(4):335–46.
Ambrose, S.H. 2001. Paleolithic technology and human evolution. *Science* 291:1748–53.
Baker, P.T. 1984. The adaptive limits of human populations. *Man* 19(1):1–14.
Balter, M. 2002. Becoming human. What made humans modern? *Science* 295:1219–25.
Balter, M. 2005. Paleoanthropology. Small but smart? Flores hominid shows signs of advanced brain. *Science* 307:1386–9.
Baumann, R., Ziprian, H., Gödicke, W., Hartrodt, W., Naumann, E. and J. Läuter. 1973. The influence of acute psychic stress situations on biochemical and vegetative parameters of essential hypertensives at the early stage of the disease. *Psychother. Psychosom.* 22:131–40.
Barrett, L. and P. Henzi. 2005. The social nature of primate cognition. *Proc. R. Soc. B* 272:1865–75.

Bongard, S., Pogge, S.F., Arslaner, H., Rohrmann, S. and V. Hodapp. 2002. Acculturation and cardiovascular reactivity of second-generation Turkish migrants in Germany. *J. Psychosomatic Res.* 53:795–803.

Bradley, B.J. 2008. Reconstructing phylogenies and phenotypes: A molecular view of human evolution. *J. Anat.* 212:337–53.

Byrne, R.W. and A. Whiten. 1988. *Machiavellian intelligence: Social expertise and the evolution of intellect in monkeys, apes and humans.* Oxford: Oxford University Press.

Cheng, L. S-C., Carmelli, D., Hunt, S.C. and R.R. Williams. 1997. Segregation analysis of cardiovascular reactivity to laboratory stressors. *Genet. Epidemiol.* 14:35–49.

Chong, R.C., Uhart, M., McCaul, M.E., Johnson, E. and G.S. Wand. 2008. Whites have a more robust hypothalamic-pituitary-adrenal axis response to a psychological stressor than blacks. *Psychoneuroendocrinol.* 33:246–54.

Cruz-Coke, R. 1987. Correlation between prevalence of hypertension and degree of acculturation. *J. Hypertens.* 5:47–50.

Dunbar, R.I.M. 1992. Neocortex size as a constraint on group size in primates. *J. Hum. Evol.* 20:469–93.

Eliasson, K., Hjemdahl, P. and T. Kahan. 1983 Circulatory and sympatho-adrenal responses to stress in borderline and established hypertension. *J. Hyperten.* 1:131–9.

Flaa, A., Eide, I.K., Kjeldsen, S.E. and M. Rostrup. 2008. Sympathoadrenal stress reactivity is a predictor of future blood pressure: An 18-year follow-up study. *Hypertension* 52(2):336–41.

Gagneux, P. and A. Varki. 2001. Genetic differences between humans and great apes. *Mol. Phylogenet. Evol.* 18:2–13.

Gregg, M.E., James, J.E., Matyas, T.A. and E.B. Thorsteinsson. 1999. Hemodynamic profile of stress-induced anticipation and recovery. *Inter. J. Psychophysiol.* 34:147–62.

Habra, M.E., Linden, W., Anderson, J.C. and J. Weinberg. 2003. Type D personality is related to cardiovascular and neuroendocrine reactivity to acute stress. *J. Psychosomatic Res.* 55:235–45.

Hackenberg, R.A., Hackenberg, B.H., Magalit, H.F., Cabral, E.I. and S.V. Guzman. 1983. Migration, modernization and hypertension: Blood pressure levels in four Philippine communities. *Med. Anthropol.* 7:45–71.

Hermans, E.J., Putman, P., Baas, J.M., Gecks, N.M., Kenemans, J.L. and J. van Honk. 2007. Exogenous testosterone attenuates the integrated central stress response in healthy young women. *Psychoneuroendocrinol.* 32:1052–61.

Holden, C. 2006. Human behavior and evolution society meeting. An evolutionary squeeze on brain size. *Science* 312(5782):1867.

Hull, D. 1979. Migration, adaptation, and illness: A review. *Soc. Sci. Med.* 13A:25–36.

IJzerman, R.G., Stehouwer, C.D.A. and D.I. Boomsma. 2000. Evidence for genetic factors explaining the birth weight-blood pressure relation analysis in twins. *Hypertension* 36:1008–12.

Iwanaga, K. 2005. The biological aspects of physiological anthropology with reference to its five keywords. *J. Physiol. Anthropol. Appl. Hum. Sci.* 24(3):231–5.

Iwanaga, K., Liu, X., Shimomura, Y. and T. Katsuura. 2005. Approach to human adaptability to stresses of city life. *J. Physiol. Anthropol. Appl. Hum. Sci.* 24(4):357–61.

Jern, S. 1982. Psychological and hemodynamic factors in borderline hypertension. *Acta Med. Scand. Suppl.* 662:1–54.

Jolly, A. 1966. Lemur social behavior and primate intelligence. *Science* 153(735):501–6.

Kane, M.J. and R.W. Engle. 2002. The role of prefrontal cortex in working-memory capacity, executive attention, and general fluid intelligence: An individual-differences perspective. *Psychonomic Bull. Rev.* 9(4):637–71.

Kasprowicz, A.L., Manuck, S.B. and S.B. Malkoff. 1990. Individual differences in behaviorally evoked cardiovascular response: Temporal stability and hemodynamic patterning. *Psychophysiol.* 27:605–19.

Kudo, H. and R.I.M. Dunbar. 2001. Neocortex size and social network size in primates. *Anim. Behav.* 62:711–22.

Lawler, K.A., Kline, K., Seabrook, E. et al. 1998. Family history of hypertension: A psychophysiological analysis. *Int. J. Psychophysiol.* 28:207–22.

Lawler, K.A., Kline, K.A. and R.F. Adlin. 2001. Psychophysiological correlates of individual differences in patterns of hemodynamic reactivity. *Int. J. Psychophysiol.* 40:93–107.

Light, K.C. and P.A. Obrist. 1980. Cardiovascular reactivity to behavioral stress in young males with and without marginally elevated casual systolic pressures. Comparison of clinic, home, and laboratory measures. *Hypertension* 2:802–8.

Lindenfors, P. 2005. Neocortex evolution in primates: The 'social brain' is for females. *Biol. Lett.* 1:407–10.

Liu, X., Iwanaga, K., Shimomura, Y. and T. Katsuura. 2007a. Comparison of stress responses between mental tasks and white noise exposure. *J. Physiol. Anthropol.* 26(2):165–71.

Liu, X., Iwanaga, K., Shimomura, Y. and T. Katsuura. 2007b. Different types of circulatory responses to mental tasks. *J. Physiol. Anthropol.* 26(3):355–64.

Markovitz, J.H., Raczynski, J.M., Wallace, D., Chettur, V. and M.A. Chesney. 1998. Cardiovascular reactivity to video game predicts subsequent blood pressure increases in young men: The CARDIA study. *Psychosom. Med.* 60(2):186–91.

Matthews, K.A., Owens, J.F., Kuller, L.H., Sutton-Tyrrell, K., Lassila, H.C. and S.K. Wolfson. 1998. Stress-induced pulse pressure change predicts women's carotid atherosclerosis. *Stroke* 29:1525–30.

Mazess, R.B. 1975. Biological adaptation: Aptitudes and acclimatization. In *Biosocial interrelations in population adaptation*, ed. E.S. Watts, F.E. Johnston and G.W. Lasker, 918–27. The Hague: Mouton Publishers.

Nestel, P.J. 1969. Blood pressure and catecholamine excretion after mental stress in labile hypertension. *Lancet* 1(7597):692–4.

Sabban, E.L. 2007. Catecholamines in stress: Molecular mechanisms of gene expression. *Endocr. Regul.* 41(2–3):61–73.

Sato, M. 2005. The development of conceptual framework in physiological anthropology. *J. Physiol. Anthropol. Appl. Hum. Sci.* 24(4):289–95.

Sawaguchi, T. 1992. The size of the neocortex in relation to ecology and social structure in moneys and apes. *Folia. Primatol.* 58:131–45.

Sherwood, A., Dolan, C.A. and K.C. Light. 1990. Hemodynamics of blood pressure responses during active and passive coping. *Psychophsiol.* 27:656–68.

Steptoe, A., Feldman, P.J., Kunz, S., Owen, N., Willemsen, G. and M. Marmot. 2002. Stress responsivity and socioeconomic status. *Eur. Heart J.* 23:1757–63.

Steptoe, A., Magid, K., Edwards, S., Brydon, L., Hong, Y. and J. Erusalimsky. 2003. The influence of psychological stress and socioeconomic status on platelet activation in men. *Atherosclerosis* 168:57–63.

Stewart, J.C. and C.R. France. 2001. Cardiovascular recovery from stress predicts longitudinal changes in blood pressure. *Biol. Psychol.* 58:105–20.

Tang, B.L. 2006. Molecular genetic determinants of human brain size. *Biochem. Biophys. Res. Commun.* 345:911–6.

Whiten, A. and R.W. Byrne. 1997. *Machiavellian Intelligence II. Extensions and Evaluations.* Cambridge: Cambridge University Press.

Widgren, B.R., Wikstrand, J., Berglund, G. and O.K. Andersson. 1992. Increased response to physical and mental stress in men with hypertensive parents. *Hypertension* 20(5):606–11.

Wilson, A.C. 1985 The molecular basis of evolution. *Sci. Am.* 253(4):164–73.

Wyles, J.S., Kunkel, J.G. and A.C. Wilson. 1983. Birds, behavior, and anatomical evolution. *Proc. Natl. Acad. Sci. U.S.A.* 80:4394–7.

9 Impact of Pollution on Physiological Systems

Taking Science from the Laboratory to the Field

*Lawrence M. Schell**
Department of Anthropology, State University
of New York, Albany, New York

CONTENTS

9.1 INTRODUCTION

Pollution affects parameters of interest to biological anthropologists. These include human reproduction, human growth and ageing, morbidity, and mortality. All these outcomes are based on physiological systems that are susceptible to disturbance by one or more pollutants. In fact, many systemic effects can be traced to effects of pollutants on endocrine pathways or on neurotransmitters which are involved in endocrine system regulation. The main exceptions are target organ effects such as damage to lung tissue by some irritating air pollutants. Correct understanding of the biological effects of pollution depends on the use and proper interpretation of both laboratory and field studies.

The basis for interest in the effects of pollutants on human biology and variation derives from a large body of literature using animal models. Experimental in vivo studies show unmistakable evidence of physiological effects. As voluminous and diverse as this literature is—encompassing effects on reproduction, development, and immunological features, as well as mortality and morbidity—acceptance of these results as appropriate models of human biology has not been unequivocal.

---

* Address all correspondence to lmschell@albany.edu.

Interpretation of some animal studies can be problematic. Until recently, most of these employed single-exposure models whereas human exposure is multifaceted. Thus, questions arise as to whether a particular pollutant has the same effect alone as when accompanied by other pollutants. In addition, until recently many laboratory studies employed a dosing regimen that increased logarithmically (for example, 0, 1, 10, 100, 1000 units) whereas human exposures are often within the lower end of that range, leaving doubt as to the applicability to humans of high dose effects in animals. Finally, the realisation that the timing of exposure is a critical variable in producing effects makes the interpretation of results from animals to humans difficult because of differences in timing of developmental stages. Nevertheless, laboratory studies have established the plausibility of certain effects of pollutants in mammals, and give substance to the epidemiological studies that find statistical associations between measures of exposure and anthropologically relevant outcomes.

9.2   KNOWING THE EFFECTS OF POLLUTANTS: SOME DIFFICULTIES AND SUCCESSES

The preference by the press to sensationalise reports of endocrine effects of pollutants has framed the scientific debate on pollution to too large a degree. Reports on hermaphroditic polar bears, for example, with scientifically unsubstantiated assertions create the impression in some uncritical readers that all studies are unsubstantiated and help to relegate the careful study of effects in humans to the same category of Internet fodder. The study of pollution and physiological effects is thus put in a defensive position. Over the past few years, several sources of supposed change in human sexual development have been proposed including phthalates, persistent organic pollutants such as polychlorinated biphenyls (PCB), and the pesticide *p-p*-dichlorodiphenyl-trichroloethane (DDT), diet (too much beef, too little zinc and folic acid), cell phone use, and even tight trousers.

Field studies of the effects of pollutants are especially susceptible to criticism due to the limitations of retrospective research designs generally, even though practitioners have laboured strenuously to reduce flaws. The basic design for the study of pollution among human populations has been retrospective because purposefully exposing humans to a suspected pollutant is unethical. Retrospective designs have flaws that are challenging. It is difficult to know the effects of pollutants from field studies because of problems in the accurate measurement of exposure, problems with the selection of biological outcomes, difficulty in knowing and measuring relevant control variables, and hindrances to the application of proper sampling techniques. Of these, only the problem of exposure assessment is discussed here. The other problems are faced in field studies of a variety of environmental variables and have been dealt with by others to varying extents.

Exposure assessment is especially difficult. Usually in field studies, the pollutant exposure took place in the past and has to be reconstructed. This may be attempted from residence histories but the kinetics of pollutants in the environment are not known precisely and proximity to a source may not be an accurate indicator of exposure itself. For example, the study of effects from the pollutants at Love Canal

TABLE 9.1

Hierarchy of Types of Measurement of Exposure or Surrogates of Exposure in Terms of Accuracy to Actual Exposure from Best (#1) to Worst (#7)

1. Quantified personal measurements (body burden).
2. Quantified measurements of the environment near the residence or other locations of activities.
3. Quantified surrogates of exposure.
4. Distance form a point source and duration of residence.
5. Distance or duration of residence.
6. Residence or employment in an area reasonably close to the location where exposure is assumed.
7. Residence or employment in defined geographic area (e.g., a postal code) of the location where exposure is assumed.

Source: Committee on Biological Markers of the National Resource Council, 1987.

in New York was plagued by uncertainty regarding who was exposed and to what extent they were exposed. Some models of exposure were based on the presence of swales, naturally occurring drainage ditches that were later filled with building materials when houses were constructed and through which pollutants were thought to have moved more freely. With this model, families near swales were thought to be more exposed (Goldman et al. 1985; Vianna 1980). Other exposure calculations used a diffusion-based model in which pollutants moved from the canal fairly uniformly in all directions. There was considerable dispute around the different models of pollutant movement from Love Canal and that discussion greatly influenced the credibility of the studies of biological outcomes there.

The National Research Council produced a hierarchy of exposure assessment techniques (Committee on Biological Markers of the National Research Council 1987) (Table 9.1). The less expensive techniques, such as using the location of residence as a surrogate measure of exposure, are rated rather low, and the most expensive method, the measurement of the pollutant in the individual, is rated the highest. However, even individual measurement, which is the gold standard, is not perfect because measurement of a pollutant in the body reflects more than exposure. Depending on the pollutant's chemical properties it can be influenced by different biological features. Lipophilic compounds are affected by body composition (Gallo et al. 2002; Schildkraut et al. 1999), while diet affects lead uptake (Schell, Denham, et al. 2003), and stage of skeletal development also affects uptake. Lead is released from bone when calcium is liberated, as it is in older individuals. Breastfeeding affects the suckled infant as well as the mother's own body burden of lipophilic compounds. This raises the question of how body burden (the retrospective assessment of exposure) reflects the biologically relevant exposure. In operational terms, this expands to how well the measurement of a pollutant, or its metabolite some years after the exposure, might reflect the original exposure, and the extent to which the original exposure created the damage as opposed to the stored material. Prospective studies may address most of these problems, but these are far from perfect. Longitudinal

TABLE 9.2
Examples of the Use of Growth Patterns to Assess Risk from Pollutants

Lead

Prenatal:

Scotland—(Moore et al. 1982)

U.S.—(Dietrich et al. 1986)

Postnatal:

Poland—(Ignasiak et al. 2006)

Belgium—(Lauwers et al. 1986)

U.S.—(Schell and Stark 1999)

U.S. —(Schwartz et al. 1986)

U.S.—(Frisancho and Ryan 1991)

Noise

Prenatal:

Finland—(Hartikainen et al. 1994; Hartikainen-Sorri et al. 1991)

U.S.—(Schell 1981)

Holland—(Knipschild et al. 1981)

Japan—(Ando 1988; Ando and Hattori 1973)

Postnatal:

Japan—(Takahashi and Kyo 1968)

Japan—(Schell and Ando 1991)

U.S.—(Schell and Hodges 1985)

Polychlorinated biphenyls and other related compounds

Prenatal:

Japan (Yusho)—(Yamashita and Hayashi 1985; Yoshimura 1974)

U.S.—(Taylor et al. 1984, 1326; Taylor et al. 1989, 760; Eskenazi 2003)

Postnatal:

Japan—(Fujisawa and Fujiwara 1972)

Japan—(Yoshimura and Ikeda 1978)

US—(Jacobson et al. 1990)

China (Yu-cheng)—(Guo et al. 1994)

studies re-examine children to monitor exposures and effects. If handled insensitively, such chronic probing may discourage long-term participation.

Despite the subtleties in the conduct of research with human populations in the field, many studies have found strong and close relationships between measures of pollutant body burden and relevant biological effects or endpoints (Table 9.2). Numerous studies of lead and growth, reproduction, and child development have produced consistent results in populations around the world (Dietrich et al. 1986; Ignasiak et al. 2006; Moore et al. 1982; Schwartz et al. 1986). Similarly, the effect of noise stress on prenatal development has been studied in France (Coblentz et al. 1990), Japan (Ando 1988; Ando and Hattori 1973), the Netherlands (Knipschild et al. 1981), Finland (Hartikainen et al. 1994; Hartikainen-Sorri et al. 1991), and the United States (Schell 1981, 1982) with similar results. PCBs, a group of lipophilic, persistent organic pollutants, have been related to disruption of a cluster of functions related to sexual maturation and sexual reproduction (Battershill 1994; Faroon et al. 2001; Yu et al. 2000).

Learning how successful field studies are conducted reveals that they employ findings from laboratory studies to guide both the selection of biological outcomes for study and the selection of exposure measures from the many available. Reviewing research with the Akwesasne Mohawk nation can reveal some of the ways to integrate methods and results from laboratory and field studies to greatest advantage.

9.3 TAKING LABORATORY-DERIVED KNOWLEDGE TO INFORM FIELD STUDIES: AKWESASNE

The study of Akwesasne Mohawk youth provides an example of use of laboratory findings to structure the study design and the statistical analysis. This study has been described (Schell, Hubicki, et al. 2003) and the information presented here concerns analyses of observations on 271 youth between the ages of ten and seventeen who were born between 1978 and 1990.

The Akwesasne Mohawk nation is a community of Haudensaunee people living near, or on, the St Regis Reservation which straddles the St Lawrence River and overlaps New York State, and Ontario and Quebec, Canada. The river has several industries along its banks. The Akwesasne community has been exposed to poly-chlorinated biphenyls from nearby manufacturing plants that have contaminated the St Lawrence River and raised pollutant levels in fish above the safety mark. Fish are a traditional component of the Mohawk diet but in the mid-1980s authorities issued advisories against eating locally caught fish and consumption declined. Despite the recent decline, several studies have documented elevated levels of PCBs in children, adolescents, and adults of both genders.

PCB molecules exist in different configurations, called congeners. There is a maximum of 209 possible forms, depending on the locations of chlorine atoms that are substituted for carbon atoms in the joined biphenyl rings (Figure 9.1). Field studies have identified around 100 of these in humans, and yet far fewer of these at noticeable levels. At Akwesasne, sixteen different congeners were found among 50% or more of the people studied. This report concentrates on the effects of those sixteen.

Whether 16 or 100, statistical testing with so many exposure variables produces biased results as the tests are not independent of one another. Furthermore, the distribution of values is highly skewed for any single congener. A practical solution is to limit testing to groups of congeners or to a small number of individual ones. Grouping congeners often produces more normally distributed values and allows conformity with assumptions of many statistical tests. The choice of which congeners to group together is based on findings from experimental studies of laboratory animals. Such studies have indicated which ones are persistent once ingested and which are short-lived in mammals. They also indicate which congeners bind to the aryl hydrocarbon receptor and which activate the phenobarbital pathway, two different pathways of effect. The aryl hydrocarbon receptor is present in tissues of the female reproductive tract (Pocar et al. 2005). Thus, laboratory studies have identified which congeners can stimulate tissues that are characteristically stimulated by estrogens and are therefore considered estrogenic. Likewise, tests of other congeners reveal which are anti-estrogenic or anti-

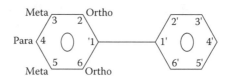

FIGURE 9.1 Polychlorinated biphenyl molecules may exist in many forms depending on the number and location of substitutions of carbon with chlorine.

androgenic. Finally, laboratory studies have established which congeners are concentrated in breast milk and which of them are stored for long periods in adipose tissue and can be released long after the exposure has occurred. The groupings of congeners reflect qualities of persistence and specific tissue-type toxicities as revealed by numerous experimental studies of laboratory animals and tissues (Table 9.3).

PCBs are not the only pollutant at Akwesasne. Like most human populations in industrialised areas, people have body burdens of DDE (*p,p'*-dichlorophenylethylene, a metabolite and marker of exposure to the pesticide DDT), lead, mercury, and HCB (hexachlorobenzene) as well as mirex. The levels of these toxicants in the Akwesasne youth have been described (Schell, Hubicki, et al. 2003). Based on results of laboratory studies of endocrine effects of these toxicants, it may be possible to identify

---

TABLE 9.3
Groupings of PCB Congeners Used in Analyses of Effects among the Akwesasne Mohawk Youth

Sum of congeners with a 50% detection rate:
IUPAC #s 52, 70, 74, 84, 87, 95, 99, 101, 105, 110, 118, 123, 138, 153, 180, 187.

Sum of 8 persistent congeners:
IUPAC #s 74, 99, 105, 118, 138, 153, 180, 187.

Sum of nonpersistent congeners:
IUPAC #s 52, 70, 84, 87, 95, 101, 110, 123.

Wolff's estrogenic/neurotoxic group:
IUPAC #s 52, 70, 101 [+90], 187

Wolff's anti-estrogenic group:
IUPAC #s 74, 105, 118, 138 [+163,164]

Wolff's enzyme-inducing group:
IUPAC #s 99, 153, 180

Note: PCB congeners are analysed in groups according to rate of detection, persistence, and certain toxicological properties. Individual congeners are identified by International Union of Pure and Applied Chemistry (IUPAC) number. Numbers in brackets are congeners that are detected in combination with another congener. (For compositions of Wolff's groups, see Wolff et al. 1997; Wolff and Toniolo 1995.)

which ones to test against specific biological outcomes in the Akwesasne community. However, a complete matrix of all toxicants by all outcomes is unwieldy and impracticable. A better strategy is to predetermine which toxicants are biologically-plausible influences on particular biological systems and then test specific outcomes related to those systems with appropriate corrections for repeated statistical tests. Thus, careful hypothesis testing with a mixture of exposures demands as much fore-knowledge of effects as possible.

Not only is the composition of the exposure important but so too is its timing. An important issue in studies of human developmental toxicity is the identification of critical windows of exposure and effect. In earlier discussions of difficulties in understanding the effects of pollutants on human populations, the matter of different life-stage timing was identified as important; from now-classic work on thalidomide, toxicologists know that the timing of exposure is critical. Thus, knowing the timing of exposure during foetal and postnatal development may be critical to being able to observe effects postnatally. However, in most instances of exposure in human populations, such as the Akwesasne Mohawk, the exposure is chronic. Often the foetus is exposed chronically from transplacental passage of toxicants that have been stored in maternal fat (such as lipophilic organic pollutants like PCBs) or bone (such as lead). Of course, extreme maternal exposure during pregnancy can produce a tempo-rary surge in foetal exposure also, but in many cases the foetal exposure is chronic. Chronic exposure makes it difficult to identify critical windows of sensitivity during the prenatal period. After parturition, a substantial change in exposure can occur through lactation. It would be worthwhile from a public health perspective to deter-mine whether postnatal exposure has similar, different, or no effects to prenatal expo-sure, because the former is easier to prevent or reduce than prenatal exposure. Prenatal exposure is based on the mother's bodily burden of toxicants from earlier exposure and is difficult to reverse. Thus, distinguishing effects of congeners or other toxicants that are present primarily through transplacental passage, lactation, or postnatal diet can provide valuable, practical information for preventive recommendations.

Exposure to lipophilic toxicants at Akwesasne has occurred by consumption of locally-caught fish and probably also by consumption of game that live in the river or that feed on river-dwelling animals. Toxicants stored in maternal adipose tissue are released into circulation where they can pass into the foetus during gestation. Postnatal exposure through breast milk can make a substantial contribution to the infant's burden because breast milk contains fat. At Akwesasne, breastfed adoles-cents have higher PCB levels than those who had not been breastfed, the difference being among the persistent congeners (Schell, Hubicki, et al. 2003). Postnatal expo-sure also occurs through consumption of foods with toxicant content.

Previous analyses using a preliminary sample from Akwesasne ($n = 115$) have shown the more common congeners (those found in 50% or more of the sample, termed here PCB50%), to be associated with changes in the levels of thyroid stimu-lating hormones (TSH) and thyrotropin (Schell et al. 2004). Recent analysis of the entire sample using multiple regression analyses ($n = 232$) show a statistically sig-nificant relationship between PCB50% and TSH with standardised beta coefficient of 0.24 ($p < 0.05$), and between PCB50% and free thyroxine with standardised beta coefficient of -0.27 ($p < 0.01$) (Schell et al. 2008). The different signs of the relationships between PCBs with thyrotropin and with thyroxine reflect the negative feedback system controlling the level of free thyroxine by altering the level of thyrotropin. However, when only the eight persistent congeners are tested against thyrotropin, the relationship is stronger, with a beta coefficient of 0.31, the relationship with thyroxine remaining unchanged. Tests of relationships with congeners that are more reflective of current exposure are all nonsignificant. This suggests that exposure much earlier in life is more influential than current exposure.

The earlier exposures are by transplacental transfer and lactation. Further testing with the sample stratified by whether they had or had not been breastfed showed the former to exhibit no relationships between PCBs and thyroid hormones, while the latter showed the strongest relationships (Schell et al. 2008). The reason for this dichotomy might be that breastfeeding is protective of PCB effects. However, the levels of PCBs in youth who had been breastfed are significantly higher than among those who had not been breastfed. According to general theories of toxicity, greater exposure is associated with more severe effects. Thus, the greater exposure of breastfed youth would be expected to produce greater effects. This militates against the possibility of a beneficial effect of breastfeeding in countering effects of PCBs on thyroid hormone regulation. When greater exposure is associated with smaller effects, it suggests the presence of other influential variables in the system.

Alternatively, the PCBs transferred through lactation may not influence thyroid hormone regulation and thus the PCBs added through lactation may not contribute to a relationship between PCB level in adolescence and thyroid hormone levels but add random variation that swamps and obscures the effect of prenatal exposure. The lack of influence could be due to two or more factors. First, the profile of PCB congeners transferred through breastfeeding could be different from those transferred across the placenta, and the different profiles could have different effects. In this explanation, it is not the stage of development of the exposed individual that influences the effects created but the character of the exposure itself. Alternatively, the profiles could be similar but the stages of development differ, such that the influence of PCBs on the development of the thyroid regulatory system is quite dependent on when the exposure occurs, being absent in postnatal exposure and present in prenatal exposure. Current knowledge of the ontogenesis of the thyroid regulatory system is not sufficient to determine whether this is so, however.

Another example of laboratory studies that are used to inform field studies is in the analysis of PCB effects on sexual maturation. Previous work by laboratory toxicologists has indicated that mechanisms affecting reproductive system endpoints were different from other types of endpoints. Accordingly, Wolff and colleagues developed a classification of PCB groups as estrogenic, antiestrogenic, and enzyme inducing (Table 9.3) for the PCB congener composition of each group (Wolff et al. 1997; Wolff and Toniolo 1995). This classification was employed in a successful study of the effects of PCBs and other toxicants on sexual maturation among Akwesasne girls (Denham et al. 2005), although some congeners comprising Wolff's groups were not detected among Akwesasne youth. The data collection methods, toxicant analysis, and sample characteristics are as described above, but for girls only ($n = 138$). Multiple logistic regression analysis has shown higher levels of the estrogenic group of PCBs to predict earlier age at menarche. The other groups of congeners were not associated with age at menarche. These associations were evident before and after controlling for covariates including age, breastfeeding, lipids, socioeconomic status of the family, and other toxicants. BMI was tested also but was not a significant predictor and was not included in the model. Forcing BMI into the model gave no change to this relationship.

9.4 CONCLUSION

The results from research with the Akwesasne Mohawk nation show how information obtained through laboratory studies can guide field studies in very important ways. The inference that earlier exposure may be a critical factor is based on laboratory studies of the structure of the PCB molecule that predicts specific biological actions. Laboratory work in this case has helped to define the exposure variable and reduce the number of possible candidates of causal variables to a manageable number. This approach can be generalised to many types of studies. Furthermore, although this review has dealt with exposure more than the specification of likely effects, laboratory results have suggested relevant outcomes to examine in field study. They have indicated that thyroid and reproductive effects were more likely to arise in humans than effects in other organ systems. Finally, field studies always derive associations between putative causes and effects and these must be substantiated in several ways. One of the usual criteria for evaluating a statistical association is the presence of a plausible biological mechanism. Laboratory studies are able to provide this essential link.

REFERENCES

Ando, Y. 1988. Effects of daily noise on fetuses and cerebral hemisphere specialization in children. *J. Sound Vib.* 127:411–7.

Ando, Y. and H. Hattori. 1973. Statistical studies on the effects of intense noise during human fetal life. *J. Sound Vib.* 27:101–10.

Battershill, J.M. 1994. Review of the safety assessment of polychlorinated biphenyls (PCBs) with particular reference to reproductive toxicity. *Hum. Exp. Toxicol.* 13:581–97.

Coblentz, A., Martel, A. and G. Ignazi. 1990. Effects of fetal exposition to aircraft noise on the birthweight of children. *Proc. Hum. Factors Soc.* 562–6.

Committee on Biological Markers of the National Research Council. 1987. Biological markers in environmental health research. *Environ. Health Perspect.* 74:3–9.

Denham, M., Schell, L.M., Deane, G., Gallo, M.V., Ravenscroft, J. and A. Decaprio. 2005. Relationship of lead, mercury, mirex, dichlorodiphenyldichloroethylene, hexachlorobenzene, and polychlorinated biphenyls to timing of menarche among Akwesasne Mohawk girls. *Pediatrics* 115:E127–34.

Dietrich, K.N., Krafft, K.M., Bier, M., Succop, P.A., Berger, O.G. and R.L. Bornschein. 1986. Early effects of fetal lead exposure: Neurobehavioral findings at 6 months. *Int. J. Biosocial Res.* 8:151–68.

Eskenazi, B., Mocarelli, P., Warner, M., Chee, W.-Y., Gerthoux, P.M., Samuels, S. et al. 2003. Maternal serum dioxin levels and birth outcomes in women of Seveso, Italy. *Environ Health Perspect.* 111:947–53.

Faroon, O.M., Keith, S., Jones, D. and C. De Rosa. 2001. Effects of polychlorinated biphenyls on development and reproduction. *Toxicol. Ind. Health* 17:63–93.

Frisancho, A.R. and A.S. Ryan. 1991. Decreased stature associated with moderate blood lead concentrations in Mexican-American children. *Am. J. Clin. Nutr.* 54:516–9.

Fujisawa, H. and B. Fujiwara. 1972. On the influence which PCB Exercises on the development of the child. *Nat. Sci.* 13:15–21.

Gallo, M.V., Ravenscroft, J., Denham, M., Schell, L.M., Decaprio, A. and the Akwesasne Task Force on the Environment. 2002. Environmental contaminants and growth of Mohawk adolescents at Akwesasne. In *Human Growth from Conception to Maturity*, ed. G. Gilli, L.M. Schell, and L. Benso, 279–87. London: Smith-Gordon.

Goldman, L.R., Paigen, B., Magnant, M.M. and J.H. Highland. 1985. Low birth weight, prematurity and birth defects in children living near the hazardous waste site, Love Canal. *Hazard Waste Hazard Mater.* 2:209–23.

Guo, Y.L., Lin, C.J., Yao, W.J., Ryan, J.J. and C.C. Hsu. 1994. Musculoskeletal changes in children prenatally exposed to polychlorinated biphenyls and related compounds (Yu-Cheng children). *J. Toxicol. Environ. Health* 41:83–93.

Hartikainen, A-L., Sorri, M., Anttonen, H., Tuimala, R. and E. Laara. 1994. Effect of occupational noise on the course and outcome of pregnancy. *Scand. J. Environ. Health* 20:444–50.

Hartikainen-Sorri, A-L., Kirkinen, P., Sorri, M., Anttonen, H. and R. Tuimala. 1991. No effect of experimental noise on human pregnancy. *Obstet. Gynecol.* 77:611–5.

Ignasiak, Z., Slawinska, T., Rozek, K., Little, B.B. and R.M. Malina. 2006. Lead and growth status of school children living in the Copper Basin of South-Western Poland: Differential effects on bone growth. *Ann. Hum. Biol.* 33:401–14.

Jacobson, J.L., Jacobson, S.W. and H.E.B. Humphrey. 1990. Effects of exposure to PCBs and related compounds on growth and activity in children. *Neurotoxicol. Teratol.* 12:319–26.

Knipschild, P., Meijer, H. and H. Sallé. 1981. Aircraft noise and birth weight. *Int. Arch. Occup. Environ. Health* 48:131–6.

Lauwers, M-C., Hauspie, R.C., Susanne, C. and J. Verheyden. 1986. Comparison of biometric data of children with high and low levels of lead in the blood. *Am. J. Phys. Anthropol.* 69:107–16.

Moore, M.R., Goldberg, A., Pocock, S. et al. 1982. Some studies of maternal and infant lead exposure in Glasgow. *Scott. Med. J.* 27:113–22.

Pocar, P., Fischer, B., Klonish, T. and S. Hombach-Klonish. 2005. Molecular interactions of the aryl hydrocarbon receptor and its biological and toxicological relevance for reproduction. *Reproduction* 129:379–89.

Schell, L.M. 1981. Environmental noise and human prenatal growth. *Am. J. Phys. Anthropol.* 56:63–70.

Schell, L.M. 1982. The effects of chronic noise exposure on human prenatal growth. In *Human Growth*, ed. J. Borms, R. Hauspie, A. Sand, C. Susanne and M. Hebbelinck, 125–9. New York: Plenum Press.

Schell, L.M. and Y. Ando. 1991. Postnatal growth of children in relation to noise from Osaka International Airport. *J. Sound. Vib.* 151:371–82.

Schell, L.M. and D.C. Hodges. 1985. Longitudinal study of growth status and airport noise exposure. *Am. J. Phys. Anthropol.* 66:383–9.

Schell, L.M. and A.D. Stark. 1999. Pollution and child health. In *Urbanism, Health and Human Biology in Industrialised Countries,* ed. L.M. Schell and S.J. Ulijaszek, 136–57. Cambridge: Cambridge University Press.

Schell, L.M., Denham, M., Stark, A.D. et al. 2003. Maternal blood lead concentration, diet during pregnancy, and anthropometry predict neonatal blood lead in a socioeconomically disadvantaged population. *Environ. Health Perspect* 111:195–200.

Schell, L.M., Hubicki, L.A., Decaprio, A.P et al. 2003. Organochlorines, lead, and mercury in Akwesasne Mohawk youth. *Environ. Health Perspect* 111:954–61.

Schell, L.M., Gallo, M.V., Decaprio, A.P. et al. 2004. Thyroid function in relation to burden of PCBs, *P,P'*-Dde, HCB, mirex and lead among Akwesasne Mohawk youth: A preliminary study. *Environ. Toxicol. Pharmacol.* 18:91–9.

Schell, L.M., Gallo, M.V., Denham, M. et al. 2008. Relationship of thyroid hormone levels to levels of polychlorinated biphenyls, lead p,p'- DDE and other toxicants in Akwesasane Mohawk youth. *Environ. Health Perspect.* 116:806–13.

Schildkraut, J.M., Demark-Wahnefried, W., Devoto, E., Hughes, C., Laseter, J.L. and B. Newman. 1999. Environmental contaminants and body fat distribution. *Cancer Epidemiol. Biomarkers Prev.* 8:179–83.

Schwartz, J., Angle, C.R. and H. Pitcher. 1986. Relationship between childhood blood lead levels and stature. *Pediatrics* 77:281–8.

Takahashi, I. and S. Kyo. 1968. Studies on the differences in adaptabilities to the noise environment in sexes and growing processes. *J. Anthropol. Soc. Nip.* 76:34–51.

Taylor, P.R., Lawrence, C.E., Hwang, H.-L. and A.S. Paulson. 1984. Polychlorinated biphenyls: influence on birthweight and gestation. *Am. J. Public Health* 74:1153–4.

Taylor, P.R., Stelma, J.M. and C.E. Lawrence. The relation of polychlorinated biphenyls to birth weight and gestational age in the offspring of occupationally exposed mothers. *Am. J. Epidem.* 129:395–406.

Vianna, N.J. 1980. Adverse pregnancy outcomes—potential endpoints of human toxicity in the Love Canal preliminary results. In *Embryonic and Fetal Death,* ed. I. Porter and E. Hook, 165–8. San Diego: Academic Press.

Wolff, M.S and P.G. Toniolo. 1995. Environmental organochlorine exposure as a potential etiologic factor in breast cancer. *Environ. Health Perspect.* 103:141–5.

Wolff, M.S., Camann, D., Gammon, M. and S.D. Stellman. 1997. Proposed PCB congener groupings for epidemiological studies. *Environ. Health Perspect.* 105:13–4.

Yamashita, F. and M. Hayashi. 1985. Fetal PCB Syndrome: Clinical features, intrauterine growth retardation and possible alteration in calcium metabolism. *Environ. Health Perspect.* 59:41–5.

Yoshimura, T. 1974. Epidemiological study on Yusho babies born to mothers who had consumed oil contaminated by PCB. *Fukuoka Igaku Zasshi* 65:74–80.

Yoshimura, T. and M. Ikeda. 1978. Growth of school Children with polychlorinated biphenyl poisoning or *Yusho. Environ. Res.* 17:416–25.

Yu, M-L., Guo, Y-Ll., Hsu, C-C. and W.J. Rogan. 2000. Menstruation and reproduction in women with polychlorinated biphenyl (PCB) poisoning: Long-Term follow-up interviews of the women from Taiwan Yucheng cohort. *Int. J. Epidemiol.* 29:672–7.

10 Bridging the Gap between Laboratory Trials and the Reality of the Human in Context

*Neil J. Mansfield**
Department of Human Sciences, Loughborough University, Leicestershire, United Kingdom

CONTENTS

10.1 INTRODUCTION

Scientific investigation of humans can be broadly categorised into 'laboratory' and 'field' studies. Results from these studies can be used to develop or test theoretical models. Laboratory studies are generally designed to be performed under controlled conditions where as many relevant variables as possible are controlled, or at least monitored. Field studies are usually designed to study the human in context, where it is generally impossible and/or undesirable to control all relevant variables, although these are often monitored.

If a particular human effect is observed in some field context, then it might be desirable to simulate the field environment to develop understanding of the effect such that it can be minimised or enhanced, as appropriate. For example, field studies of drivers of mining machines (such as large bulldozers) might show 'effects' such as elevated risks of low back pain, evidence of fatigue, and occasional lapses in concentration. Field studies might also show several possible 'causes' of adverse effects such as lack of postural variation, regular twisting, and exposure to whole-body vibration

---

* Address all correspondence to n.j.mansfield@lboro.ac.uk.

(see, for example, Newell et al. 2006). Based on these observations, a researcher might then take a laboratory study approach in order to use sophisticated biomechanical, physiological, or psychological methods to indicate the main contributors to risks and to provide practical advice to the machine manufacturer, site manager, and machine driver (Newell and Mansfield 2008). In this example, there is a clear link between the field study, the laboratory study, and the ultimate benefactors.

Humans in context are diverse, having varied social, cultural, and physical attributes, skills, and experiences. Similarly, environments are diverse, including factors such as temperature, humidity, air quality, noise, vibration, physical layouts, and accommodation. The interactions between humans and their environments are complex, and unless diversity is considered, recommendations from field or laboratory trials might be misleading. For example, when setting limits for environmental noise, the nature of the noise, the affected population, and location interact such that it is not possible to predict the community response from pure objective measures such as dB(A), even though dB(A) can be shown to correlate well with annoyance under controlled laboratory conditions (see, for example, Job and Hatfield 2001; Flindell 1998; Kuwano and Namba 1985).

It is possible to perform laboratory experiments which are repeatable and consistent, yet the results might not be applicable in the field. This can be thought of as analogous to the Müller-Lyer illusion (Figure 10.1; Müller-Lyer 1889). If experimental subjects are asked to judge the size of the two items in Figure 10.1a, it would be correct for them to state that they were of identical length. If these two items were then placed into context, as being part of a sketch of a partitioned office (Figure 10.1b), they would then likely state that they were of different size, despite them being identical length (Figure 10.1c). Thus the context of observation can alter the results, since what might be considered important in one context might be considered trivial in another.

Utmost care is needed to ensure that any study framework is appropriate and fits the purpose. The framework might alter depending on the purpose of the work, while practical constraints might require that compromises be made to make a study feasible. A problem arises when the data from a 'compromised' study is applied into a context for which it is unintended and unsuitable. Even within studies, compromises might be necessary. During a repeated-measures design laboratory study with several independent variables, one might need to choose between obtaining several repeat measures of the same combination of conditions (thus increasing confidence in the data), increasing the number of conditions tested (thus increasing resolution in the data), or risking participant fatigue (thus questioning the validity of any of the data). In order to ensure a strong framework for experimental studies, sound use of standard scientific approaches is essential and might include power analyses, pilot studies, careful randomisation, and reporting limitations. Before embarking on any study, the scientist must have confidence that the framework is robust enough to support any conclusions which might be sought.

This chapter highlights some difficulties in transferring findings from the laboratory to the field in order to assist those designing studies in the future to maximise scientific validity. Although the chapter takes its examples from environmental ergonomics, the principles can be applied in many contexts; indeed, the chapter itself is compromised by seeking to provide general guidance and inspiration by using several examples, but at the cost of lack of detailed reporting of case study limitations.

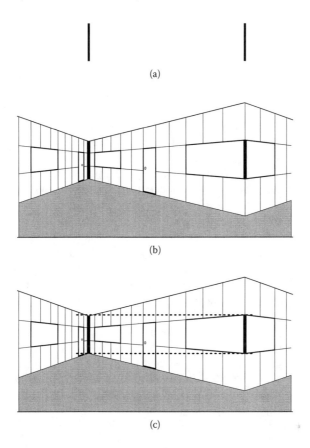

FIGURE 10.1   Effect of context of judgement of line length. In (a) the two lines are perceived as having identical length; in (b) the two lines are perceived as having different lengths, despite them being identical as shown in (c).

Therefore, the reader is encouraged to study primary information sources if intending to use data reported here.

10.2   THE RISKS OF REPEATABILITY

One of the main reasons to transfer research from the field to the laboratory is to enable control of environmental variables. If one is attempting to understand the effects of one particular variable, then other variables can be held constant, and the variance in the data due to those other variables will be reduced or eliminated. Although this improvement in repeatability might appear desirable, if this occurs at the cost of obtaining realistic data which is comparable to what is observed in the field, it could be counterproductive. To illustrate this effect, risk assessments for workers exposed to hand–arm vibration emitted from handheld power tools will be considered.

In 2005, the European Physical Agents (Vibration) Directive was implemented across the European Union. The purpose of the directive is to protect workers from risks arising from vibration, in particular musculo-skeletal, neurological, and vascular disorders. Amongst other requirements, the directive requires assessment of risk for workers exposed to hand–arm vibration and whole–body vibration. In Europe, machinery suppliers have a requirement to report the vibration emission of their products. To ensure fair comparisons are made between suppliers, each tool must be tested according to the appropriate test code which defines the conditions under which it must be used during the vibration measurements (Mansfield 2005). Central to the test codes is a principle that the same result should be obtained for any test house performing the same standard test on that particular model of tool. As there is great deal of variability due to the work being completed, the test conditions are often artificial. For example, in the test codes, breakers do not break concrete but are operated on steel ball bearings; rammers are used on neoprene foam rather than ramming to compact loose material. Although these codes are under revision, there remains the problem that in the interests of repeatability, only a single test condition is required, and this test condition is rarely representative of the task for which the tool is used by operators on site.

Results reported in the manufacturer's handbooks are derived from tool test codes and can provide little indication of the vibration experienced by the tool user. For example, for a combi-hammer (a popular type of heavy-duty power drill used in construction), the vibration emission depends on the type of drill bit used (Figure 10.2) and has a great deal of variability (coefficients of variation being typically 10% for back-to-back repeat measures). As illustrated in Figure 10.2, manufacturers' reported data can greatly underestimate the true vibration emission. A large discrepancy between reported vibration emissions and measurements of exposures in the field has led to end-users loss of trust in manufacturers' data and they have looked elsewhere for guidance on risk management.

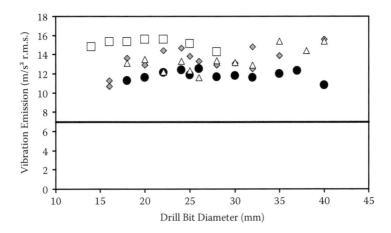

FIGURE 10.2 Vibration emission of an electric power tool (a combi-hammer) measured whilst drilling 40 N concrete using different types of drill bit (denoted by different marker shapes) of different diameter. Solid line denotes the declared value reported in the manufacturer's handbook as 'typical', which is lower than any value measured on site.

The situation for power tools parallels the requirements of testing passive safety performance of automobiles in Europe in the early 1990s. European law required safety testing to occur under standardised conditions which, although designed to be highly repeatable, were not representative of the speeds and types of crashes which cause injury to vehicle occupants and pedestrians. The European New Car Assessment Programme (EuroNCAP) was established in 1995 by research and motoring consumer organisations, which recognised the unsuitability of the legally-required tests and embarked on a programme to test independently a range of popular new automobiles under more representative conditions (EuroNCAP 2005). This scheme has grown to such an extent that, although vehicles are still required to pass the basic safety tests, consumers are mostly unaware of these and take great interest in EuroNCAP test results. Manufacturers have a clear incentive to ensure that their products perform better than their competitors' and this has contributed to the rapid improvements in secondary safety. As the test codes remain independent, they can be changed rapidly and in response to innovation. In the case of vehicle safety then, laboratory tests were considered unsatisfactory by the end-users who effectively took things into their own hands by defining new tests which were more meaningful, albeit less repeatable.

Following the example of EuroNCAP, building construction trade organisations, professional bodies, and individual companies looked to obtain tool vibration-emission data from an independent source. Thus, the trade association OPERC and Loughborough University worked together to establish a database of hand–arm vibration measurements for tools measured in simulated work-site conditions (www.operc.com). HAVTEC (Loughborough University's Hand–Arm Vibration Test Centre) has since measured many thousands of tool/consumable/workpiece combinations and provided these data in an easily searchable format to industry, free of charge. This has followed the principles of EuroNCAP, providing freely and easily accessible data, independent tests, optimised test methods, realistic test scenarios, and use of established measurement techniques (Mansfield 2006; Rimell, Notini et al. 2008).

For both examples, those with an interest in health and safety are able to select products and to distinguish between the 'best' and 'worst' in class. Manufacturers have a desire to perform well in the rankings and have, as a result, improved their technologies to keep up with their best-in-class competitors. This means that the products are safer for all end-users, whether or not they consider safety a selling point.

At the heart of the problems in testing tool vibration and vehicle passive safety has been the desire to optimise repeatability. The consequence of this has been end-users establishing their own tests, through HAVTEC and EuroNCAP, as repeatability was obtained at the expense of realism ('ecological validity'). When considering any form of human study, the requirement for ecological validity must be considered. In some cases, it is of no importance; in others, it is essential.

10.3   CROSS-MODAL INFLUENCES

In psychophysics, modality (or channel) refers to the sense being used to detect a stimulus. In most cases, a full cognitive model of an environment is achieved through many modalities simultaneously, although one particular channel could dominate. For qualitative research, such effects can be important, and some types of physiological research are also affected by cross-modal influences. As a simple example, if one is interested in changes in heart rate due to physical loading from high-g manoeuvres in high-performance automobiles or aircraft, experimental subjects would need to be carefully selected such that the response is not purely due to fear of a novel environment (although in some cases, this might be the objective of the work itself).

Cross-modal effects can be subtle and have been exploited by some industries to enhance the perceived quality of their products. One of the most commonly experienced cross-modal effects is combined auditory and tactile stimuli in motor vehicles. When an automobile traverses an irregularity in the road, a small shock is transferred through the floor, seat, and steering wheel (tactile channels) whilst a sound simultaneously occurs (auditory channel). Although the driver's attention might only be drawn to the tactile channel, the auditory signal can add to the overall sensation of the event. This means that if one is performing laboratory-based research on just one of these factors (such as perception of the vibration/shock at the hands for drivers or motorcyclists), then unless the auditory channel is controlled, the results could be misleading.

In a laboratory experiment to study the tactile–auditory cross-modal interaction, subjects were required to judge the intensity of a sequence of impulses of different magnitudes replayed through a steering wheel, typical of shocks experienced by the hands whilst driving over road irregularities (Mansfield et al. 2007). Each of these impulses was accompanied by the sound of driving over a bump in a road. For 75% of the stimuli, the sound experienced was identical [93 dB(A)$_{peak}$], but for the remainder, it was either 5 dB(A) higher or lower. Whilst subjects were instructed to rate the intensity of the shock as felt by the hands, there was no reference to the auditory stimulus. In this way the true purpose of the experiment was withheld from the subjects. Subjective ratings of the shocks increased with shock magnitude for those conditions where there was no change in sound; however, an effect of sound level was also observed, where ratings of shock vibrations were significantly higher with the higher intensity sound, and significantly lower for the lower intensity sound. This demonstrates that tactile sensation can be manipulated by the auditory channel. Whilst this might be exploited in the automotive industry to improve perceptions of 'quality', it could cause misleading results in experimental research, where one requires subjects to express opinions of one detail of an environment which cannot be isolated.

Subjective opinion of stimuli can also be manipulated by diverting attention to one modality. Rimell, Mansfield et al. (2008) showed that if subjects were shown digitally processed 'talking head' material on a screen, similar to watching a newsreader, subjective ratings of the quality of the audio and video could not be separated. As the audio quality was reduced whilst the video signal did not change, subjects perceived that the quality of both the audio and video channels was degraded. Likewise, as the video quality was reduced with no change in the audio signal, subjects perceived that the quality of both channels was degraded. In a second experiment, subjects were given a task to complete, designed to direct attention to either the auditory or video channel. After each short talking-head clip, subjects were asked their opinion of either the audio quality or the video quality. When no task was performed, the subjective ratings for either audio or video quality reduced with degradations in either, or both, modalities, as seen before (Figure 10.3C and Figure 10.3F). This effect is a classic cross-modal interaction. If the attention of subjects was drawn to one modality, then

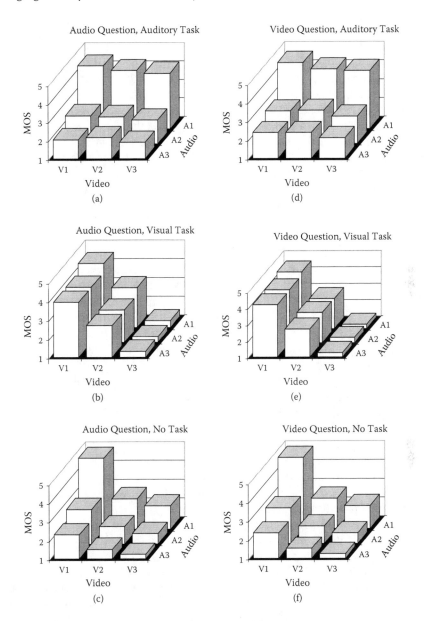

FIGURE 10.3 Effect of focussed attention of perceptions of quality of 'talking head' video and audio material. (A) Audio-quality question and auditory task, (B) Audio-quality question and visual task, (C) Audio-quality question and no task (control condition), (D) Video-quality question and auditory task, (E) Video-quality question and visual task, and (F) Video-quality question and no task (control condition). The z axis denotes the mean opinion score (MOS) using the ITU quality scale (ITU, 1996) where 1 = Bad, 2 = Poor, 3 = Fair, 4 = Good, 5 = Excellent. The three different levels of video are labelled V1 to V3 and the three different levels of audio are labelled A1 to A3, where higher numbers denote more distortion. (Reprinted with permission from Rimell, Mansfield et al. 2008).

reports of the quality of any channel would be dominated by the quality of the modality of attention. If subjects had their attention drawn to the audio channel (Figure 10.3A and Figure 10.3D), and this channel was of high quality (A1), then opinions of the audio would be high but opinions of the video would also be high, irrespective of the true quality of the video channel. If subjects had their attention drawn to the video channel (Figure 10.3B and Figure 10.3C), and this channel was of high quality (V1), then opinions of the video would be high but opinions of the audio would also be high, irrespective of the true quality of the audio channel. This demonstrates that subjects do not necessarily answer the question which is posed by the experimenter, but their opinions can be driven by what is considered to be important information to them.

Thus, unless the whole environment is carefully controlled, subjective experimental data can give misleading results.

10.4   MULTIFACTORIAL SCENARIOS

Cross-modal interactions occur when sensation from one psychophysical channel interferes with outputs from a second psychophysical channel. This could be considered as one 'input' interfering with multiple 'outputs'. An alternative problem occurs when one 'output' is affected by several 'inputs'. In the human sciences, there is rarely a simple single-input, single-output relationship, and there is a danger of oversimplification to ensure compatibility with previous research.

Low back pain for forklift truck drivers is an example of a multifactorial scenario (multiple input, one output). Such drivers often are required to manually handle some of their loads, are exposed to high magnitudes of whole-body vibration, and are required to drive in twisted postures. A manual handling expert might associate a high prevalence of low back pain with the requirements of the lifting aspect of the job, a vibration specialist might associate the pain with the vibration, and a postural specialist might consider the twisted posture the cause of the low back pain. A social scientist might also find issues with psychosocial aspects of the job. In this example there is unlikely to be a single identifiable cause of the problem, and it is possible that some or all of the risk factors interact.

Another example of a multifactorial scenario which has been studied in the laboratory is that of driver discomfort. Laboratory and field trials have demonstrated that discomfort increases with extended sitting time and that the rate of increase in discomfort varies between seats (Porter et al. 2003). It has been recommended that at least two hours of sitting are required in order to differentiate between different car-seat designs (Gyi and Porter 1999). These trials have shown that it is insufficient to simply sit in a seat and rate the comfort immediately (a 'showroom' evaluation), since fatigue is an additional important consideration. Ebe and Griffin (2000a, 2000b) have also shown that showroom evaluations are insufficient and demonstrated the importance of the dynamic properties of the seat. This means that for smooth roads, the dynamic aspects are less important than for rough roads, where vibration isolation properties of the seat can become an important aspect of overall seat discomfort. Combining these two dimensions allows for a conceptual model of overall seat discomfort to be developed, including dynamic, static, and temporal factors (Figure 10.4).

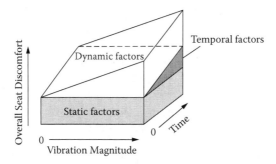

FIGURE 10.4 Model of vehicle seat discomfort including static, dynamic, and temporal factors. (Reprinted with permission from Mansfield 2005.)

Testing this model in the laboratory using an independent-measures design showed that for no vibration, discomfort increased with time (as per Porter et al. 2003). It was also shown that at each time interval the discomfort was greater with increased vibration magnitude (as per Ebe and Griffin 2000a, 2000b). However, there was also an interaction between vibration and time, where the vibration accelerated the onset of discomfort (Figure 10.5). This means that although the presence of vibration

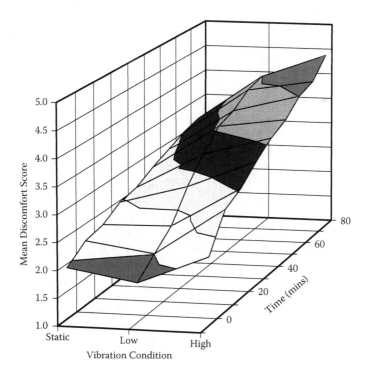

FIGURE 10.5 Mean discomfort score against vibration magnitude and exposure time for subjects sitting in an automobile in the laboratory. Discomfort increases with time and with vibration magnitude.

increased the discomfort at the start of the user trials, the magnitude of the difference became greater as time passed. Thus, long-term sitting whilst driving on rough roads produces more discomfort than would be predicted from user trials considering just long-term sitting, or user trials considering just the dynamic performance of the seat and its efficacy at reducing the vibration.

In multifactorial scenarios then, it might be possible to produce results which are internally valid and repeatable but miss other important factors. This means that if an end-user wishes to implement the recommendations, the improvements in context could be disappointing.

10.5 COMMUNICATION OF RESULTS

Whilst there might be difficulties in ensuring that laboratory-based research is valid, even sound research can fail to meet its ultimate objectives if it is not communicated effectively to the target audience. Both the message and the medium must be carefully selected to be appropriate.

If the target audience consists of other researchers, then publication of results in peer-reviewed journals might be appropriate, as one should have confidence that researchers proactively seek out information in their area of interest. In many cases, though, the ultimate beneficiaries from research findings will not have access to academic library resources and might even have no knowledge that the topic area exists. Since these individuals will not seek appropriate data, they will not find it. In such circumstances, findings should also be disseminated through conduits such as popular press, television, radio, trade magazines, direct to policymakers, conference presentations, workshops, community posters, and the Internet.

The HAVTEC tool-vibration database is an example where the tool user should be the ultimate beneficiary such that risks of occupational disease can be minimised. These users need to be trained to understand that vibration is a potential risk, otherwise they will never seek out the database; therefore, research results are targeted at trainers, including health and safety managers, through media such as trade magazines as well as free workshops. Once trained, users still need to have access to appropriate information which is quick and easy to understand. In the case of HAVTEC, a traffic light system was developed such that, in collaboration with tool hire companies, each tool can be tagged according to how long it can be used and how much work can be done before reaching thresholds for risk. This means that even those distant from the academic laboratory and with no access to the database at the point of sourcing the tool can use the information in a simple and effective way. As it is also important to communicate best practise to the research community, outcomes have also been published in the academic literature.

The problem of communication of results is nothing novel; in 1884 the physicist Lord Rayleigh stated at a meeting of the British Association for the Advancement of Science:

> At the present time and in some departments, the accumulation of material is so rapid that there is a danger of indigestion. By a fiction as remarkable as any to be found in law, what has once been published, even though it be in the Russian language, is usually

spoken of as 'known', and it is often forgotten that the rediscovery in the library may be a more difficult and uncertain process than the first discovery in the laboratory. (Lord Rayleigh 1884)

There has been little improvement in the intervening years (see, for example, Vickery 1963). For information to be 'rediscovered' in a library, an individual needs to have the inclination and means to search for it. Findings need to be publicised for the nonscientist in a form which can be easily understood, and published for the scientist in a form which can be easily sourced. The task of scanning the library for relevant articles has become easier with automated electronic searches and databases, although care is needed to ensure that, as authors, the appropriate keywords are used, and as readers, the searches are appropriate. Communication of results is an integral part of research programme design, with care to ensure that the target audience is reached effectively.

10.6 CLOSING COMMENTS

This chapter has shown that unless care is taken when conducting research in the laboratory, there is a risk that it can become compromised such that it is no longer relevant to the context of application. Many of the studies cited here have actively sought out the limitations of experimental method and have been designed to establish the thresholds of external applicability, despite results being internally valid.

There are compromises which need to be made in the laboratory, and researchers are ultimately faced with the challenge of pragmatism, ensuring that studies can proceed with available resources. There are recommendations in this chapter which might appear contradictory: for example, Section 10.2 warns that simplification can compromise applicability, whereas Section 10.3 demonstrates how simplification might be required in order to remove cross-modal influences. Researchers must balance these contrasting risks to perform their work where their findings are reliable.

Research represents a method of traversing a gap from what is known to what is desired to become known. The metaphorical bridge between these points needs to be strong. This can be achieved by consideration of

1. Assumptions which must be made due to the limitations of method
2. Complexity of human responses to their environment
3. Interaction of the human with complex environments
4. Confounding factors outside the immediate research focus
5. Targeting reporting of results, in terms of media and message

If these issues are ignored, then the bridge which represents the findings might not be robust enough to be of practical benefit.

Taking such a cautious approach to research can incur costs in terms of resource allocation, but if the applicability is increased, the cost represents good value. One rewarding way of minimising the types of risks addressed here is to create cross-disciplinary teams, potentially including advisers in areas of application, policymakers, and/or those recognised as experts in methods used for the work.

Humans are unpredictable and complex, but despite this, human scientists seek to understand and explain their observed behaviour and response. Human and environmental diversity mean that many of the details studied with human biology only follow deterministic patterns at a population level, rather than at the individual level (for example, although it might be possible to predict disease incidence and risk factors, it is usually not possible to predict exactly which individuals will be affected). One of the side effects of this is that individuals within the population of those protected from adverse effects are unaware of the extent to which their lives had been improved by the application of research.

REFERENCES

Ebe, K. and M.J. Griffin. 2000a. Qualitative models of seat discomfort including static and dynamic factors. *Ergonomics* 43(6):771–90.

Ebe K. and M.J. Griffin. 2000b. Quantitative prediction of overall seat discomfort. *Ergonomics* 43(6):791–806.

EuroNCAP. 2005. Creating a market for safety. http://www.euroncap.com.

European Commission. 2002. Directive 2002/44/EC of the European Parliament and of the Council of 25 June 2002 on the minimum health and safety requirements regarding exposure of workers to the risks arising from physical agents (vibration). *Official Journal of the European Communities* L177, 13–9.

Flindell, I.H. 1998. Fundamentals of human response to sound. In *Fundamentals of Noise and Vibration*, ed. F.J. Fahy and J.G. Walker. London: E & FN Spon.

Gyi, D.E. and J.M. Porter. 1999. Interface pressure and the prediction of car seat discomfort. *Appl. Ergon.* 30(2):99–107.

International Telecommunications Union. 1996. Subjective video quality assessment methods for multimedia applications. ITU-T Recommendation, p. 910.

Job, R.F.S. and J. Hatfield. 2001. Responses to noise from combined sources and regulation against background noise levels. In *Proceeding of InterNoise 2001*. The Hague, The Netherlands.

Kuwano, S. and S. Namba. 1985. Continuous judgment of level-fluctuating sounds and the relationship between overall loudness and instantaneous loudness. *Psychol. Res.* 47(1):27–37.

Mansfield, N.J. 2005. *Human Response to Vibration.* Boca Raton: CRC Press.

Mansfield, N.J. 2006. Filling the gap between manufacturers' declared values and hand-arm vibration emission under real working conditions. *Proc. Inst. Acoust.* 28(3):1–10.

Mansfield, N.J., Ashley, J. and A.N. Rimell. 2007. Changes in subjective ratings of impulsive steering wheel vibration due to changes in noise level: A cross-modal interaction. *Int. J. Vehicle Inf. Commun. Syst.* 3(2):185–96.

Müller-Lyer, F.C. 1889. Optische Urteilstauschungen. Archiv fur Anatomie und Physiologie. *Physiologische Abteilung* 2:263–70.

Newell, G.S. and N.J. Mansfield. 2008. Evaluation of reaction time performance and subjective workload during whole-body vibration exposure while seated in upright and twisted postures with and without armrests. *Int. J. Ind. Ergon.* 38:499–508.

Newell, G.S., Mansfield, N.J. and L. Notini. 2006. Inter-cycle variation in whole-body vibration exposures of drivers operating track-type machines. *J. Sound Vib.* 298:563–79.

Porter, J.M., Gyi, D.E. and H.A. Tait. 2003. Interface pressure data and the prediction of driver discomfort in road trials. *Appl. Ergon.* 34(3):207–14.

Rayleigh. 1884. Report of the 54th meeting of the British Association for the Advancement of Science held in Montreal, August and September 1884. London: John Murray, 1885.

Rimell, A.N., Mansfield, N.J. and D. Hands. 2008. The influence of content, task and sensory interaction on multimedia quality perception. *Ergonomics* 51:85–97.

Rimell, A.N., Notini, L., Mansfield, N.J. and D.J. Edwards. 2008. Variation between manufacturers' declared vibration emission values and those measured under simulated workplace conditions for a range of hand-held power tools typically found in the construction industry. *Int. J. Ind. Ergon.* 38:661–75.

Vickery, B.C. 1963. Scientific information: Problems and prospects. *Minerva* 2(1):21–38.

11 Geography, Migration, Climate, and Environmental Plasticity as Contributors to Human Variation

*Michael A. Little**
Department of Anthropology, Binghamton University,
State University of New York, Binghamton, New York

CONTENTS

11.1 INTRODUCTION

Human prehistory and history identify our species, modern *Homo sapiens*, and our antecedent species, as capable of migrating over vast distances across the globe. While hominin populations were evolving large complex brains in Africa and developing

---

* Address all correspondence to mlittle@binghamton.edu.

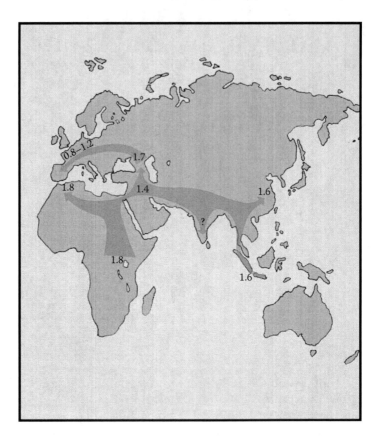

FIGURE 11.1   Migration of early *Homo erectus* from Africa to other parts of the Old World; dates are in millions of years before the present. (Data from Anton and Swisher 2004.)

new kinds of cultural behaviour, the Pleistocene epoch was displaying dramatic climatic fluctuations that imposed selective pressures on these hominin populations. There were two major migrations out of Africa that contributed to the widespread distribution of humans throughout the Old World. The first occurred sometime around 1.8 million years ago when a small-brained hominin (either *Homo habilis* or an early *Homo erectus*) moved into the Caucasus region and then as far as Southeast and East Asia and Western Europe (Anton and Swisher 2004; see Figure 11.1). Whether this migration was a demic/colonial expansion or a series of episodic population movements is not yet known (Rightmire 2001). The second major migration was of a relatively modern *Homo sapiens* who moved out of Africa around 100,000 to 150,000 years ago (see Figure 11.2) and either totally displaced, or assimilated to varying degrees, the more primitive hominin residents (Cann et al. 1987; Stringer and McKie 1996). Following the habitation of Europe, Asia, and probably the far reaches of Africa, *Homo sapiens* successively moved to Australia, the New World, and into the distant islands of the Pacific Ocean (see Figure 11.3) (Merriwether and Ferrel 1996; Kirch 2000; Schurr 2004). By roughly a thousand years ago, with the exception of

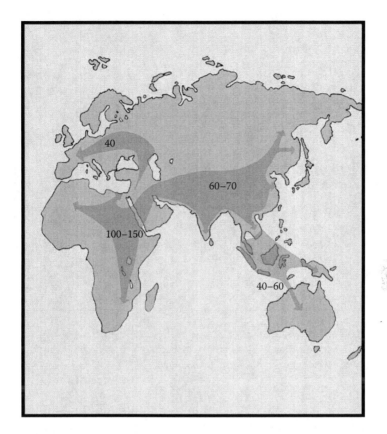

FIGURE 11.2 Migration of modern *Homo sapiens* from Africa to other parts of the Old World; dates are in thousands of years before the present. (Data from Rightmire 2001.)

some Arctic zones, Antarctica, and a few Pacific Islands, much of the land surface of the world was inhabited by fully modern humans, all intelligent, all with a remarkable array of cultures, and all with an astounding range of biological variation.

During the modern, historical era of migration, beginning about 500 years ago, Europeans moved into the New World, displacing and overwhelming Native American populations, and later colonisations moved Europeans, Asians, and Africans to different parts of the world, sometimes voluntarily, sometimes not. Populations that had a recent evolutionary history of indigenous residence migrated to new cultural and physical environments over relatively short periods of time. Many of these migrations during the modern era were stimulated by 'push' factors such as population pressure and ethnic and political repression. Others were stimulated by 'pull' factors such as perceived open territories, or lands inhabited by technologically less advanced peoples, or by a variety of economic opportunities. Human migrations continue in present times at high levels with economic, political, and ethnic considerations acting as both push and pull causes of migrant flows. The question, however,

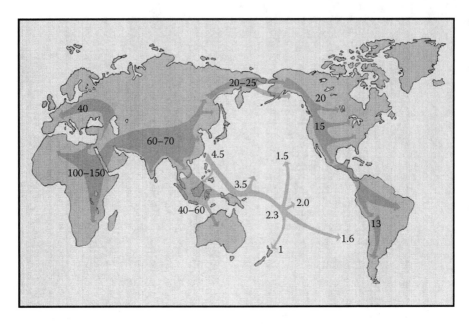

FIGURE 11.3 Precolonial migrations of modern *Homo sapiens* from Africa to other parts of the Old World and the New World; dates are in thousands of years before the present. (Data from Merriwether and Ferrel 1996, Kirch 2000, Schurr 2004.)

of migration analogies from the present to the past is similar to the danger of ethnographic analogies from the present to the past (Fix 1999, 149–50). The migratory colonial expansion in prehistoric times into uninhabited territory is quite different from the historical patterns of migration during the past 1,000 years.

The early migrations of hominins and later *Homo sapiens* exposed individuals to new environments where evolutionary processes such as selection, genetic adaptation, gene flow, and genetic drift could operate because of sufficient time. Plasticity, including behavioural and developmental adaptation, was important within these evolutionary processes. Human plasticity and adaptability arising from culture has been one of the hallmarks of human evolution. For recent migrations over the past 500 years, genetic, biological, and cultural plasticity has played an increasingly important role along with fundamental evolutionary processes.

An understanding of these complex evolutionary and adaptive processes requires information from fields that include anthropology, palaeontology, prehistory, ecology, demography, climatology, and a host of others. Climate is but one variable in a multitude of environmental variables that act on humans and serve as selective pressures, provide opportunities or impose stresses in the quest to procure food, establish the need to avoid predators and disease, and set up conditions for survival and reproduction. However, climate is an all-pervasive and dominant variable that governs a significant part of human behaviour. Global climatic conditions today, in even the most benign environments and with the most sophisticated technology, suggest that human controls over our ambient conditions are limited.

11.2 MIGRATION AND CLIMATE

There are three fundamental demographic variables that characterise how population numbers change through time in a given territory: through births (fertility, natality), deaths (mortality), and through movement (migration). Changes in population numbers have evolutionary as well as demographic significance for a variety of reasons, especially the action of selection on differential fertility and mortality, gene flow arising from mating practices and movement of people, and the founder effect from differential migration. Migration has evolutionary significance as an adaptive means to respond to environmental change in the home environment and also as a means of exposing humans to stresses, resources, and selective pressures in new environments. It is certainly the case throughout human prehistory and history that human movement has been driven by need, but there is also what might be called a 'peripatetic drive' that is associated with the mammalian characteristic of curiosity to explore its environment, a motivation that in humans can be intense. During the course of such movements, which began in antiquity in early hominins, humans have been exposed to new environments. Hence, migration can be seen as an adaptive response to a changing environment while at the same time as producing a need to develop new adaptive patterns in new environments. Human need and mammalian curiosity have led to human movement, while it is culture that has enabled humans to move into diverse, and often stressful, environments. Finally, human migration can serve as a natural experiment (Garruto et al. 1999), where migrants who have been exposed to new environments can be tested to determine how they are able to adapt to these new environments over lifetimes and over generations.

Migration patterns have been very useful as models for studies of human adaptation to the environment (Boas 1912; Harrison 1966; Baker 1976; Little and Leslie 1993). They are consistent with the use of time-honoured practices of 'comparative studies' to solve anthropological, adaptational, and evolutionary problems when manipulative and experimental research designs are not feasible.

11.2.1 EARLY RESEARCH

Early twentieth-century studies of migrants were conducted in New York City by Fishberg (1905) and then Boas (1912). Boas' motivations were to dispel the myths associated with the fixity of race and race categories, and the influence of plasticity, particularly developmental plasticity, on human size and proportions (Mascie-Taylor and Little 2004). Prior to this massive study of immigrants from Europe to the United States, Boas had conducted extensive research on growth of children, and he was convinced that their growth and adult physical characteristics were highly adaptable or plastic responses to the environment. His results on cranial dimensions were controversial then and now (Sparks and Jantz 2002; Gravlee et al. 2003; Relethford 2004), but it is generally agreed that Boas' pioneering research demonstrated the joint action of genetics and the environment on the human body and paved the way for later studies of human environmental plasticity (Lasker 1969). A number of anthropological studies of Mexican, Puerto Rican, Japanese, and Chinese migrants to Hawai'i and the United States in the 1930s, 1940s, and

1950s were stimulated by Boas' work (Shapiro and Hulse 1939; Ito 1942; Goldstein 1943; Lasker 1946, 1952, 1954; Greulich 1957; Thieme 1957). They demonstrated and reaffirmed many of the effects of migration that had been shown by or suggested in Boas' early study.

11.2.2 WORLD WAR TWO STUDIES OF CLIMATIC TOLERANCE

The next major series of studies was stimulated by the Second World War from the 1930s to the mid-1940s. The migrants, in this case, were military personnel transported throughout the world: Japanese in the tropical Pacific, northern Asia with its cold winters, and in tropical Southeast Asia; Germans in the dry deserts and Sahel of North Africa, and the cold winters of Scandinavia and the Soviet Union; Italians in Ethiopia; and other Europeans and Americans in cold zones and both hot arid and hot moist zones. Research was conducted to measure the extremes of tolerance to heat and cold by scientists from most nations involved in the global conflict. Although academic research generally was curtailed during this period, some of the applied research on human tolerance to climatic extremes was invaluable to the understanding of human adaptation to the environment. Some studies were concerned with rationing of food for military personnel, such as the research on climate and diet funded by the U.S. Quartermaster Corps and conducted at the University of Illinois (Mitchell and Edman 1951). Other research that was notable in the United States included the Rochester Desert unit which was charged with the research task of determining water and food requirements, sweat rates, energy balance, heat tolerance, work capacity, and the probability of survival while living under the hot-dry conditions of desert environments. Much of the desert research was conducted on military personnel on manoeuvres in southern California. Other studies were conducted of men on life rafts without water, and some limited tests were conducted under hot-wet conditions in Florida to simulate troop conditions in tropical forests. In the preface to the volume that reported this research, the authors identified this work as a part of the 'recently developing field of environmental physiology' (Adolph et al. 1947, vii). What was not known then was that this work was to stand as one of the first major works in climatic stress physiology, and one that would serve as the basis for heat stress studies both in environmental physiology and human biology.

In Japan, as with other nations, there were concerns about frostbite and other cold damage to military personnel. Yoshimura conducted pioneering studies of peripheral cold tolerance by comparing Japanese and a number of ethnic groups from Manchuria in a standardised finger exposure to stirred ice water. As with much of this research, it was conducted and reported during the war but published in scientific journals afterwards (Yoshimura 1960; Yoshimura and Iida 1950, 1952). A brief history of these and other early studies was provided by Yoshimura and Kuno (1966) at a major conference on human adaptability in Kyoto in 1965. Yoshimura and Iida (1950, 1952) were interested in the reactivity of blood vessels in the context of cold-induced vasodilation (Lewis waves) and its time/temperature parameters, particularly with reference to frostbite resistance. In addition to establishing a 'Resistance Index' (based on mean temperature after five to thirty minutes, time at first temperature rise, and temperature at first rise; the index ranged from 3 [poor response] to 9 [good response]), they recorded effects of time of day (circadian temperature rhythms), a hot bath, a hot meal, fasting, sleep deprivation, exercise, and season on finger responses to cold. They also found considerable individual variation in peripheral cold resistance. When residents and sojourners in Manchuria were tested, Japanese sojourners showed the poorest reaction to cold, Chinese residents and Mongol natives were next, and Orochon reindeer nomads of the Kingan Mountains showed the best reaction. Differences were attributed to developmental acclimatisation to cold rather than population differences. Many of these same relationships also were found for the fingers and toes of cold-exposed Peruvian Andean natives in studies conducted in the 1960s (Little 1969; Little et al. 1971, 1973).

11.2.3 The International Biological Programme and the Man and the Biosphere Programme

An upsurge in migration and thermal tolerance research began with the organisation of the International Biological Programme (IBP) and its human studies. The IBP was a global research programme initiated by the International Council of Scientific Unions (ICSU) (now the International Council for Science) with a planning phase from 1964 to 1967, a research phase from 1967 to 1972, and a synthesis phase that followed from 1972 to 1976. The orientation of this worldwide programme was ecological, with the theme 'The Biological Basis of Productivity and Human Welfare'. A number of sections of the IBP were designed to cover various components of ecology and ecosystems studies. A separate section, called Human Adaptability (HA), was designed to cover 'the ecology of mankind' from a variety of perspectives including health and welfare, environmental physiology, population genetics, child growth, anthropology, and demography (Weiner 1965). Joseph S. Weiner, who was international convener (director) of the Human Adaptability Section of the IBP, encouraged consistent methods (Yoshimura and Weiner 1966), prepared a handbook of standardised methods (Weiner and Lourie 1969), and at the end of the IBP published a compendium of the completed research (Collins and Weiner 1977). The planning and research that followed resulted in the participation of forty nations, the completion of more than 230 projects, and several thousand publications under the Human Adaptability umbrella. In a retrospective review of the human adaptability research, there has been some criticism based on topical omissions, but the contributions far outweigh the shortcomings (Ulijaszek and Huss-Ashmore 1997). Harrison (1997, 25) noted, in a contribution to the same review: 'Notwithstanding its limitations, it played a major part in converting the old defunct physical anthropology into the vibrant and exciting component of biological anthropology as it is today'.

The IBP/HA projects were particularly important because one of the major themes was identified as 'Migration and Hybrid Populations' (Weiner 1977, 13) and many individual research projects employed migration models and designs to explore human variation and adaptation (Collins and Weiner 1977). Thermal tolerance studies were conducted of migrants in comparative designs by the British,

Japanese, Americans, Israelis, Canadians, Indians, and Soviets. Joint British and Israeli research focused on Kurdish, Yemenite, and Ashkenazi Jewish immigrants, and on Bedouins who were tested under hot-dry conditions in the Negev Desert (Edholm and Samueloff 1973). This design was ideal for comparing populations of different ethnicity and origins and testing them under the same conditions. A consortium of nations (France, Canada, the United States, Denmark) studied Eskimos from Alaska, Canada, and Greenland. This design employed similar ethnic populations in widely dispersed territories. Much cold-tolerance research had been conducted on Eskimos prior to the IBP, so this work was deemphasised. However, the Canadian Igloolik project did find a higher peripheral reactivity to cold in Eskimo women than in men (Hughes 1977, 70).

There were seven Japanese IBP (JIBP) Human Adaptability Projects (JHA): the first dealt with 'Analysis of tolerance to heat and cold and its methodology', the fifth was on 'Studies on adaptabilities of underwater workers' with a major component on thermal regulation, and the seventh was on 'U.S.–Japan cooperative projects of "Comparative studies on human adaptability based on contrasting living habits and habitats"' (JIBP 1972). These human adaptability studies were reported on in considerable detail in the four JIBP (1975) synthesis volumes which were reviewed in the *American Anthropologist* (Little 1976). Some of the principal researchers were H. Yoshimura, H. Matsui, and K. Ohara from Nagoya, S. Kondo, and S. M. Horvath from Santa Barbara. Yoshimura, who played an important role in the Japanese IBP, was the Human Adaptability convener for Japan. Comparative studies were conducted in Nagoya, Los Angeles, San Francisco, and Santa Barbara of Japanese in Japan, mixed Japanese-Americans in Japan, Japanese-Americans in Santa Barbara and Los Angeles, and non-Japanese in the United States. Also, some comparisons of finger cooling of Okinawa natives indicated that they responded less favourably (lower reaction index) than mainland Japanese (Nakamura 1975). Other studies of thermoregulation included those of Ama divers chronically exposed to cold sea water, Ainu natives from Hokkaido, and Japanese (Wajin) migrants to Hokkaido. Comparing Ainu and Wajin in Hokkaido, the Ainu were less well-nourished and shorter in stature than Wajin, but they responded more favourably in the finger cold test (Koishi et al. 1975). There were a variety of results from the thermal investigations, with most of the variation attributed to health status of subjects, growth processes, body composition, acclimatisation level, seasonality, secular trends, and slight variations in methodology. These JIBP volumes are an important 'state-of-knowledge' for the human biology of Japanese and migrants up to 1975, particularly because of the review of Japanese-language articles previously published and their extensive reference materials.

Migration research designs were productive in studies of human adaptability to high altitude both during the IBP years and up to the present. Altitudes of 2,500–3,000 metres above sea level or higher are stressful environments because of the combined effects of low temperatures and low atmospheric oxygen. However, although technology can ameliorate somewhat the effects of cold, there are limits to the influence that traditional-society technology can have on the effects of high-altitude hypoxia. The Soviet, Indian, and American IBP research focussed on the

TABLE 11.1

Harrison's (1966) Migration Research Design for Comparisons among High-Altitude and Low-Altitude Native Groups

Comparisons	Bases for Differences (Assuming Controls for Other Variables)
HAN vs. LAN	Differences caused by altitude.
HAN vs. LAN↑	Differences based on acclimatization, developmental, or genetic adaptation.
LAN vs. LAN↑	Acclimatization or detrimental effects from the hypoxia of high altitude?
HAN vs. HAN↓	Detrimental effects from the hyperoxia of low altitude?
LAN vs. HAN↓	Differences based on acclimatization, developmental, or genetic adaptation.

High-altitude natives (HAN), low-altitude natives (LAN), high-altitude natives who migrated to low altitude (HAN↓), and low-altitude natives who migrated to high altitude (LAN↑)

Source: Modified from Baker (1976).

climatic effects of high altitude in the Tien Shan and Pamirs, the Himalayas, and the Andes, respectively. These studies were reported in the synthesis volumes edited by Baker (1977; Baker and Little 1976), and the cold studies conducted at high altitudes were summarised by Little and Hanna (1977).

High-altitude studies employed migration research designs as an important component of the research. Monge (1948, 1953) documented some of the problems at high altitude encountered by the first Europeans migrants (Spaniards) in their conquest of the Inca Empire. Monge's review of the Spanish archives recorded problems with health and reproduction by these early Spanish migrants. Early in the IBP planning phase, Harrison (1966) designed a simple but effective migration model to explore patterns of adaptation to high altitude, principally for British Himalayan research (see Table 11.1). He identified four groups for investigation: high-altitude natives (HAN), low-altitude natives (LAN), high-altitude natives who migrated to low altitude (HAN↓), and low-altitude natives who migrated to high altitude (LAN↑). Comparisons of these four groups and conclusions that might be drawn from differences are shown in Table 11.1. Baker (1976; Thomas et al. 1977) applied this migration model to Andean Quechua-speaking populations during the IBP and continued this work as a part of the UNESCO Man and the Biosphere (MAB) Programme.

A complex matrix of migrant flows characterised (and continues to characterise) the central Andes as typified by movements within three biogeographic zones in Peru

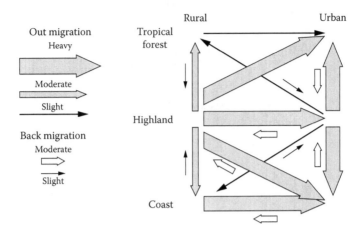

FIGURE 11.4   Migration patterns in Peru across the three major biogeographic zones of the central Andes. (Redrawn from Thomas et al. 1977.)

and the two major zones in Bolivia (Ives and Stites 1975; Glaser and Celecia 1981). The biogeographic zones of the Peruvian Andes are (1) the arid coast and western slopes of the central Andes, (2) the highland Andes and *altiplano* (grassland *puna*), and (3) the tropical wet eastern montaña and lowland forests. Bolivia has no coastal biogeographic zone. Figure 11.4 shows a model of the patterns of migration during the late 1970s in the central Andes (Thomas et al. 1977). Population pressure in the rural highlands and the coast led to large donor migrant flows to urban centres in the highlands, the tropical forest lowlands, and the coast. Very few migrants were moving into or returning to the highlands from the lowlands; hence, Harrison's (1966) model could not be used as effectively as he envisioned for the Andes. Baker and his colleagues (Dutt and Baker 1978; Baker and Beall 1982) were able to conduct extensive research on downward migrants from the highlands to the southern Peruvian coast, so comparisons could be made between sedentary lowland Quechua-speaking natives (LAN) and highland natives who had migrated to the coast from different elevations (HAN↓). Sorting out the reduced stress of high-altitude hypoxia and cold in the downward migrants was difficult in the context of many changes in nutrition and social factors. However, the HAN↓ did show higher fertility, increased child growth, earlier maturation, and reduced cardiovascular risk than high-altitude natives (HAN) (Baker and Beall 1982).

    In one of the few studies of low-altitude natives who migrated to high altitude (LAN↑), Frisancho and his colleagues (1973) found a remarkable effect of age at migration on work capacity (maximal oxygen intake during exercise). Figure 11.5 shows a decreased ability to perform work at high altitude with later ages of migration between the ages of two and sixteen years. Children who move to high altitude in infancy achieve levels of adaptation to high altitude as adults that approach native residents. This provides strong evidence that developmental adaptation is a significant contributor to overall adaptation to high-altitude hypoxia.

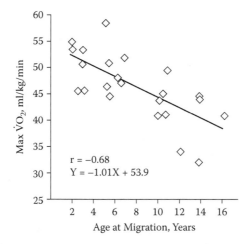

FIGURE 11.5 Relationship between age of migration from the lowlands to high altitude in the Peruvian Andes and maximal oxygen intake (aerobic exercise capacity) in adults. (Redrawn from Frisancho et al. 1973.)

11.3 THERMAL ADAPTATION AND HUMAN BIOGEOGRAPHY

Adaptation to temperature and other associated climatic variables (humidity, wind flow, hypoxia, solar radiation) can occur through three principal components: (1) behaviour and culture—to buffer or protect the body from climatic extremes; (2) active function—metabolic and physiological processes to generate heat, transfer heat within the body, deliver oxygen to tissues, and dissipate heat to the external environment; and (3) passive structure—body morphology, size, shape, composition, and pigmentation. The effectiveness of behaviour depends on knowledge of the environment and how to deal with it. It also depends on technology and ways that climatic extremes can be ameliorated by cultural practices. The dynamics of physiological functions with intricate feedback control mechanisms are complex indeed. In the case of cold stress or heat stress, blood, as a transporting medium of heat, is either shunted from the skin (to slow down heat loss) or to the skin (to enhance heat loss). At the same time, heat-generating organs and tissues are powering the basic furnace to maintain body heat content and other organs are participating in regulatory processes. The passive component of body structure and size is less flexible than either the behavioural or the physiological components but is equally likely to be moulded by climate, either over the short term, the long term, or transgenerationally.

11.3.1 BEHAVIOUR AND CULTURE

It is generally agreed that humans are physiologically well adapted to heat, largely because of their tropical evolutionary heritage (Hanna et al. 1989, Beall and Steegmann 2000). Behavioural responses to heat generally include seeking shelter from the sun and limiting activity during hot periods of the daily thermal cycle.

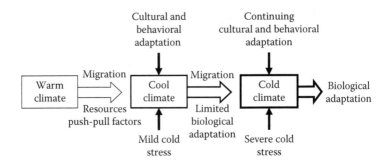

FIGURE 11.6   A diagram illustrating adaptation to a cold climate as it may have occurred with the Neanderthals in Europe during the Riss and Würm glaciations or with modern *Homo sapiens* in Europe and Asia or in migrants entering the New World during the last glacial maximum of the Würm/Wisconsin glaciation.

Cultural responses to heat centre on construction of shelter and dwellings that facilitate air flow, limiting decorative clothing to minimal insulation, and establishing work patterns during cooler times of the day to regulate heat production from muscular activity. Acclimatisation to heat can be achieved in most healthy, young individuals over a period of several weeks while working or exercising in the heat (Frisancho 1993, 55–56).

When humans moved out of their African tropical environment to subtropical, temperate, boreal, and tundra environments, cold stress, which began to be experienced, required a variety of adaptations in behaviour and biology. Figure 11.6 is a simple schematic diagram to illustrate the process of adaptation to a cold climate as may have occurred with the Neanderthals in Europe during the Riss and Würm glaciations or with modern *Homo sapiens* in Europe and Asia or in migrants entering the New World during the last glacial maximum of the Würm/Wisconsin glaciation (Steegmann et al. 2002; Hoffecker 2005; Snodgrass et al. 2007). An important point to be made is that whereas rudimentary clothing and caves or rock shelters were possible at the glacial margins, cold arctic residence required more advanced material culture in the form of tailored clothing and insulated dwellings. Snodgrass et al. (2007, 167) noted that only modern humans were able to move into the subarctic and Arctic regions of Siberia and Europe at latitudes at or above 55° north latitude. Hence, it is the case that sophisticated material culture probably preceded any major biological adaptations to cold, because material culture protecting against cold was required before chronic exposure to cold climates could take place.

11.3.2   ACTIVE FUNCTION: METABOLISM

Metabolic heat output from internal organs and muscles maintains appropriate body heat content for homeothermic mammals, including humans. Ito (1980) and, later, Little and Steegmann (2006) surveyed basal metabolic rates (BMR—minimum energy for body maintenance during rest and after fasting) for Alaskan and Canadian Inuit (Eskimos) and found consistently high values even under a variety of conditions that included controlling for dietary intake and anxiety. This suggests a metabolic acclimatisation to cold in Eskimos and, perhaps, other Arctic residents. There is some supporting evidence from studies of the diving women of Korea, where seasonal variations in BMR were correlated with cold stress from water temperature variation (Kang et al. 1963; Hong 1963). Also, Roberts (1952, 1973, 1978, 14–15) found a significant negative correlation ($r = -0.74$) between BMR and mean annual environmental temperature in a sample of 160 populations around the world, providing additional support for an adaptation of BMR to cold. The Native American subset of these populations showed a significant correlation as well in Roberts' survey.

The most persuasive evidence for a metabolic adaptation in cold-climate dwellers was compiled by Leonard in several studies of Arctic New World and native Siberian populations (Leonard et al. 1999, 2002, 2005). Leonard and his colleagues (2002) surveyed data from the literature on sixteen Inuit, Chippewa, Athapascan, and native Siberian samples from their studies and the literature. These data pooled a total of 109 indigenous men, 122 indigenous women, and smaller samples of nonindigenous men and women. What is different about this study from other literature reviews is that the authors considered the three reference standards used for BMR: body surface area (m²), body mass (kg), and fat-free mass or lean-body mass (kg). Controlling for each of these referents, indigenous Arctic residents showed persistent elevated BMRs, particularly when fat-free mass (which controls for age and body fatness) was used. Indigenous BMR values ranged from 3% to 19% above predicted values for these samples of males and females.

In a synopsis of the Siberian research, Leonard et al. (2005) pooled data on BMR from several indigenous Siberian populations—Evenki, Ket, Buryat, Yakut, Chukchi, and Nganasan. All were reindeer, horse, or cattle herders, hunters and fishers, or coastal sea mammal hunters. Again, controlling for fat-free mass, BMR values for three populations in pooled samples of men and women ranged between 14% and 23% above expected values. They state, 'In sum, indigenous Siberian populations show substantial elevation in BMR relative to standard reference values based on body weight and FFM' (Leonard et al. 2005, 458). The authors proposed three hypotheses to explain the elevated BMRs: (1) thermogenesis from high consumption of protein, (2) acclimatisation to cold stress, and (3) genetic adaptation to cold stress. The first was discounted because, although protein intake is about double that recommended for indigenous Siberians, the protein in their diets is no greater than U.S. intakes, and Americans do not show elevated BMR values. The second and third mechanisms were explored in the context of thyroid function, because thyroid hormones promote oxidative metabolism in cells at the mitochondrial level. The evidence provided was based on the sensitivity of thyroid function to seasonal factors (Hong et al. 1966) and other climatic effects and higher levels of thyroxine (T_4) in Evenki when compared to Russians, but also to the presence of seasonal variations in Russians and indigenous Siberians. The authors concluded that 'thyroid hormones appear to play a central role in regulating metabolic responses among indigenous Arctic groups' (Leonard et al. 2005, 462), and that these adaptations may be both acclimatisational (seasonal) and genetic (either linked to high thyroxin production or increased sensitivity to the hormone).

11.3.3 Active Function: Physiology

Physiological thermoregulatory systems are largely controlled by feedback at the level of the hypothalamus or under local circulatory and nervous controls. Recent work (Romanovsky 2006) has suggested that concepts of thermoregulation have dramatically changed in recent years with the concept of a central controller and a 'set point' no longer correct, but rather a 'balance point' idea to reflect a multiple feedback system of thermal balance is suggested. Basically, however, temperature regulation in humans maintains core body temperature (T_b) within a narrow range to avoid hypothermia and hyperthermia which can alter tissue and organ function, especially the brain. Other mechanisms control local temperatures, especially peripheral temperatures, to both regulate heat loss and prevent tissue damage, as in frostbite, through vasodilation and vasoconstriction of superficial blood vessels. It is presumed that maintenance of manual dexterity during cold is also an important adaptive function. The 'core-shell' concept of human temperature regulation of Carlson (Carlson and Hsieh 1970, 54; as described most recently by Steegmann 2007) contracts blood flow from the surface to conserve heat under cold conditions (reduces the heat core and expands the shell) and dilates blood flow to the surface from the core to dissipate heat under hot conditions (expands the heat core and reduces the shell) (see Figure 11.7). These physiological effects depend, to some degree, on body composition which influences tissue conductance of heat from the core to the shell or surface. This will be discussed in Section 11.3.4.

A good example of individual and population variation in thermal responses is in peripheral blood flow and temperature with cold exposure. Early studies of responses to cold of the extremities began with work by Grant (1930; Grant and Bland 1931), and Lewis (1930), who discovered cyclic cold-induced vasodilation (CIVD) and

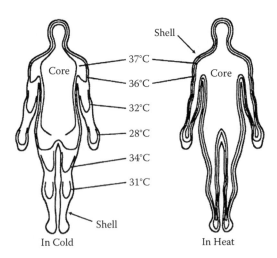

FIGURE 11.7 The core-shell concept of Loren Carlson illustrates vascular responses in cold and heat. The core contracts (to conserve) or expands (to dissipate) heat. (Redrawn from Carlson and Hsieh 1970.)

the arteriovenous anastomoses (AVA) that dilate to enhance blood flow to cooled fingers and toes. Following these discoveries, a number of studies were conducted of peripheral responses to cold in the extremities of cold-exposed and non-cold-exposed native and non-native groups, including Inuit (Eskimos), Canadian Arctic Indians, Andean Quechua Indians, Ainu, Canadians, Americans, Chinese, Japanese, Orochons, Mongols, and Sami (Lapps). Maintaining warm extremities has the advantage of allowing for mobility (feet) and manual dexterity (hands), and preventing tissue damage from frostbite or chronic nonfreezing cold. These attributes are crucial for people whose subsistence requires working out-of-doors in the cold. On the other hand, warm extremities require a high metabolic rate and contribute to loss of body heat content, which can lead to a dangerous state of hypothermia. Steegmann (1975), who did a thorough review of the literature on extremity cooling, concluded that Asians and Asian-derived Native American populations respond best to extremity cold exposure and that African-derived populations show limited responses to cold. Although he suggested that these results implied genetic adaptation to extremity cold responses, developmental acclimatisation was not discounted, especially in light of studies demonstrating developmental responses to cold in other studies (Little 1969; Little et al. 1971; Nakamura 1975). A latitudinal gradient among Asian populations for the finger-cooling Reaction Index is shown in Figure 11.8, based on calculations by Steegmann (Beall and Steegmann 2000) from Yoshimura and Iida (1952), Kondo (1969), and Toda (1975).

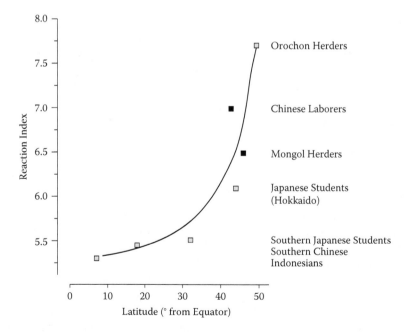

FIGURE 11.8 Relationship between Yoshimura's Reaction Index and geographic latitude for several populations tested in the finger immersion cold test. (Data from Beall and Steegmann 2000; Yoshimura and Iida 1952; Kondo 1969; and Toda 1975.)

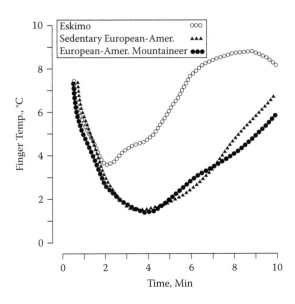

FIGURE 11.9   Finger cooling in young Inuit (Eskimo) men after nine months at an urban Oregon boarding school. (Redrawn from Eagan 1967.)

Sorting out developmental acclimatisation from any genetic or genetic/developmental effect is a difficult task, despite the likelihood of intense selective pressure for physiological adaptations in Arctic dwellers. Eagan (1967) did an interesting study of a small group of young Eskimos from Alaska who had been resident at an Oregon boarding school for nine months and had just returned to Alaska. They were compared with sedentary Americans and well-trained American mountaineers (see Figure 11.9) in a finger immersion test in water at 0°C for ten minutes. The Eskimos were able to maintain remarkably warm finger temperatures after an initial drop during the first two minutes of the test. Eagan (1967) attributed the difference to a genetic cold tolerance in the Eskimos, but there is really no way to rule out developmental acclimatisation.

11.3.4   Passive Structure: Size, Morphology, and Composition

Bergmann's (1847) and Allen's (1877) biogeographic or ecological rules reveal the evolutionary outcome of passive exchange of heat from an object to its external environment in homeothermic mammals and birds. Large objects/animals minimise the relationship between surface area and volume to slow heat dissipation. Small objects/ animals maximise surface area to volume to facilitate heat transfer to the exterior. Round objects also minimise heat dissipation, whereas linear objects enhance heat loss. These are fundamental laws of passive heat exchange that are a function of surface dimensions and volume or mass, and that work in concert with active physiology to transfer heat throughout the body and to the skin surface. Because humans show a remarkable variation in size, shape, and composition, they manifest a broad capacity to show adaptive features to a variety of thermal environments.

Coon et al. (1950) first introduced body size and shape as important variables in climatic adaptation of peoples around the world. Around that time and shortly after, several human biologists began to explore relationships of surface area, weight, and climate (Schreider 1950, 1951; Newman 1953; Newman and Munro 1955; Roberts 1953; Baker 1960). They found that some of the same climatic patterns for nonhuman homeotherms also applied to humans. Newman (1953) demonstrated clines in stature, sitting height, and nasal index for native New World populations that he attributed to climatic variation, and Roberts (1953) found a high negative correlation between body weight and mean annual temperature in 116 populations from around the world. Both papers cited their results as supporting Bergmann's and Allen's rules. Shortly after this, the distinguished environmental physiologist Per Scholander (1955) published a devastating critique of Bergmann's and Allen's rules in their application to both animals and humans. His principal argument was that only culture (in the case of humans) and physiological mechanisms (for all warm-blooded animals) have allowed homeotherms to adapt to environmental temperature extremes, and he expressed a surprising ignorance of evolutionary process. Responses were published by Mayr (1956) and Newman (1956) with evolutionary arguments, but it remained for several years before biogeographic correlations demonstrated the values of surface area/weight variation in adaptation to climatic diversity.

Baker (1960) explored the question of human adaptation to the thermal environment via the surface area to weight ratio (SA/wt) in humans. He noted that from experimental studies of hot-dry (desert) stress, a high SA/wt ratio gave no advantage to workers in the heat since evaporative cooling is maximally effective under these conditions (Baker 1958). Baker (1960) then suggested that a high SA/wt ratio *would* play a role in hot-wet conditions since evaporative cooling would be less effective, although he noted that there was no experimental support for the prediction at that time. He then continued to argue for the validity of the SA/wt ratio under conditions of cold, noting the importance of subcutaneous fat as an insulator and an additional variable. Baker (1960) noted from earlier studies that 'when subcutaneous fat was held constant [in experimental cold studies], individuals with lower [SA/wt] ratios maintained higher [body] core temperatures at lower weight-adjusted metabolic rates' (Baker and Daniels 1956; Baker 1959). He noted that since body weight and stature were both under some degree of genetic control, and the thermal value of the SA/wt ratio had been demonstrated experimentally, then selection has probably played a role in establishing these relationships for human populations. A later study by Shvartz et al. (1973) experimentally supported Baker's (1960) prediction for the value of a high SA/wt ratio in humid heat. They tested men whose SA/wt ratios ranged from 0.0232 to 0.0307 m^2/kg under hot-dry (25 subjects at 50°C dry bulb, 28°C wet bulb) and hot-wet (eight subjects at 30°C dry bulb, saturated) conditions. The rate of heat storage was negatively related to SA/wt at a low level of correlation ($r = -0.22$) for the hot-dry conditions ($Y = -37.7X + 215$), whereas the rate of heat storage was negatively related to SA/wt at a high level of correlation ($r = -0.93$) for the hot-wet conditions ($Y = -87.3X + 262$): that is, under hot-wet conditions, body heat storage was less with high SA/wt ratios.

Roberts built on an earlier study of body weight and climate (Roberts 1953) to provide regressions and correlations between mean annual temperature and a variety of human linear, dimensions, weight (body mass), and indices (Roberts 1973, 1978). He found high correlations and relatively steep regressions for body size vs. mean annual temperature in comparisons of between 100 and 160 populations around the world. These results confirmed the operation of both Bergmann's and Allen's rule for size, relative limb lengths, and trunk dimensions in humans. Katzmarzyk and Leonard (1998) restudied human biogeographical relationships in humans by using substantial numbers of populations (223 male samples, 195 female samples) that were gathered during the period after Roberts' 1953 study. They found that humans still conform to Bergmann's and Allen's rules, but that the strength of the associations has markedly declined—there has been a *secular trend* in these relationships. This may be attributed to a worldwide increase in body weights, especially in tropical populations, and 'This trend likely reflects the impact of acculturation and lifestyle change and the associated improvements in health care and nutrition' (Katzmarzyk and Leonard 1998). For example, in the case of the correlation between SA/wt ratio and mean annual temperature, Roberts' correlation for males of $r = 0.59$ dropped to $r = 0.29$ and the regression slope of the modern sample was half the earlier one.

Finally, Ruff (1991, 1994) explored the SA/wt ratio by modelling the human body as a cylinder, where the important variable is the trunk diameter (equivalent to bi-iliac or hip breadth) and height becomes less important. In his studies, he compared seventy-one living populations and several hominins from skeletal remains for SA/wt vs. latitude of residence in absolute degrees from the equator. SA/wt in the living populations gave a correlation of $r = -0.65$ with latitude, while bi-iliac breadth give a substantially higher correlation of $r = 0.87$. This comparative work with living and early hominin populations confirms the relationships between SA/wt and climate and projects these relationships back in time to our antecedents.

11.4 WHAT HAVE WE LEARNED ABOUT CLIMATIC ADAPTATION FROM MIGRATION STUDIES?

Summary statements outlining our present state of knowledge on human adaptation to climatic variation follow:

- Humans have evolved in the tropics and migrated to other climatic zones.
- This worldwide migration has exposed populations to diverse climatic stress.
- Material culture (clothing, shelter, hunting techniques) has enabled humans to migrate to climates with extreme cold.
- On theoretical grounds, selection has operated on the SA/wt ratio to produce global variation in this value for both females and males.
- This theory has been confirmed by biogeographic studies that demonstrate relationships between body sizes and SA/wt with mean annual temperature or latitude from the equator.
- Experimental studies have confirmed the importance of SA/wt under hot-wet and cold conditions.

- Body composition, particularly adipose tissue but also muscle, can protect against cold.
- SA/wt ratios can also vary according to developmental causes.
- Developmental acclimatisation can mimic genetic adaptation to cold or altitude and makes it difficult to distinguish between the two modes of adaptation.
- Basal metabolism tends to be elevated in populations resident in cold climates and this may result from variations in thyroid function.
- Adaptation to high-altitude hypoxia, which is multifactorial, shows a genetic adaptation in Himalayan populations that favours oxygen transport.
- Migration itself serves as an excellent model or research design in which to compare populations exposed to different environmental circumstances.
- Today, migration continues unabated and can serve as a rich opportunity for scientific designs of human population variation.

11.5 PROSPECTS FOR THE FUTURE: GLOBAL CLIMATE CHANGE

Over the next century and more, global climate changes will profoundly affect the geographic features of the Earth and its biosphere, including human populations. Figure 11.10 shows some of the relationships in this complex scenario of anthropogenic change in our planet. A number of changes will impact human health and

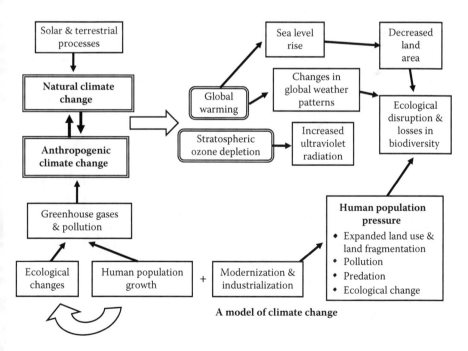

FIGURE 11.10 Model of some of the relationships associated with global climate change. (Reprinted with permission from Little and Garruto 2007.)

well-being and lead to new migrant flows stimulated by climate change. (1) Global warming and stratospheric ozone depletion will enhance sunlight and ultraviolet B radiation. One result will be higher levels of skin cancer in those with minimal melanin pigmentation. Other documented effects of high UVB radiation are immune suppression and damage to the eye. (2) Global warming will produce increases in heat waves with urban centres most susceptible. Heat stress in plants and domestic animals will reduce productivity and affect food availability. Climatic extremes, such as drought, floods, and storms, are expected to increase with temperature increases. (3) Sea-level rises over the next century from melting continental and mountain glaciers will reduce habitable land areas and contribute to salinisation of coastal fresh water areas. Areas most vulnerable are Pacific atolls, the European low countries, Bangladesh, coastal China, and many coastal cities such as New York City. The movement of people from flooded areas will be exacerbated by the continuing increases in population growth during the twenty-first century and will lead to massive out-migration from flooded coastal zones to inland zones. This will further concentrate inland population numbers and lead to competition for resources and conflict. (4) Migration from coastal areas will also lead to a rise in opportunistic, emerging infections that will appear because of population and ecological disruptions, and the transmission of disease with population movement. Tropical diseases will move into more temperate zones with global warming.

It is generally agreed by the scientific community that even if immediate drastic action were taken to reduce CO_2 and chlorofluorocarbon emissions, many of the global climate changes predicted would take place. However, anticipation of the problems and widespread planning may reduce the catastrophic effects of global climate change. Certainly these new environmental and social stresses will present novel adaptive challenges to our species.

ACKNOWLEDGEMENTS

I wish to thank the organisers of the Joint Meeting of the Society for the Study of Human Biology (SSHB) and the Japan Society for the Study of Physiological Anthropology (JSPA) for inviting me to present a draft of this paper. I particularly wish to express my gratitude to our Cambridge University hosts, Nicholas Mascie-Taylor and Rie Goto, whose warm hospitality made my introduction to Cambridge a memorable one. I am grateful to my Japanese, European, and American colleagues who made the meeting a pleasure to attend. Finally I offer my thanks to Anne D. Hull, who expertly drafted several of the figures.

DEDICATION

This paper is dedicated to the memory of Paul T. Baker (1927–2007), whose pioneering twentieth-century studies of climate, human adaptation, and migration served as a framework for studies that followed.

REFERENCES

Adolph, E.F. et al. 1947. *Physiology of Man in the Desert.* New York: Wiley.

Allen, J.A. 1877. The influence of physical conditions in the genesis of species. *Radic. Rev.* 1:108–40.

Anton, S.C. and C.C. Swisher III. 2004. Early dispersals of *Homo* from Africa. *Annu. Rev. Anthropol.* 33:271–96.

Baker, P.T. 1958. The biological adaptation of man to hot deserts. *Am. Nat.* 42:337–57.

Baker, P.T. 1959. American Negro-White differences in the thermal insulative aspects of human fat. *Hum. Biol.* 31:316–24.

Baker, P.T. 1960. Climate, culture and evolution. *Hum. Biol.* 32:3–16.

Baker, P.T. 1976. Research strategies in population biology and environmental stress. In *The Measures of Man: Methodologies in Biological Anthropology*, ed. E. Giles and J.S. Friedlaender, 230–59. Cambridge, Massachusetts: Peabody Museum Press.

Baker, P.T. 1977. *The Biology of High-Altitude Peoples.* International Biological Programme 14, Cambridge: Cambridge University Press.

Baker, P.T. and C.M. Beall. 1982. The biology and health of Andean migrants: A case study of south coastal Peru. In *Human Population and Biosphere Interactions in the Central Andes*, ed. P.T. Baker, special issue of *Mt. Res. Dev.* 2(1):81–95.

Baker, P.T. and F. Daniels Jr. 1956. Relationship between skinfold thickness and body cooling for two hours at 15°C. *J. Appl. Physiol.* 8:409–16.

Baker, P.T. and M.A. Little. 1976. *Man in the Andes: A Multidisciplinary Study of High Altitude Quechua.* Stroudsburg, Pennsylvania: Dowden, Hutchison & Ross.

Beall, C.M. and A.T. Steegmann Jr. 2000. Human adaptation to climate: Temperature, ultraviolet radiation, and altitude. In *Human Biology: An Evolutionary and Biocultural Approach*, ed. S. Stinson, B. Bogin, R. Huss-Ashmore, and D. O'Rourke, 163–224. New York: Wiley-Liss.

Bergmann, C. 1847. Uber die Verhaltnisse der Wärmeökonomie des Thiere zu ihrer Grösse. *Göttinger Studien.* 3:595–708.

Boas, F. 1912. *Changes in Bodily Form of Descendants of Immigrants.* New York: Columbia University Press.

Cann, R.L, Stoneking, M. and A.C. Wilson. 1987. Mitochondrial DNA and human evolution. *Nature* 325:32–36.

Carlson, L.D. and A.C.L. Hsieh. 1970. *Control of Energy Exchange.* New York: Macmillan.

Collins, K.J. and J.S. Weiner. 1977. *Human Adaptability: A History and Compendium of Research.* London: Taylor and Francis.

Coon, C.S., Garn, S.M. and J.B. Birdsell. 1950. *Race: A Study of the Problems of Race Formation in Man.* Springfield, Illinois: C.C. Thomas.

Dutt, J.S. and P.T. Baker. 1978. Environment, migration, and health in southern Peru. *Soc. Sci. Med.* 12:29–38.

Eagan, C.J. 1967. The responses to finger cooling of Alaskan Eskimos after nine months of urban life in a temperate climate. *Biometeorology* 2(2):822–30.

Edholm, O.G. and S. Samueloff. 1973. Biological studies of Yemenite and Kurdish Jews in Israel and other groups in South West Asia. *Philos. Trans. R. Soc. Lond.* 266:83–224.

Fishberg, M. 1905. Materials for the physical anthropology of the Eastern European Jews. *Ann. N. Y. Acad. Sci.* 16:155–297.

Fix, A.G. 1999. *Migration and Colonization in Human Microevolution.* Cambridge: Cambridge University Press.

Frisancho, A.R. 1993. *Human Adaptation and Accommodation.* Ann Arbor: University of Michigan Press.

Frisancho, A.R., Martinez, C., Velásquez, T., Sanchez, J. and H. Montoye. 1973. Influence of developmental adaptation on aerobic capacity at high altitude. *J. Appl. Physiol.* 34:176–80.

Garruto, R.M., Little, M.A., James, G.D. and D.E. Brown. 1999. Natural experimental models: The global search for biomedical paradigms among traditional, modernizing, and modern populations. *Proc. Natl. Acad. Sci. U.S.A.* 96:10536–43.

Glaser, G. and J. Celecia. 1981. Guidelines for integrated ecological research in the Andean region. Special Issue of *Mt Res. Dev.* 1(2):171–86.

Goldstein, M.S. 1943. *Demographic and Bodily Changes in Descendants of Mexican Immigrants.* Austin: Institute of Latin American Studies, University of Texas.

Grant, R.T. 1930. Observations on direct communications between arteries and veins in the rabbit's ear. *Heart* 15:281–303.

Grant, R.T. and E.F. Bland. 1931. Observations on arteriovenous anastomoses in human skin and in the bird's foot with special reference to the reaction to cold. *Heart* 15:365–407.

Gravlee, C.C., Bernard, H.R. and W.R. Leonard. 2003. Heredity, environment, and cranial form: A reanalysis of Boas's immigrant data. *Am. Anthropol.* 105:125–38.

Greulich, W.W. 1957. A comparison of the physical growth and development of American-born and native Japanese children. *Am. J. Phys. Anthropol.* 15:489–515.

Hanna, J.M., Little, M.A. and D.M. Austin. 1989. Climatic physiology. In *Human Population Biology: A Transdisciplinary Science*, ed. M.A. Little and J.D. Haas, 132–51. New York: Oxford University Press.

Harrison, G.A. 1966. Human adaptability with reference to the IBP proposals for high altitude research. In *The Biology of Human Adaptability*, ed. P.T. Baker and J.S. Weiner, 109–19. Oxford: Clarendon Press.

Harrison, G.A. 1997. The role of the Human Adaptability International Biological Programme in the development of human population biology. In *Human Adaptability: Past, Present, and Future*, ed. S.J. Ulijaszek and R. Huss-Ashmore, 17–25. Oxford: Oxford University Press.

Hoffecker, J.F. 2005. *A Prehistory of the North: Human Settlement of the Higher Latitudes.* New Brunswick: Rutgers University Press.

Hong, S.K. 1963. Comparison of diving and nondiving women of Korea. *Fed. Proc.* 22:831–33.

Hong, S.K. 1966. Seasonal studies of thyroid function in Ama. In *Human adaptability and its methodology*, ed. H. Yoshimura and J.S. Weiner, 78–82. Papers given at a symposium held in Kyoto, Japan, under the auspices of the International Union of Physiological Sciences and the International Biological Programme. Tokyo: Japan Society for the Promotion of Sciences.

Hughes, D.R. 1977. International study of Eskimos. In *Human Adaptability: A History and compendium of Research*, ed. K.J. Collins and J.S. Weiner, 69–71. London: Taylor and Francis.

Ito, P.K. 1942. Comparative biometrical study of physique of Japanese women born and reared under different environments. *Hum. Biol.* 14:279–351.

Itoh, S. 1980. Physiology of Circumpolar peoples. In *The Human Biology of Circumpolar Populations*, ed. F.A. Milan, 285–303. Cambridge: Cambridge University Press.

Ives, J.D. and A. Stites. 1975. *Project 6: Impact of Human Activities on Mountain and Tundra Ecosystems.* Proceedings of the Boulder Workshop, July, 1974, United States MAB (Man and the Biosphere) Report, Boulder, Colorado: INSTAAR Special Publication.

JIBP. 1972. *Japanese Contribution to IBP in 1971.* Report No. 8 of The Japanese National Committee for IBP (JIBP). Tokyo: The Science Council of Japan.

JIBP. 1975. *Japanese International Biological Program Synthesis: Human Adaptability,* Vols. 1–4, numerous editors. Japanese Committee for the IBP, Science Council of Japan, Tokyo: University of Tokyo Press.

Kang, B.S., Song, J.H, Suh, C.S. and S.K. Hong. 1963. Changes in body temperature and basal metabolic rate of the Ama. *J. Appl. Physiol.* 18:483–88.

Katzmarzyk, P.T. and W.R. Leonard. 1998. Climatic influences on human body size and proportions: ecological adaptations and secular trends. *Am. J. Phys. Anthropol.* 106: 483–503.

Kirch, P.V. 2000. *On the Road of the Winds: An Archaeological History of the Pacific Islands before European Contact.* Berkeley: University of California Press.

Koishi, H., Okuda, T., Matsudaira, T., Takaya, S. and Y. Kōhara. 1975. Physiological functions of the Ainu: Nutrition and cold tolerance. In *JIBP Synthesis, Human Adaptability*, Vol. 2, *Anthropological and Genetic Studies on the Japanese*, ed. S. Watanabe, S. Kondo, and E. Matsunaga, 309–19. Japanese Committee for the International Biological Program, Tokyo: University of Tokyo Press.

Kondo, S. 1969. A study on the acclimatization of the Ainu and the Japanese with reference to the hunting temperature reaction. *Journal of the Faculty of Science* (Univ. of Tokyo), III, 4:254–65.

Lasker, G.W. 1946. Migration and physical differentiation. A comparison of immigrant and American-born Chinese. *Am. J. Phys. Anthropol.* 4:273–300.

Lasker, G.W. 1952. Environmental growth factors and selective migration. *Hum. Biol.* 24:262–89.

Lasker, G.W. 1954. The question of physical selection of Mexican migrants to the USA. *Hum. Biol.* 26:52–8.

Lasker, G.W. 1969. Human biological adaptability. *Science* 166:1480–6.

Leonard, W.R., Galloway, V.A., Ivakine, F., Osipova, L. and M. Kazakovtseva. 1999. Nutrition, thyroid function and basal metabolism of the Evenki of Central Siberia. *Int. J. Circumpolar Health* 58:281–95.

Leonard, W.R., Sorensen, M.V., Galloway, V.A. et al. 2002. Climatic influences on basal metabolic rates among Circumpolar populations. *Am. J. Hum. Biol.* 14:609–20.

Leonard, W.R., Snodgrass, J.J. and M.V. Sorensen. 2005. Metabolic adaptation in Indigenous Siberian populations. *Annu. Rev. Anthropol.* 34:451–71.

Lewis, T. 1930. Observations upon the reaction of the vessels of the human skin to cold. *Heart* 15:177–208.

Little, M.A. 1969. Temperature regulation at high altitude: Quechua Indians and U.S. Whites during foot exposure to cold water and cold air. *Hum. Biol.* 41:519–35.

Little, M.A. 1976. *Review of the Japanese International Biological Program Synthesis, Volumes 1–4*. Tokyo: University of Tokyo Press, *Am. Anthropol.* 78(3):700–2.

Little, M.A. and Garruto, R.M. 2007. Chapter 5: Global impacts of anthropogenic climate change on human health and adaptability. In *The Anthropologist* (Special Issue 3)— *Anthropology Today: Trends, Scope and Applications*, ed. V. Bhasin and M.K. Bhasin, Delhi, India: Kamla-Raj Enterprises.

Little, M.A. and J.M. Hanna. 1977. The responses of high-altitude populations to cold and other stresses. In *The Biology of High-Altitude Peoples*, International Biological Programme 14, ed. P.T. Baker, 251–98. Cambridge: Cambridge University Press.

Little, M.A. and P.W. Leslie. 1993. 'Migration'. In *Research Strategies in Human Biology*, ed. G.W. Lasker and C.G.N. Mascie-Taylor, 62–91. Cambridge: Cambridge University Press.

Little, M.A. and A.T. Steegmann Jr. 2006. Acclimatization and adaptation: responses to cold. In *Handbook of North American Indians*, Vol. 3: *Environment, Origins and Population*, ed. D.H. Ubelaker, 740–7. Washington D.C.: Smithsonian Institution.

Little, M.A., Thomas, R.B., Mazess, R.B. and P.T. Baker. 1971. Population differences and developmental changes in extremity temperature responses to cold among Andean Indians, *Hum. Biol.* 43:70–91.

Little, M.A., Thomas, R.B. and J.W. Larrick. 1973. Skin temperature and cold pressor responses of Andean Indians during hand immersion in water at 4°C. *Hum. Biol.* 45:643–62.

Mascie-Taylor, C.G.N. and M.A. Little. 2004. History of migration studies in biological anthropology. *Am. J. Hum. Biol.* 16:365–78.

Mayr, E. 1956. Geographical character gradients and climatic adaptation. *Evolution* 10:105–8.

Merriwether, D.A. and R.E. Ferrel. 1996. The four founding lineage hypothesis for the New World: A critical reevaluation. *Mol. Phylogenet. Evol.* 5:241–6.

Mitchell, H.H. and M. Edman. 1951. *Nutrition and Climatic Stress: With Particular Reference to Man.* Springfield, Illinois: C.C. Thomas.

Monge, M.C. 1948. *Acclimatization in the Andes.* Baltimore: Johns Hopkins Press.

Monge, M.C. 1953. Biological basis of human behavior. In *Anthropology Today: An Encyclopedic Inventory*, ed. A.L. Kroeber, 127–44. Chicago: University of Chicago Press.

Nakamura, M. 1975. Peripheral cold tolerance, In *JIBP Synthesis, Human Adaptability*, Vol. 3, *Physiological Adaptability and Nutritional Status of the Japanese, A. Thermal Adaptability of the Japanese and Physiology of the Ama*, ed. H. Yoshimura and S. Kobayashi, 27–35. Japanese Committee for the International Biological Program, Tokyo: University of Tokyo Press.

Newman, M.T. 1953. The application of ecological rules to the racial anthropology of the aboriginal New World. *Am. Anthropol.* 55:311–27.

Newman, M.T. 1956. Adaptation of man to cold climates, *Evolution* 10:101–4.

Newman, R.W. and E.H. Munro. 1955. The relation of climate and body size in U.S. males. *Am. J. Phys. Anthropol.* 13:1–17.

Relethford, J.H. 2004. Boas and beyond: migration and craniometric variation. *Am. J. Hum. Biol.* 16:379–86.

Rightmire, G.P. 2001. Patterns of hominid evolution and dispersal in the Middle Pleistocene. *Quat. Int.* 75:77–84.

Roberts, D.F. 1952. Basal metabolism, race, and climate. *J. R. Anthropol. Inst.* 82:169–83.

Roberts, D.F. 1953. Body weight, race and climate. *Am. J. Phys. Anthropol.* 11:533–58.

Roberts, D.F. 1973. *Climate and human variability, Addison-Wesley Module in Anthropology No. 34.* Reading, Massachusetts: Addison-Wesley.

Roberts, D.F. 1978. *Climate and Human Variability.* 2nd Edition, Menlo Park, California: Cummings.

Romanovsky, A.A. 2006. Thermoregulation: Some concepts have changed: Functional architecture of the thermoregulatory system. *Am. J. Physiol. Regul. Integr. Comp. Physiol.* 292:R37–46.

Ruff, C.B. 1991. Climate, body size and body shape in hominid evolution: The thermoregulatory imperative. *J. Hum. Evol.* 21:81–105.

Ruff, C.B. 1994. Morphological adaptation to climate in modern and fossil hominids. *Yearb. Phys. Anthropol.* 37:65–107.

Scholander, P.F. 1955. Evolution of climatic adaptation in homeotherms. *Evolution* 9:15–26.

Schreider, E. 1950. Geographical distribution of the body-weight/body-surface ratio. *Nature* 165:286.

Schreider, E. 1951. Anatomical factors of body-heat regulation. *Nature* 167:823–4.

Schurr, T.G. 2004. The peopling of the New World: Perspectives from molecular anthropology. *Annu. Rev. Anthropol.* 33:551–83.

Shapiro, H.L. (with the assistance of F.S. Hulse). 1939. *Migration and Environment: A Study of the Physical Characteristics of the Japanese Immigrants to Hawaii and the Effects of Environment on Their Descendants.* Oxford: Oxford University Press.

Shvartz, E., Saavand, E. and D. Benor. 1973. Physique and heat tolerance in hot-dry and hot-humid enviornments. *J. Appl. Physiol.* 34:799–803.

Snodgrass, J.J., Sorensen, M.V., Tarskaia, L.A. and W.R. Leonard. 2007. Adaptive dimensions of health research among indigenous Siberians. *Am. J. Hum. Biol.* 19:165–80.

Sparks, C.S. and R.L. Jantz. 2002. A reassessment of human cranial plasticity: Boas revisited. *Proc. Natl. Acad. Sci. U.S.A.* 99:14636–9.

Steegmann, A.T. Jr. 1975. Human adaptation to cold. In *Physiological Anthropology*, ed. A. Damon, 130–66. Oxford: Oxford University Press.

Steegmann, A.T. Jr. 2007. Human cold adaptation: An unfinished agenda. *Am. J. Hum. Biol.* 19:218–27.

Steegmann, A.T. Jr., Cerny, F.C. and T.W. Holliday. 2002. Neanderthal cold adaptation: Physiological and energetic factors. *Am. J. Hum. Biol.* 14:566–83.

Stringer, C. and R. McKie. 1996. *African Exodus: The Origins of Modern Humanity*. New York: Henry Holt.

Thieme, F.P. 1957. A comparison of Puerto Rico migrants and *sedentes*. *Michigan Academy of Science, Arts, and Letters*, 42:249–67.

Thomas, R.B., Baker, P.T. and J.D. Haas. 1977. Man and the environment in the central Andes: A research prospectus. In *Human population Problems in the Biosphere: Some Research Strategies and Designs*, ed. P.T. Baker, 33–52. MaB (Man and the Biosphere) Technical Notes 3, Paris: UNESCO.

Toda, Y. 1975. Thermal adaptability of Indonesians. In *JIBP Synthesis, Human Adaptability Volume 3, Physiological Adaptability and Nutritional Status of the Japanese, A. Thermal Adaptability of the Japanese and Physiology of the Ama*, ed. H. Yoshimura and S. Kobayashi, 147–56. Japanese Committee for the International Biological Program, Tokyo: University of Tokyo Press.

Ulijaszek, S.J. and R. Huss-Ashmore. 1997. *Human Adaptability: Past, Present, and Future*. Oxford: Oxford University Press.

Weiner, J.S. 1965. *International Biological Programme Guide to the Human Adaptability Proposals*. International Council of Scientific Unions, London: Special Committee for the IBP.

Weiner, J.S. 1977. The history of the Human Adaptability Section. In *Human Adaptability: A History and Compendium of Research*, ed. K.J. Collins and J.S. Weiner, 1–23. London: Taylor and Francis.

Weiner, J.S. and J.A. Lourie. 1969. *Human Biology: A Guide to Field Methods*. IBP Handbook No. 9, Philadelphia: Davis.

Yoshimura, H. 1960. Acclimatization to cold, In *Essential Problems of Climatic Physiology*, ed. H. Toshimura, 61–106. Tokyo: Nankodo Ltd.

Yoshimura, H. and T. Iida. 1950. Studies on the reactivity of skin vessels to extreme cold: Part I, A point test on the resistance against frost bite. *Jpn. J. Physiol.* 1(2):147–59.

Yoshimura, H. and T. Iida. 1952. Studies on the reactivity of skin vessels to extreme cold: Part II, Factors governing the individual difference of the reactivity, or the resistance against frostbite. *Jpn. J. Physiol.* 2(3):177–88.

Yoshimura, H. and Y. Kuno. 1966. Review of Japanese research on adaptation to heat and cold, In *Human Adaptability and Its Methodology*, In H. Yoshimura and J.S. Weiner, 1–4. Papers given at a symposium held in Kyoto, Japan, under the auspices of the International Union of Physiological Sciences and the International Biological Programme. Tokyo: Japan Society for the Promotion of Sciences.

Yoshimura, H. and J.S. Weiner. 1966. *Human Adaptability and Its Methodology*. Papers given at a symposium held in Kyoto, Japan, under the auspices of the International Union of Physiological Sciences and the International Biological Programme. Tokyo: Japan Society for the Promotion of Sciences.

12 From a History of Anthropometry to Anthropometric History

Stanley Ulijaszek[*1] *and John Komlos*[2]
[1]Institute of Social and Cultural Anthropology,
University of Oxford, Oxford, United Kingdom
[2]Department of Economics, University
of Munich, Munich, Germany

CONTENTS

12.1 INTRODUCTION

Anthropometry is the longest-used measure of human variation and, since it measures surface morphology, is intuitively understood at the elementary level. By standing in any major junction of any major city on Earth, one can easily marvel at the range of human physical diversity: short, tall, thin, fat, long-legged, stumpy; native wit provides the face validity for the study of anthropometric variation and its application (Ulijaszek and Mascie-Taylor 2005). Ideas of biological difference between human populations are of great antiquity but only became quantitatively formalised in the nineteenth century, with early attempts at doing so having taken place in the eighteenth century. Prior to innovations that could identify variation at microscopic levels, including physiological, biochemical, endocrinological, and genetic ones, morphology was the prime means of classification of nature.

The strength of anthropometry as a measure of human variation is its relative cheapness and simplicity of application, although accuracy is another issue. It is no surprise that anthropometric methods have changed little since the nineteenth century. However,

---

[*] Address all correspondence to stanley.ulijaszek@anthro.ox.ac.uk.

the meaning and interpretation of anthropometric variation has changed with new understandings of how human variation is generated and maintained. With changing meaning has come change in usage: from racial classification to international public health and anthropometric history. Morphology may be measured in many ways, and different anthropometric measures have gained and lost primacy as the pressing issues of the day have dictated the measures of importance. These have ranged from head and body breadth and length dimensions informing studies of racial affinities in the nineteenth and early twentieth century to stature affecting likely physical productivity in army, navy, and slave recruitment during the eighteenth and nineteenth centuries, and from use in public health in the nineteenth and twentieth centuries to measures of weight, stature, circumference, and skinfold thickness in identifying environmental influences in child growth and nutritional status in the second half of the twentieth century and into the twenty-first.

In the nineteenth century, anthropometry was used in the creation and validation of racial typologies. Changing understandings of human diversity and its production led to ideas of typology and classification being challenged across the first half of the twentieth century and overturned in the second half of it. By the mid-twentieth century, biological anthropology embraced the idea of anthropometry as a measure of physiological and developmental plasticity in relation to environmental quality, and in most quarters rejected its use in taxonomy. The reframing of anthropometry for use in adaptability research reflected these newer understandings, placing human adaptation as determined by physical human growth patterns in evolutionary context.

From the eighteenth century, a parallel stream of anthropometric use was the determination of healthy or strong physique by means of stature, whose application featured in the assessment of the physical quality of slaves to the Americas and of recruits to the armies of Europe and North America. This was extended to the investigation of human welfare by anthropometric description of both adults and children in general populations from the nineteenth century onwards (Tanner 1981). Its use in this context accelerated in the second half of the twentieth century, becoming embedded in epidemiological practice, most usually as a proxy for nutritional status (Jelliffe 1966). However, crossover between anthropometric practice in anthropology and in public health has been maintained.

Most recently, this has been extended to the understanding of human welfare in the past by the investigation of historic anthropometric datasets, such as those developed from investigations of slave and military recruitment records. This work has led to debate among economists concerning the biological quality of life and how it might be factored into economic analysis more broadly. This chapter traces changes in the use of anthropometric methods and the interpretative lenses used for analysis and understanding of anthropometric data from the eighteenth century to the present. Anthropometric practice in New Guinea across the twentieth century is detailed, to demonstrate ways in which the use of this method in anthropology and in public health, respectively, have emerged and changed across the past hundred years or so.

12.2 ANTHROPOMETRY AND ANTHROPOLOGY

While interest in biological and cultural variation among human populations has great time depth, this became of particular interest during the eighteenth and nineteenth centuries when colonial expansions were bringing Europeans into contact with a wide range of non-European humanity. Linnaeus included classification of human groups within his taxonomy of species (Linne 1767). He proposed subdivisions of humanity into four 'stocks', each with particular characteristics. According to him, *Homo sapiens americanus* were reddish, stubborn, and easily angered; *Homo sapiens asiaticus* were yellowish, greedy, and easily distracted; *Homo sapiens africanus* were black, relaxed, and negligent; and *Homo sapiens europeanus* were white, gentle, and inventive. The characters attributed to each stock placed each grouping according to the perceptions of the dominant European imperial powers of the time, and the science that served them. Although Darwin (1874) proposed mechanisms of natural selection to have operated in the generation of human biological variation, ideas of biological difference between human populations had become formalised into notions of race by the nineteenth century, usually employing some or all of the dominant human taxonomic categories proposed by Linnaeus. The use of anthropometry was central to the creation and reification of racial typologies, forming the basis of scientific racism and eugenics in the late nineteenth century. The use of morphology in human classification continued to be central to physical anthropology into the twentieth century, the concerns in the first half of the twentieth century continuing to be dominated by concerns about racial origins, typologies, and affinities (Harrison 1997).

Ideas of typology and classification were challenged and overturned in the second half of the twentieth century with the empirical testing of evolutionary and ecological mechanisms for human biological variation (Harrison 1997). According to Weiner (1966),

> It is true that anthropologists (of the older schools) have collected data in great abundance but with the avowed aim of making 'racial' and taxonomic comparisons. Consequently (and despite the praiseworthy standardisation of technique and the development of statistical method) the data, in their hands, has proved of rather limited biological value. Undue concentration was paid to cranial and facial dimensions; body weight, circumferences, bone, fat, and muscle components were much less often measured. For the assessment of physique, body composition, or growth patterns and their relation to working capacity and to nutritional, climatic, and other environmental factors—in fact on the general questions of selection and adaptation—the older material is badly deficient.

From the 1960s, human population biology sought to document and explain processes that contribute to human biological variability. At this time, the Human Adaptability section of the International Biological Programme (HAIBP) was founded to extend ecological understandings from plant and animal communities to human populations. The majority of work involved attempts to describe and understand adaptation and adaptability as the ecological processes by which natural selection takes place among human beings (Collins and Weiner 1977). Plasticity of human form was identified by Boas (1912), but it was not until the 1960s and various HAIBP studies that it was used as an explanatory framework in studies of human adaptation (Lasker 1969). The reframing of anthropometric practice for adaptability research reflected newer understandings of human physical plasticity and health, as determined by physical human growth patterns understood in evolutionary context (Bogin 1999).

TABLE 12.1
Major Racial Typology and Ethnic Classifications for the Purposes of Characterizing Anthropometric Differences among Populations

Linnaeus	Eveleth and Tanner	Martorell et al.	Eveleth and Tanner
1767	1976	1988	1990
Homo americanus		Latin American	
Homo asiaticus	Asiatic	Asiatic	Asiatic: Asian
			Asiatic: Amerind and Eskimo
Homo africanus	African	African	African
	African ancestry		African ancestry
Homo europeanus	European	European	European
	European ancestry		European ancestry
	Indo Mediterranean	Indo Mediterranean	Indo Mediterranean
	Australian aboriginal and Pacific Islander		Australian aboriginal and Pacific Islander
	Inter-racial crosses		

The HAIBP instigated studies in ninety-three nations between 1964 and 1974 (Collins and Weiner 1977) and, while different measures were employed among different populations in different places, most collected anthropometry, usually of adults, but often of children too. The major anthropometric outcome was the publication of the volume *Worldwide Variation in Human Growth* (Eveleth and Tanner 1976). The success of this publication stimulated a rapid expansion in the study of child growth and adult morphology. The second edition of *Worldwide Variation in Human Growth* (Eveleth and Tanner 1990) reflected this, containing summary data of studies carried out and published since the first edition of 1976. From the extensive range of studies reported in Eveleth and Tanner (1976, 1990), it was possible to identify the key features of between-population variation in adult morphology and child growth in weight, height, and skinfold thicknesses. However, such variation was classified into population typologies which retained the major classes put forward by Linnaeus (1767) (Table 12.1). Although Eveleth and Tanner (1976, 1990) were careful not to use the term 'ethnic group' to inform their interpretation of between-population differences in human growth patterns and adult morphology, their comparisons took such grouping as implicit to human migration and genetic histories.

The use of such classifications continues to inform the use of anthropometry in public health practice to the present day, despite such classification obscuring as much as it clarifies. The classification of 'ethnic group' was put into the anthropological mainstream by Seligmann (1909) as a descriptor of New Guinea groups that lie at a level intermediate between individual 'tribes', distinguished mostly through cultural characteristics, and larger 'stocks', which could be characterised in racial terms. Weber (1921/1968) put forward a definition of ethnic group that was broadly accepted by anthropologists and sociologists:

> those human groups that entertain a subjective belief in their common descent because of similarities of physical type or of customs or both, or because of memories of colonisation and migration; this belief must be important for group formation; furthermore it does not matter whether an objective blood relationship exists.

This more flexible category allowed the use of the term more generally in relation to populations across the world in ways that have allowed boundaries around particular ethnic groups to be narrowed or broadened according to political circumstances (Cohen 1978). The extent to which ethnic groups map onto patterns of genetic diversity remains controversial, although the demonstration of gradients of human genetic diversity both within and among continents rather than of discrete clusters of genetic similarity argue against it (Serre and Pääbo 2004). This opens the possibility that variation in growth patterns across human groups arise primarily from differences in environmental quality and secondarily from biocultural variation in the past.

12.3 ANTHROPOMETRY FOR THE COMMON GOOD

From the eighteenth century, measures of stature had been used to identify and reject adult males of poor productive quality from the army, navy or slavery, implicitly acknowledging a link between body size, health, and productivity. This was anthropometry for the good of Empire and private enterprise. From the political concern for the quality of the fighting and working stock of imperial nations emerged the welfare-oriented anthropometry of reformers such as Villerme, Chadwick, and Bowditch (Tanner 1981). The theoretical basis for this work remained eugenic: Individuals and groups of poor physical quality might reproduce a cycle of inferior 'stock', not fit for the purposes of Empire or enterprise. It was not appreciated until the surveys of the physical characteristics of immigrants to the United States, undertaken between 1908 and 1911 (Boas 1912), that inferior stock might be improved by living in improved environmental circumstances. In the cities of the United States, reform of social conditions of the poor had been underway since the 1890s (Burnstein 2006), and American-born Bohemians and Hebrews were shown to be taller than their foreign-born counterparts (Table 12.2). In respect of the physical characteristics of children, Boas comments that 'the change in stature and weight increases with the time elapsed between the arrival of the mother and the birth of the child'. This opened the door for a plastic, rather than typological, understanding of human morphology, and the application of anthropometric practice for the common good.

The transition from anthropometric practice for racial typology to that for welfare was not neat, however. In Britain and the United States, eugenic discourse continued well into the first three decades of the twentieth century, as evidenced by the three International Eugenics Conferences, held in London (1912) and New York (1921 and 1932). World War Two saw the discrediting of anthropometry as an instrument of eugenics, although its use in racial typology persists to the present day in some quarters. Post–World War Two, broader public health campaigns were developed, having

TABLE 12.2

Stature of Immigrants and Their Descendants to the United States (Adults)

Bohemians							
	Male				Female		
	Foreign-born		American-born		Foreign-born		American-born	
Age (years)	N	Mean	N	Mean	N	Mean	N	Mean
25 and less	14	170.1	11	174.5	11	157.7	12	160.9
26 and over	448	167.5	58	171.1	598	156.9	71	159.3

Hebrews							
	Male				Female		
	Foreign-born		American-born		Foreign-born		American-born	
Age (years)	N	Mean	N	Mean	N	Mean	N	Mean
All ages	762	164.2	38	167.2	896	155.0	67	158.4

Source: Data from Boas (1912).

emerged in the context of decolonisation and the world health agenda set by the United Nations in 1947 and resulting in the formation of the World Health Organisation in 1948. An acceptance of plasticity in child growth and morphology formed the basis of the use of anthropometry in what Tanner (1981) called auxological epidemiology. This saw the start of the present era of epidemiology that uses statistically-derived growth standards based on large-scale data collection in the surveillance, screening, and monitoring of populations towards improved social welfare (Tomkins 2005).

Known environmental factors that influence growth, body size, and body composition of children include nutrition, infection, interactions between the two, psychosocial stress, food contaminants, pollution, and hypoxia (Ulijaszek 2006). Most of these factors are conditioned by poverty and socioeconomic status, making body size a sensitive marker of the quality of life (Tanner 1986) and anthropometric practice an instrument of social welfare. These factors are also conditioned historically, culturally, and politically, making anthropometric practice a political economic instrument. Diet, nutrition, disease, hypoxia, pollution, contamination, behavioural toxicants, deprivation, and psychosocial stress are clustered as proximate environmental agents that influence child growth (Figure 12.1). They vary in importance according to circumstance and the age and stage in infancy, childhood, and adolescence. Culture, society, behaviour, socioeconomic status, poverty, and political economy are also clustered as structurally powerful but distal agents in the production of growth and body size outcomes, at all ages and stages of childhood and adolescence. In the developing world, poverty is associated with poor food security, inadequate health infrastructure, and environmental hazards, while in the industrialised world, it is associated with single parenthood, overcrowding, paternal ill health, dependence on social welfare, and parental abuse of alcohol and drugs.

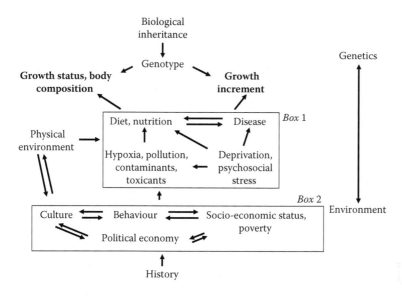

FIGURE 12.1 Proximate and distal agents influencing child growth. (Adapted from Ulijaszek 2006.)

In both developed and developing worlds, there are similar associations between stature and socioeconomic status, height correlating positively with wealth (Komlos and Lauderdale 2007). Weight, however, does not always relate positively with wealth, with overweight and obesity being the provenance of low socioeconomic status in most industrialised nations (Pickett et al. 2005) and becoming increasingly so among emerging nations undergoing the health transition (Amuna and Zotor 2008).

Since human growth and body size is sensitive to changing environmental quality, the generational changes observed by Boas (1912) among immigrants to the United States was observed across the world during the twentieth century. The secular trend describes marked changes in growth and development of successive generations of human populations living in the same territories. Positive secular trends in increased stature and weight, and earlier timing of the adolescent growth spurt have been documented across the world (Ulijaszek 2001). These have been attributed universally to improvements in the quality of life. Negative secular trends have been identified among some populations in Africa, Papua New Guinea, and Central and Latin America. These are usually seen as outcomes of environmental, social, or political deterioration. The mapping by historians of social, economic, and political factors onto such secular trends forms the basis of the discipline of anthropometric history.

12.4 CHANGING RELATIONS AMONG ANTHROPOMETRY, ANTHROPOLOGY, AND WELFARE IN NEW GUINEA

New Guinea was the last major land mass to be colonised by Europeans and was seen by anthropologists of the day as a new field for testing theories of human variation. However, it was primarily seen by its colonisers as a site for economic exploitation of both natural and human resources. Anthropometry was initially employed by scientists in New Guinea to test Wallace's theory concerning migrations in Southeast Asia and the Pacific (Keane 1880; Brown 1887), where a division was made between Malays and Polynesians along a line that ran parallel to, and east of, the present-day 'Wallace Line', which divides Asian and Oceanic flora and fauna. To this end, use of anthropometry to classify the populations of New Guinea led to numerous local surveys, as well as considerable debate. Early on, Haddon (1900) noted the difficulty of identifying 'races' and instead spoke of 'peoples'.

There is no evidence that anthropometry was used to identify and reject adult males of poor productive quality from plantation and other work. This may be because British, and subsequently Australian, administration from the 1890s to the 1940s recruited large numbers of adult males in usually indiscriminate manner, often to the point of depopulating villages of males of productive age (Maher 1961). Anthropometry in physical anthropology was largely discontinued after around 1910, despite opportunities for such surveys or occasional measurements by administrators and medical officers. Anthropologists more generally turned their attention to social and cultural issues.

In public health, the disease patterns observed by colonists in New Guinea in the first half of the twentieth century were seen through racist spectacles, obscuring the social and environmental conditions which gave rise to most of the infections of the region (Denoon 1989). It was not until the late 1920s that high infant mortality rates were seen as a serious issue, after the publication of Rivers' (1922) concern about the decline of the Melanesian population, which continued into the 1950s (Maher 1961). These concerns were replaced with ones about nutrition and infection, post–World War Two, when anthropometry as a measure of child health and nutritional status was first used in 1947 by the New Guinea Nutrition Survey Expedition (Hipsley and Clements 1951). The purpose of this expedition was to identify appropriate ration scales for labourers in New Guinea, based on indigenous foods. A second aim was the ascertainment of the nutritional health of the people.

Anthropometric practice in anthropology reemerged in the 1960s with the HAIBP. Globally, the aim of the HAIBP was to elucidate the interaction of nature and nurture on physiological, morphological, and developmental characters. Researchers included physiologists and nutritionists, for whom anthropometry was useful for standardisation of measurement, assessment of nutritional status, and the study of developmental plasticity. A new synergy in anthropometric practice emerged from the outset. While the Papua New Guinea Institute of Human Biology undertook local coordination of this project as well as medical examination, socio-demography, fertility, and physical growth and development, the Department of Health provided administrative assistance and personnel to help in nutritional and physiological studies (Walsh 1974). By the mid 1970s, the Institute of Human Biology was renamed the Institute of Medical Research, reflecting its emphasis on problem-derived research, most commonly in relation to nutrition and infection, where anthropometry retained importance both in studies of plasticity and in studies of nutritional physiology that required control for body size.

Although initially ignored, the recommendations for nutritional improvement made by Hipsley and Clements (1951) were taken up in the 1970s with the adoption of primary health care as the governing principle of health improvement enshrined in the 1974–1978 National Health Plan (Denoon 1989). Embedded in this plan was simple anthropometrically-based nutrition monitoring and surveillance, using weight for age as the measure. National Nutrition Surveys were carried out in 1975, 1978, and 1982–1983. In the first two of these surveys, weight-for-age data was collected from maternal and child health clinics across the country and used in the development of the National Nutrition Plan. The 1982–1983 survey used a sampling framework to obtain an accurate picture of anthropometric nutritional status of rural children, using weight and height measures. Nutrition programmes were initiated in all provinces from the late 1970s, using anthropometric data for surveillance, and special surveys among groups seen to be of particular nutritional risk.

Studies of human adaptability from the 1970s echoed concerns about nutritional health, with nutritional status, health, physique, body composition, and demography becoming the overriding foci, and studies relating physique to nutrition proliferating (Norgan 1997). Most recently, the emphasis has shifted to relationships between child growth, embodiment, nutrition, and infection with political economy, political ecology, and public health (Ulijaszek 2003).

Thus, anthropometric practice with anthropological focus increasingly converged with that of public health as medical and ecological anthropology became increasingly political and economic in outlook (Goodman and Leatherman 1998). This mirrors similar patterns elsewhere in the world, where anthropometric practice has been reframed within biocultural approaches in anthropology (Ulijaszek 2007). One outcome of such reframing in New Guinea has been the reanalysis of the anthropometry carried out by anthropologists at the turn of the twentieth century in relation to more data collected more recently among the same groups, to examine the changing welfare of New Guinea populations across the twentieth century (Ulijaszek 1993). While not yet in place, the mapping of social, economic, and political change onto secular changes in body size would open New Guinea to investigation through anthropometric history, a field that emerged in the 1970s among cliometricians, who are concerned with analyses of secular changes and cross-sectional patterns in physical growth and development to understand the effects of economic development on biological welfare.

12.5 ANTHROPOMETRIC HISTORY

The concern of economic historians in the 1970s was to extend the existing indices of living standards back in time, to illuminate debates about living conditions of workers during the Industrial Revolution, and to provide indices where none existed. For example, conventional measures of money income do not exist for subgroups of society such as housewives, children, or self-sufficient peasants, and anthropometry can shed light on the welfare of such groups. Economic historians have need for alternative measures of welfare when conventional indicators are unavailable (Steckel 1995), and cliometricians introduced biological measures, including physical stature, as complements to conventional indicators of well-being (Fogel 1994; Steckel 1995).

The analysis of historic anthropometric data created new understandings of the impact of economic processes on humans. This discipline has identified the existence

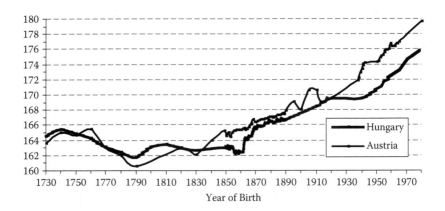

FIGURE 12.2   Height (cm) of Austrian and Hungarian men, age twenty-one.

of long cycles in physical stature, brought about by demographic growth, urbanisation, changes in relative prices, market structure, income, inequality, and climate, as for example, in the Austro-Hungarian Empire between 1730 and 1910 (Figure 12.2) (Baten 2002; Baten and Murray 2000; Steckel 1995). It has also revealed shorter cycles in height associated with business cycles (Brabec 2005; Woitek 2003). Only in the twentieth century have these cycles become attenuated due to improvements in medicine, increases in labour productivity, and a substantial decline in the relative price of food.

Historical anthropometry has been useful in identifying changing welfare during industrialisation. In the preindustrial world, remoteness from markets tended to confer a distinct advantage for human welfare, because farmers or peasants not able to sell their products tended to consume all they produced (Cuff 2005; Craig and Weiss 1998; Haines 1998). Thus, Irishmen were taller than Englishmen, and Northerners in America were shorter than Southerners. Once this circumstance changed, such people traded away their nutritional benefits. Thus, when self-sufficiency declined with industrialisation, heights decreased as well.

Some socioeconomic groups were exempt from anthropometric perturbations during industrialisation and modernisation. These included high-status groups such as European aristocrats and middle-class students whose income was less variable, and either large enough, or rising fast enough, to keep the increase in the relative price of nutrients from influencing their food consumption bundle substantially. Also isolated from the effects of structural change, ironically, were male slaves in the United States, whose value increased so much that their owners had sufficient incentives to maintain and sometimes increase their food allotments to maintain their productivity. Slaves remained well-nourished relative to the European lower classes (Figure 12.3), even if they were shorter than their masters (Figure 12.4).

That income is protective of anthropometric nutritional status is confirmed for past populations by anthropometric history. For example, the height of the upper class in America did not decline in the mid-nineteenth century. Rather, the difference in the height of upper-class men and the average increased from one to three centimetres between 1830 and 1840, a value much lower than those found in Europe.

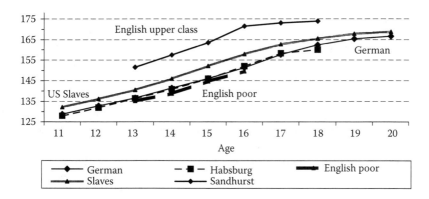

FIGURE 12.3   Height (cm) of youth during the Industrial Revolution.

The greatest social gradient in height ever recorded was found in early-industrial England, where the difference between upper and lower class fifteen-year-olds reached twenty centimetres (Figure 12.3). These exceptions to the general downward trend in physical stature show that the decline in heights was not brought about by an increase in disease incidence alone, because that would have affected the rest of the population.

The secular increases in heights that lasted until the twenty-first century began in the birth cohorts of the 1860s. Several technological breakthroughs were important for the environmental improvements that fuelled them. The decline in the cost of long-distance ocean shipping brought the productivity of the American prairies within the reach of European consumers. The invention of refrigeration enabled perishable agricultural products to be shipped over longer distances. Improvements in public health and sanitation, associated with running water and sewer systems, also made a contribution, as did improvements in biomedicine. Moreover, the increased productivity of

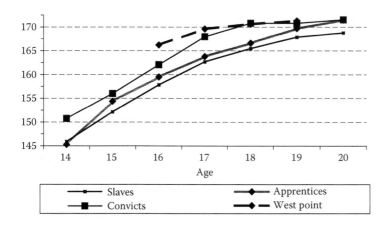

FIGURE 12.4   Height (cm) of U.S. youth, early nineteenth century.

the agricultural sector brought about by mechanisation allowed biological well-being in all societies in the Western Hemisphere to increase by the twentieth century.

Anthropometry has become useful more recently in the understanding of population obesity. The past half century has seen a great change in the size and shape of the population of the United States. While being the tallest of world populations prior to World War Two, they have become among the most overweight at the onset of the twenty-first century. The abundant natural resources of the New World combined with the low population density conferred considerable biological advantages on its inhabitants prior to World War Two. Yet, citizens of the United States have increased in height by only a few centimetres since then. In contrast, many European populations increased by as much as fifteen centimetres in the meanwhile. The American height advantage over Western European populations in the middle of the nineteenth century was as much as three to six centimetres. In contrast, they are now considerably shorter than Western and Northern Europeans, with the Dutch, Swedes, and Norwegians being the tallest, and the Danes, British, and Germans also being taller than U.S. citizens (Figure 12.5). Inasmuch as the United States is a high-income country with advanced medical services that has enjoyed a practically continuous

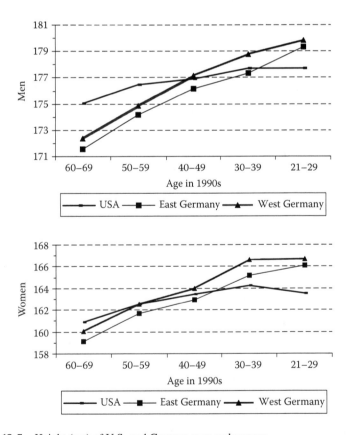

FIGURE 12.5   Height (cm) of U.S. and German men and women.

expansion in economic activity since World War Two, the fact that heights in particular and health in general have not kept pace with Western and Northern European developments is a conundrum.

Human stature is a useful complementary indicator of well being. With the development of the concept of the 'Biological Standard of Living' as distinct from conventional indicators of well-being, human biology, usually by way of anthropometry, is being integrated into mainstream economics and economic history.

12.6 CONCLUSIONS

Anthropometry is the longest-used measure of human variation, one which has undergone considerable change in understanding and interpretation since the nineteenth century. With changing meaning has come change in usage: from racial classification to international public health and, most recently, anthropometric history. Thus, anthropometry has been used in the creation and validation of racial typologies, in the determination of healthy or strong physique, as a measure of physiological and developmental plasticity in relation to environmental quality, and in the investigation of human welfare in relation to both health and economics. The illustration of anthropometric practice in New Guinea across the twentieth century shows how multiple uses of this method have been able to coexist during pivotal periods of history, and how global issues have been reflected in local practice from the European colonisation of the nation. Thus, anthropometric practice in New Guinea became focussed on welfare of the general population after World War Two, when the notions of world health became enshrined in a decolonising world, and was reinforced with the incorporation of primary health care into the National Health Plan. Perhaps the fastest growing area of anthropometric practice is that of anthropometric history. Anthropometric indicators may not be substitutes for, but complements of, conventional measures of living standards, but they dramatically extend the power of historical economic analysis.

REFERENCES

Amuna, P. and F. Zotor. 2008. Epidemiological and nutrition transition in developing countries: Impact on human health and development. *Proc. Nutr. Soc.* 67:82–90.

Baten, J. 2002. Climate, grain production and nutritional status in eighteenth-century Southern Germany. *J. Eur. Econ. Hist.* 30(1):9–47.

Baten, J. and J. Murray. 2000. Heights of men and women in nineteenth-century Bavaria: Economic, nutritional, and disease influences. *Explor. Econ. Hist.* 37:351–69.

Boas, F. 1912. *Changes in Bodily Form of Descendants of Immigrants.* New York: Columbia University Press.

Bogin, B. 1999. Evolutionary perspective on human growth. *Annu. Rev. Anthropol.* 28:109–53.

Brabec, M. 2005. Analysis of periodic fluctuations of the height of Swedish soldiers in the eighteenth and nineteenth centuries. *Econ. Hum. Biol.* 3(1):1–16.

Brown, G. 1887. Papuans and Polynesians. *J. Anthropol. Inst. Great Britain and Ireland.* 16: 311–27.

Burnstein, D. E. 2006. *Next to Godliness: Confronting Dirt and Despair in Progressive Era New York City*. Urbana: University of Illinois Press.

Cohen, R. 1978. Ethnicity: Problem and focus in anthropology. *Annu. Rev. Anthropol.* 7:383.

Collins, K.J. and J.S. Weiner. 1977. *Human Adaptability: A History and Compendium of Research*. London: Taylor and Francis.

Craig, L. and T. Weiss. 1998. Nutritional status and agricultural surpluses in the antebellum United States. In *The Biological Standard of Living in Comparative Perspective*, ed. J. Komlos and J. Baten, 190–207. Stuttgart: Franz Steiner.

Cuff, T. 2005. *The Hidden Cost of Economic Development: The Biological Standard of Living in Antebellum Pennsylvania*. Aldershot, Hants, England: Ashgate.

Darwin, C. 1874. *The Descent of Man, and Selection in Relation to Sex*. London: John Murray.

Denoon, D. 1989. *Public Health in Papua New Guinea: Medical Possibility and Social Constraint, 1884–1984*. Cambridge: Cambridge University Press.

Engerman, S. 1976. The height of U.S. slaves. *Local Popul. Stud.* 16(1):45–9.

Eveleth, P.B. and J.M. Tanner. 1976. *Worldwide Variation in Human Growth*, 1st edition, Cambridge: Cambridge University Press.

Eveleth, P.B. and J.M. Tanner. 1990. *Worldwide Variation in Human Growth*, 2nd edition, Cambridge: Cambridge University Press.

Fogel, R. 1994. Economic growth, population theory, and physiology: The bearing of long-term processes on the making of economic policy. *Am. Econ. Rev.* 84:369–94.

Goodman, A.H. and Leatherman, T.L. 1998. *Building a New Biocultural Synthesis: Political-Economic Perspectives on Human Biology*. Ann Arbor: University of Michigan Press.

Haddon, A.C. 1900. Studies in the anthropogeography of British New Guinea. *Geogr. J.* 16(265–91):414–41.

Haines, M. 1998. Health, height, nutrition, and mortality: Evidence on the 'antebellum puzzle' from Union army recruits for the New York State and the United States. In *The Biological Standard of Living in Comparative Perspective*, ed. J. Komlos and J. Baten. 155–80. Stuttgart: Franz Steiner.

Harrison, G.A. 1997. The role of the Human Adaptability International Biological Programme in the development of human population biology. In *Human Adaptability: Past, Present and Future*, ed. S.J. Ulijaszek and R.A. Huss-Ashmore, 17–25. Oxford: Oxford University Press.

Hipsley, E.H. and F.W. Clements. 1951. *Report of the New Guinea Nutrition Survey Expedition, 1947*. Sydney: Government Printer.

Jelliffe, D.B. 1966. *Assessment of Nutritional Status of the Community*. Geneva: World Health Organization.

Keane, A.H. 1880. Monograph on the relations of the Indo-Chinese and interoceanic races and languages. *J. Anthropol. Inst.* 9:254–89.

Komlos, J. and M. Baur. 2004. From the tallest to (one of) the fattest: The enigmatic fate of the American population in the 20th century *Econ. Hum. Biol.* 2(1):57–74.

Komlos, J. and B. Lauderdale. 2007. Underperformance in affluence: The remarkable relative decline in U.S. heights in the second half of the 20th century. *Soc. Sci. Q.* 88:283–305.

Lasker, G.W. 1969. Human biological adaptability. *Science* 166:1480–6.

Von Linne, C. 1767. *Systema naturae, per regna tria naturae: Secundum classes, ordines, genera, species cum characteribus, differentiis, synonymis, locis*. Vindobonae [Vienna]: Typis Ioannis Thomae.

Maher, R.F. 1961. *New Men of Papua*. Madison: University of Wisconsin Press.

Norgan, N.G. 1997. Human adaptability in Papua New Guinea: Original aims, successes, and failures. In *Human Adaptability: Past, Present and Future*, ed. S.J. Ulijaszek and R.A. Huss-Ashmore, 102–25. Oxford: Oxford University Press.

Pickett, K., Kelly, S., Brunner, E., Lobstein, T. and R.G. Wilkinson. 2005. Wider income gaps, wider waistbands? An ecological study of obesity and income inequality. *J. Epidemiol. Community Health* 59:670–4.

Rivers, W.H.R. 1922. *Essays on the Depopulation of Melanesia*. Cambridge: Cambridge University Press.

Seligmann, C.G. 1909. A classification of the natives of British New Guinea. *J. R. Anthropol. Inst.* 39(247–75):314–33.

Serre, D. and Pääbo, S. 2004. Evidence for gradients of human genetic diversity within and among continents. *Genome Res.* 14:1679–85.

Steckel, R.H. 1995. Stature and the Standard of Living. *J. Econ. Lit.* 33:1903–40.

Tanner, J.M. 1981. *A History of the Study of Human Growth*. Cambridge: Cambridge University Press.

Tanner, J.M. 1986. Growth as a mirror of the condition of society: Secular trends and class distinctions. In *Human Growth. A Multidisciplinary Review*, ed. A. Demirjian, 3–34. London: Taylor and Francis.

Tomkins, A. 2005. Growth monitoring, screening and surveillance in developing countries. In *Anthropometry: The Individual and the Population*, ed. S.J. Ulijaszek and C.G.N. Mascie-Taylor, 108–16. Cambridge: Cambridge University Press.

Ulijaszek, S.J. 1993. Evidence for a secular trend in heights and weights of adults in Papua New Guinea. *Ann. Hum. Biol.* 20:349–55.

Ulijaszek, S.J. 2001. Ethnic differences in patterns of human growth in stature. In *Nutrition and Growth*, ed. R. Martorell and F. Haschke, 1–15. Philadelphia: Lippincott Williams and Wilkins.

Ulijaszek, S.J. 2003. Socioeconomic factors associated with physique of adults of the Purari delta of the Gulf Province, Papua New Guinea. *Ann. Hum. Biol.* 30:316–28.

Ulijaszek, S.J. 2006. The international growth reference for children and adolescents project: Environmental influences on preadolescent and adolescent growth in weight and height. *Food Nutr. Bull.* 27(Suppl):S279–94.

Ulijaszek, S.J. 2007. Bioculturalism. In *Holistic Anthropology: Emergence and Convergence*, ed. D. Parkin and S. Ulijaszek, 21–51. Oxford: Berghahn Books.

Ulijaszek, S.J. and C.G.N. Mascie-Taylor 2005. *Anthropometry: The Individual and the Population*. Cambridge: Cambridge University Press.

Walsh, R.J. 1974. Geographical, historical and social background of the peoples studied in the I.B.P. *Philos. Trans. R. Soc. Lond. B Biol. Sci.* 268:223–8.

Weber, M. 1921/1968. *Max Weber on Law in Economy and Society*, ed. M. Rheinstein, Translated by E. Shils and M. Rheinstein. New York: Simon and Schuster.

Weiner, J.S. 1966. Major problems in human population biology. In *The Biology of Human Adaptability*, ed. P.T. Baker and J.S. Weiner, 1–24. Oxford: Clarendon Press.

Woitek, U. 2003. Height cycles in the eighteenth and nineteenth centuries. *Econ. Hum. Biol.* 1(2):243–57.

13 Growth and Maturation
Interactions and Sources of Variation

*Robert M. Malina**
Department of Kinesiology and Health Education,
University of Texas at Austin, Austin, Texas;
Department of Health and Physical Education,
Tarleton State University, Stephenville, Texas

CONTENTS

13.1 INTRODUCTION

All children and adolescents have three primary tasks: to grow (increase in the size of the body as a whole and of its parts and systems), to mature (progress towards the biologically mature state, which is an operational concept because maturity varies with the body system), and to develop (learn the appropriate cognitive, social, affective, moral, motor, and other behaviours expected by society). Growth and maturation are biological processes, while development is a behavioural process, often subsumed in the term *socialisation*, specific to a culture. The three processes are distinct, though related and interacting, tasks that dominate the daily lives of youth for approximately the first two decades of life (Malina, Bouchard, et al. 2004). Interactions among growth, maturation, and development vary during childhood and adolescence, among individuals, and within and between cultural groups.

The focus of this discussion is biological growth and maturation, two processes that occur in context. The processes occur in, and interact with, the specific environments

---

* Address all correspondence to rmalina@skyconnect.net.

within which the individual is reared: the natural environment, human modifications of the environment, and the cultural environment. Though definitions of culture vary, it can be viewed as a system of meanings and associated feelings in which individuals and groups are embedded; they are rooted to some extent in the past and interact with the present. The system of meanings is an amalgam of symbols, values and behaviours that characterise a population (Kroeber et al. 1952). Meanings influence beliefs, sanctions, and definitions, and ultimately human behaviours, including values, perceptions, and behaviours related to children, puberty, family, eating, physical activity and inactivity, obesity, and wellness and illness, among others.

It is generally assumed that interactions among genes, hormones, nutrients, and energy are the primary determinants of variation in the growth and maturation of healthy children and adolescents. Many genes have been identified with these processes and many more will be identified. The key, of course, is identification of the mechanisms whereby genetic variants exert their effects on the processes of growth and maturation, and interact with the environments in which children grow and mature.

The shared cultural environment also affects growth and maturation. These include present lifestyle characteristics and those transmitted from parents to children through modelling, education, and socioeconomic status. These are potentially important sources of variation in growth and maturation, especially transgenerational secular changes, which may be either positive or negative. A key issue is the identification of the mechanisms whereby cultural practices exert their effects on and interact with the processes of biological growth and maturation.

13.2   COMMON STUDY THEMES

Studies of growth and maturation can be sorted, to some extent, into several themes. These have traditionally included normative studies of body size, proportions, and composition from infancy through adolescence into young adulthood; interactions of growth and maturity status during childhood and adolescence; modelling of the growth curve, specifically for height and the adolescent spurt; consequences of nutritional stress in populations living in diverse environmental settings and specifically interactions between nutrition status and environmental conditions; social factors affecting the processes, specifically socioeconomic status, family size, and urban/rural residence; and secular trends, their presence, absence, and reversals. More recent study themes include emphasis on infancy and early childhood, especially in the context of the association between birth weight and/or early growth and later health outcomes; the recent secular increase in obesity and the metabolic consequences of unhealthy weight gain; consequences of environmental pollution; and internal and external migration in developing/lesser developed countries and associated marginalization, social inequality, and rapid growth of urban centres. The subsequent discussion highlights several of these themes.

13.3   SECULAR TRENDS

Evaluation of secular changes is a dominant theme in the study of growth and maturation. The primary focus of such analyses is height, especially with the acceptance of height as an index of well-being and of conditions in a society. Another major focus is age at menarche. In the context of variation, secular trends can be short term, spanning a decade or two, or long term, spanning several generations. Secular trends vary among populations and also in magnitude and tempo. They have not occurred in some populations and have been reversed in others. Evidence from national surveys of the United States and the Netherlands provide some insights and raise several questions. Heights of children and adolescents show, on average, no consistent differences among five surveys between 1963–1970 and 1999–2002 (Figure 13.1), while pubertal onset in both sexes and age at menarche show a relatively small decline over this interval (Sun et al. 2005; Himes 2006). The preceding suggests that the secular trend towards larger size and earlier maturation of children and adolescents has stopped in the United States. However, detailed analysis of U.S. data for American White and Black youth across national surveys indicates secular gains in height for those born after 1970 (Komlos and Breitfelder 2008). Heights of youth were expressed as age- and sex-specific z-scores relative to U.S. growth charts (Kuczmarski et al. 2000). Heights of children born in the 1940s increased slightly, those of children born in the 1950s and 1960s were generally constant (that is, no secular change), while heights of children born after 1970s indicated significant secular gains. In addition to the gain in height, the analysis of z-scores also suggested acceleration in tempo of growth (Komlos and Breitfelder 2008).

Analysis of heights of U.S. adults by year of birth suggested 'stagnation' in attained size in 1955–1974 birth cohorts but an increase in size in 1975–1983 birth cohorts (Table 13.1). If the data are placed in the context of those for youth cited above, the following is suggested: 1) there was no change in heights of children born between around 1950 and 1970 and a secular increase in heights of children born after 1970, and 2) the stagnation in heights of adults born between 1955 and 1974 and the secular increase in heights of adults born between 1975 and 1983 began, respectively, after the corresponding trends in children. By inference, it is likely that the secular trend in heights of American White and Black youth and adults will continue at least into the near future (Komlos and Lauderdale 2007; Komlos and Breitfelder 2008). The detailed analyses of heights of the American White and Black population also highlight periodicity in attained height in the nationally-representative samples.

In contrast to the United States, four national surveys of the Dutch population of the Netherlands between 1955 and 1997 (Fredriks et al. 2000; Mul et al. 2001) show a continued secular increase in height, especially late adolescent growth, which appears seemingly prolonged compared to other populations. On the other hand, there is no change in age at pubertal onset between 1980 and 1997, while age at menarche has declined by about 0.5 year over this same interval.

A question of relevance is what underlies the variation between these two well-off countries. Discussions generally focus on the availability of universal health care and lack thereof, and social welfare network systems and lack thereof. The situation in the United States is confounded to some extent by ethnic variation and variation in sampling strategies among national surveys. Estimated median ages at menarche declined from 12.8 to 12.6 years in American White girls and from 12.5 to 12.1 years in American Black girls (MacMahon 1973; Chumlea et al. 2003). The increase in adult height in the 1975–1983 birth cohorts was evident in American Whites but not in American Blacks (Komlos and Lauderdale 2007). On the other hand, the late adolescent growth of Dutch youth appears in contrast to the observed cessation of the secular trend in

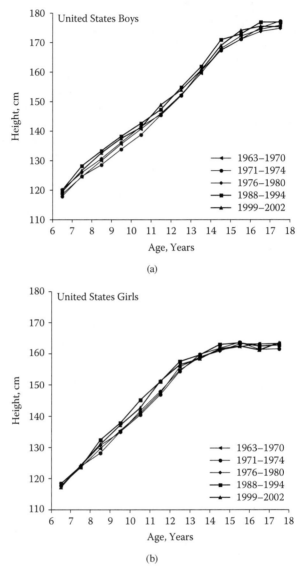

FIGURE 13.1   Mean heights of U.S. boys (a) and girls (b) in five national surveys between 1963–1970 and 1999–2004. [Drawn from data reported in Hamill et al. (1970, 1973, 1977), Najjar and Rowland (1987), National Center for Health Statistics (2005), and McDowell et al. (2005).]

many European countries (Bodzsar and Susanne 1998), prompting the question of what specifically contributes to the prolonged adolescent growth observed in the Dutch.

In contrast to the preceding, there has been a lack of secular change in some parts of the world. As an example, evidence for mean adult heights among indigenous Zapotec adults in Oaxaca, southern Mexico, between 1899 and 2000 is summarised

TABLE 13.1
Mean Heights of U.S. White
Males and Females in Five
National Surveys, 1959–1962
through 1999–2004, by Birth
Cohort

Birth Year	Males	Females
1910–14	174.6	160.5
1915–19	174.1	160.9
1920–24	175.1	161.7
1925–29	176.1	162.7
1930–34	176.9	162.7
1935–39	176.9	163.0
1940–44	176.5	163.5
1945–49	176.8	163.2
1950–54	177.6	163.3
1955–59	177.6	164.4
1960–64	177.9	163.7
1965–69	177.9	163.7
1970–74	177.7	163.6
1975–79	178.6	164.6
1980–83	179.3	165.1

Source: Adapted from Komlos and
Lauderdale (2007).
Note: Means are standardized for
income and education levels.

in Table 13.2. The data for males suggests, perhaps, a decline in adult height during the first half of the twentieth century. Unfortunately, ages of earlier samples are not reported, which potentially confounds interpretations of the trend, or lack thereof. Nevertheless, the results are consistent with the marginal and inequitable position of the state of Oaxaca compared to other federal entities in the country. The rank of the state of Oaxaca among the thirty-two federal entities in Mexico was thirtieth in 1950 and declined to thirty-second in 1960; subsequently, the rank of the state of Oaxaca has changed little: thirty-second in 1960, 1970, and 1980, and thirty-first in 1990 and 2000 (Programa de las Naciones Unidas para el Desarrollo 2003). Conditions have recently improved in the latter part of the twentieth century as evident in a secular increase in heights of indigenous young adults 19–29 years and especially in children 6–13 years and adolescents 13–17 years between surveys conducted in 1978 and 2000 in a rural Zapotec community in Oaxaca (Malina, Peña Reyes, Tan, et al. 2004). The magnitude of the secular increase in heights of children and adolescents is significantly greater within each sex compared to young adults in the indigenous community.

TABLE 13.2

Heights of Zapotec Adults in Oaxaca, Southern Mexico, 1899 to 2000, by Year of Survey

Year of Survey	n	Age (yrs)	Height, cm Mean	SD	Range	Adjusted Height, cm Mean	SD
Males							
1899	100	–	158.6		143.2–177.2		
1933	50	–	155.4	4.9	143.5–164.3		
1941	236	–	156.0		133.0–173.0		
1950s	–	~30.0	156.1				
1971	178	36.2	157.8	5.4	141.0–172.2	158.3	5.2
1978	127	33.7	157.2	4.9	142.6–167.2	157.7	4.8
2000	155	48.4	157.1	5.7	142.0–169.2	158.2	5.4
Females							
1899	25	–	147.5		140.3–157.0		
1971	112	37.4	146.2	4.4	135.3–156.1	146.8	4.4
1978	125	32.7	145.7	5.2	134.5–159.2	146.1	5.1
2000	255	44.4	145.7	5.1	130.5–160.4	146.5	4.8

Source: Adapted from Malina et al. (1983, 2010).

Note: Adjusted height refers to measured height adjusted for estimated age-associated height loss.

13.4   SECULAR TREND IN OBESITY

The prevalence of obesity in children and adolescents has increased in many populations within the past generation. The trend appears to be worldwide and is associated with the emergence of metabolic complications early in life (Cook et al. 2003). Evidence for the increasing prevalence of obesity in the United States across national surveys between the 1960s and 2004 is summarised in Table 13.3. The increasing prevalence emerged during the 1980s and has continued to increase, albeit at a slower rate, since. Moreover, within each age group, there appears to have been a greater increase in the upper percentiles of the BMI, producing an effect of increasing skewness in the distribution over time (Flegal and Troiano 2000).

Two components of the environment are often implicated in the trend: eating behavior (increased energy intake) and physical inactivity (reduced energy expenditure). Though far from conclusive, presently available data imply both increased energy intake and reduced physical activity in the obesity epidemic. Food energy available per capita (kcal/capita/day) in the U.S. population from 1909 to 2000 declined from 1920–1929 to 1960–1969, increased to a small extent in 1970–1979, and then increased to a larger extent in 1980–1989 and subsequent years (Gerrior et al. 2004). After correction for spoilage and wastage, food energy available per capita increased by 460 kcal/day between 1970 and 1997. While these figures refer to food energy available and not to actual energy intake, it seems likely that much of this increase is reflected in intake.

TABLE 13.3

Secular Trend in the Prevalence of Obesity in U.S. Youth,
National Surveys 1963–1970 through 2003–2004

| Year of Survey | 6–11 Years | | | | 12–19 Years | | | |
| | Males | | Females | | Males | | Females | |
	%	SE	%	SE	%	SE	%	SE
'63–70	4.0	0.4	4.5	0.6	4.5	0.4	4.7	0.3
'71–74	4.3	0.8	3.6	0.6	6.1	0.8	6.2	0.8
'76–80	6.6	0.8	6.4	1.0	4.8	0.5	5.3	0.8
'88–94	11.6	1.3	11.0	1.4	11.3	1.3	9.7	1.1
'99–00	15.7	1.8	14.3	2.1	14.8	1.3	14.8	1.0
'01–02	17.5	1.9	14.9	2.4	17.6	1.3	15.7	1.9
'03–04	19.9	2.0	17.6	1.3	18.3	1.9	16.4	2.3

Source: Adapted from Ogden et al. (2002, 2006).

Note: BMI > P95, age, sex-specific.

Estimated energy intakes of adolescents in the United Kingdom have declined, on average, from the 1930s to the 1980s, 2640 to 1880 kcal/day in girls (–29%) and 3065 to 2490 kcal/day in boys (–19%) 14–15 years of age (Durnin 1992). Corresponding data for the United States over this interval are not available, but mean daily energy intakes in four national surveys between 1971 and 2000 indicate that, with the exception of adolescent females 16–19 years, mean daily energy intakes of children and adolescents did not change appreciably among surveys, including the interval in which the prevalence of obesity doubled, 1976–1980 to 1988–1994 (Briefel and Johnson 2004).

An important issue is estimated energy intakes relative to changes in body mass. Among adolescents in the United Kingdom, body weights were, on average, similar in boys and girls from the 1930s to the 1980s (Durnin 1992). Similarly, in the United States, mean body weights and heights showed little variation among three national surveys between the 1960s and 1976–1980 (Figure 13.2). Over these intervals in the United Kingdom and the United States, mean energy intakes of children and adolescents declined; by inference, the reduced energy intake likely reflected reduced energy expenditure (decreased physical activity). However, in the 1988–1994 and 1999–2002 national surveys of the U.S. population, mean weights of children and adolescents increased compared to the earlier surveys while mean heights did not appreciably change (Figure 13.2), which suggests that increased energy intake and reduced energy expenditure (reduced physical activity) contributed to the weight gain.

The increase in food energy available to the American population coincided with an increased per capita consumption of high fructose corn syrup (HFCS) products in the United States. Calories from HFCS consumed per day in the United States were negligible in 1970 and increased slightly through the 1970s. However, estimated HFCS calories consumed per day increased markedly from about 30 calories per day in 1978

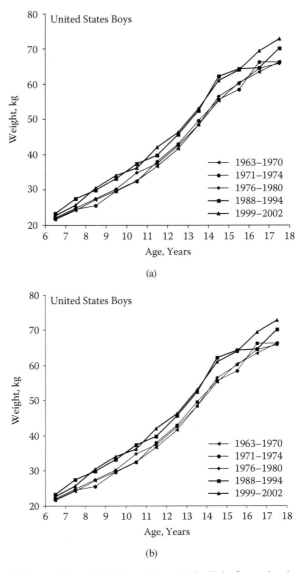

FIGURE 13.2 Median weights of U.S. boys (a) and girls (b) in five national surveys between 1963–1970 and 1999–2004. [Drawn from data reported in Hamill et al. (1970, 1973, 1977), (Najjar and Rowland 1987), National Center for Health Statistics (2005), and McDowell et al. (2005).]

to about 150 calories per day in 1985; HFCS calories consumed per day has since increased more gradually through the late 1980s and 1990s, reaching more than 200 calories per day in 2000 (Schoonover and Muller 2006). The increased consumption of calories from HFCS coincided with the increased prevalence of obesity in the United States in the 1988–1994 national health survey, which continued through 2003–2004.

It is possible that human metabolism is not adapted to handle HFCS, which is a synthetically-produced sugar using a synthetic enzyme. Fructose is metabolised in the liver, whereas glucose is metabolised in all cells, and the digestive and absorptive processes for the two sugars are different (Bray et al. 2004). Humans may not be able to completely metabolise this specific sugar species. The enzymatic handicap, in turn, may lead to an accumulation of glycemic excesses that ultimately results in storage of high levels of fat in many people with a prevalent variant combination of a complex of polymorphic genes.

A decline in levels of habitual physical activity and energy expenditure of youth from developed countries and segments of developing countries is generally accepted. It is not clear, however, when the decline set in and whether it coincides with the increased prevalence of obesity. General demographic information suggests a decline in habitual physical activity over the past forty to fifty years or so. Common examples include forms of physical activity and opportunities for physical inactivity: reduction in active transport (walking, bicycling) to work and school; increase in number of cars per household and in hours of television viewing; increased time in school (a major socially sanctioned form of physical activity) and academic demands (homework); reduction in free time and time in unstructured activities (play); and others (Malina and Katzmarzyk 2006). Many of the trends suggest an increase in opportunities for physically inactive behaviours in contrast to a single sedentary behavior: video games, personal computer activities, DVDs, homework, extracurricular classes (tutoring, art, music), other organised activities (youth groups), motorised transport to school, and probably others.

Short-term physical activity trends based on questionnaires and telephone surveys of youth are inconsistent. Canadian youth 12–19 years indicated an increase in self-reported leisure-time physical activity between 1981 and 1988, but there was no change between 1988 and 1998 (Eisenmann et al. 2004). The percentage of American high school students who attended physical education daily declined from 42% in 1991 to 28% in 2003, while the percentage of those enrolled in physical education classes who were physically active more than twenty minutes in a class, three to five days per week, varied only slightly between 32% in 1991 and 39% in 2003 (Lowry et al. 2004). The percentage of high school students reporting participation in physical activities which made them sweat and breathe hard for more than twenty minutes on three of the past seven days also changed only slightly, 66% in 1993 and 63% in 2003 (Centers for Disease Control 2005).

A more direct indicator of physical activity is the physical activity level (PAL), the ratio of total energy expenditure (TEE) to basal metabolic rate (BMR). PAL provides an approximation of activity-related energy expenditure over twenty-four hours; it is not an indicator of activity energy expenditure. PAL increases with age during childhood and into adolescence (Black et al. 1996; Torun et al. 1996). Estimates of PALs based on the doubly-labelled water (DLW) method for measuring total energy expenditure (TEE) have been available since the 1980s. Two summaries of PALs in children and adolescents based on DLW data from the 1980s through the mid-1990s provide a recent baseline (Black et al. 1996; Torun et al. 1996). PALs based on the DLW method for children and adolescents in studies done since the mid-1990s were collated (Malina and Little 2008). Mean PALs weighted for sample sizes in individual studies have declined in recent samples of boys and girls 6–13 years of age, while corresponding PALs for adolescents 14–18 years have not changed appreciably, though data for adolescents are less extensive than for children. Data for adolescents based on heart rate monitoring and indirect calorimetry are more available.

Allowing for differences in methods, the data for adolescents overlap between samples from the 1970s through the early 1990s and more recent estimates.

Estimates of absolute and relative TEE (kcal/day and kcal/kg/day) for studies of children since the 1970s and adolescents since the mid-1960s were also collated (Malina and Little 2008). Data are based on heart rate monitoring, indirect calorimetry, and activity diaries, singly and in combination, and the doubly-labelled water method. In addition to intra- and interindividual variation in TEE among free-living youth, i.e., children and adolescents carrying out their regular daily activities, potential variation associated with the methods should be noted. Although estimated TEE based on the different methods agree reasonably well, the diary method tends to underestimate TEE of older adolescents (Torun et al. 1996). Absolute TEE shows no trend over time in children 7–13 years but tends to increase over time among adolescents 14–18 years, which likely reflects the secular increase in body mass. Relative TEE, on the other hand, tends to decline over time in children 7–13 years of age but increases slightly over time in adolescents 14–18 years of age (Malina and Little 2008). Sample sizes in studies of TEE are relatively small and probably do not include obese youth. Absolute TEE is greater in obese youth given the energy cost of moving a larger body size, but per unit body weight, TEE is less in obese youth (Bandini et al. 1990; Treuth et al. 1998; Lazzer et al. 2003).

The issue of 'unhealthy weight gain' often surfaces in discussions of the need to reduce the risk of overweight and obesity in children and adolescents. Given the individuality of growth rate and the timing and tempo of growth and maturation during the adolescent growth spurt, it may be difficult to specify 'unhealthy weight gain'. Physical activity is often indicated as important in limiting weight gain. Nonobese youth who are relatively high in physical activity tend to have less adiposity (BMI, skinfolds, and estimated fatness), but enhanced activity programmes in normal-weight youth have a minimal effect on adiposity (Malina et al. 2007). It is possible that normal-weight youth may require a greater activity volume to bring about changes in adiposity. On the other hand, intervention studies with overweight/obese youth indicate reductions in overall and visceral adiposity with programmes of moderate/vigorous intensity activity (Malina et al. 2007). Unfortunately, the studies do not address the issue of the amount of activity needed to maintain the beneficial effects of activity programmes on adiposity.

Physical activity is associated with smaller gains in the BMI in normal-weight youth (Berkey et al. 2003). Thus, maintenance of smaller gains in the BMI through physical activity over time may prevent unhealthy weight gain and in turn reduce risk of overweight/obesity. Two longitudinal studies suggest a potentially important role of physical activity in limiting the accumulation of fatness in different phases of growth: more active children between 4 and 11 years have less fatness in early adolescence and may also have a later adiposity rebound (Moore et al. 2003), and an increase in physical activity during adolescence may limit the accrual of fat mass in males but not females (Mundt et al. 2006).

13.5 MIGRATION AND GROWTH

Relatively large-scale migration is a current sociopolitical reality fuelled to a large extent by economic inequities. Both internal (from rural areas to cities within a country) and external (international) migration are realities in many parts of the world. Migration of a family member or of families has the potential to impact on the growth status of children. A comparison of children 5–12 years of age of Guatemalan migrants to the United States, 93% of whom were born in the United States, with children resident in rural and agricultural communities in Guatemala shows differences in size and proportions (Figure 13.3). Children of Mayan ancestry born in the United States were taller than the corresponding sample resident in rural Guatemala (age-adjusted difference in height 11.5 ± 0.4 cm) and also had absolutely and proportionally longer legs (Bogin et al. 2002).

It is an increasingly common practice in rural areas of many developing regions of the world for a family member to emigrate for the purpose of improving the economic well-being of the family. Remittances from migrants are a major source of income which is used to improve housing and overall living conditions of the family. In rural Mexico, as in other areas of the world, the decision to emigrate is a family matter. In this context, the growth status of children from households with a family member

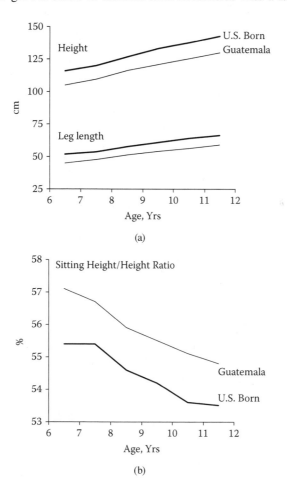

FIGURE 13.3   Height and estimated leg length A) and proportions B) of children of Mayan ancestry born and resident in the United States and in rural Guatemala, sexes combined. (Drawn from data reported by Bogin et al. 2002.)

working outside a rural indigenous community in Oaxaca, southern Mexico, was compared with children from households without emigrants. Children from households with emigrants were significantly shorter and lighter (Malina, Peña Reyes, Tan, et al. 2004). The data may be interpreted from two perspectives. The smaller size of children from households with a family member working outside the community may be less well-off economically, as suggested by the growth status of the children. An earlier study in the community, carried out in 1978, indicated socioeconomic variation in the growth status of children, significantly so in boys but not in girls (Malina et al. 1985). Socioeconomic status (SES) was based on land holdings, household goods, and occupation. Boys from low-SES households were significantly shorter and lighter than those from high-SES households, while girls from low- and high-SES households did not differ (Malina et al. 1985). Longitudinal observations over a one-year interval indicated no SES differences in estimated growth rates, suggesting that SES-associated effects on growth occur early in life prior to school entry at six years of age (Little et al. 1988). Alternatively, household financial gains associated with emigration are not reflected in, or require more time, to influence the growth status of school-age children. A confounding factor in this interpretation is lack of control of three related factors: phase of the migration cycle, duration of the migration, and age of the household. Ethnographic interviews in the community indicate a cycle of approximately three years. A major portion of income during the first year a migrant is away from home is used to pay expenses associated with migration (usually a loan), while a major portion of income during the second year is used to establish the migrant in the new residence. Funds begin to regularly flow back to the family during the third year.

Internal migration within many Latin American countries has contributed to the expansion of cities. These urban developments are often irregular neighbourhoods on the edge of a city which are variously labelled shanty towns or squatter settlements (*colonias populares, favelas, barriadas, pueblos jovenes, villa miseria*). In some instances, established communities adjacent to a city are recipients of a large influx of migrants from rural areas and take on the characteristics of these neighbourhoods. Residents of such communities in Mexico (*colonias populares*) are often described as a heterogeneous mix of classes and ethnic groups with some urban advantages; however, they generally do not show improved health and nutritional status compared with economically better-off segments of the urban population (Selby and Murphy 1979; Murphy and Stepick 1991).

Social class variation in the growth status of school children 6–13 years resident in a *colonia* within the urban limit of the city of Oaxaca de Juarez in southern Mexico was considered in 1972 and 2000. Household SES was based on occupation of the head of the household; age of the child, place of birth of the child, and household size were statistically controlled. Socioeconomic contrasts in growth status in 1972 and 2000 are summarised in Table 13.4 (Malina et al. 2009). Growth status did not differ significantly among SES groups of boys in 1972 and in 2000 and among SES groups of girls in 1972, but differed significantly among SES groups of girls in 2000. Post hoc comparisons indicated no difference between middle- and low-middle-SES girls, while both groups were significantly taller and heavier than low-SES girls in 2000. Overall, the SES contrasts among urban school children in Oaxaca were not as clearly defined as in some Western countries. The differences in height z-scores within SES between the 1972 and 2000 samples suggest that conditions

TABLE 13.4

Variation in Growth Status by Socioeconomic Status of the Household in School Children 6–13 Years of Age Resident in an Urban *Colonia* in Oaxaca, Southern Mexico, in 1972 and 2000: Adjusted Means[a]

	Males			Females		
	Low	Low-Middle	Middle	Low	Low-Middle	Middle
1972, *n*	59	128	31	44	121	26
Height z-score	−2.06	−1.93	−1.93	−2.25	−2.09	−2.17
Height, cm	125.4	125.9	126.0	125.3	126.2	125.5
Weight, kg	25.7	25.9	25.7	26.0	26.4	26.0
BMI, kg/m²	16.1	16.1	15.8	16.2	16.3	16.2
2000, *n*	19	76	71	26	86	51
Height z-score	−1.22	−1.14	−0.93	−1.46	−0.83	−0.80
Height, cm	126.2	126.8	128.0	125.4	129.1	129.1
Weight, kg	28.1	28.5	29.7	26.4	29.5	30.1
BMI, kg/m²	17.3	17.4	17.6	16.6	17.4	17.6

[a] Means in both years are adjusted for age and place of birth of the child and household size (Malina et al. 2009).

have improved across time in the three groups. With sexes combined, the prevalence of growth stunting did not differ by SES in 1972, but the prevalence declined over time within SES, from 53% to 31% in the low SES, from 53% to 17% in the low-middle SES, and from 56% to 7% in the middle SES.

13.6 SOCIAL PROCESSES AND GROWTH

Social processes within a community or within a family may be potential source of variation in the growth and maturity status of children. For example, mate choice is largely an economic strategy, for example, in patrilocal vs. matrilocal postnuptial labour contracts in Mesoamerican subsistence agriculture peasant communities (Selby 1966, 1974; Nutini 1967). The influence of social processes related to mate choice on growth status was analysed in a rural indigenous community in Oaxaca, southern Mexico (Little and Malina, under review). Z-scores for the heights of school children and adult height were compared by postnuptial matrilocal, patrilocal, neolocal, and current matrilocal, patrilocal, and nuclear residence. Household economic status was estimated from household resources related to production and demography. Results are summarised in Figure 13.4. Matrilocal strategy was used by males from poorer households (reflected in their shorter stature) for socioeconomic mobility, resulting in improved growth status of their offspring (taller stature). Mate choice, associated labour contract (an economic strategy), and the inheritance cycle driven by social processes had a significant effect on child growth status, adult

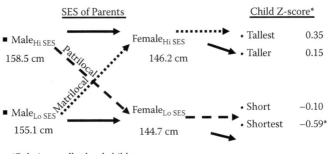

FIGURE 13.4 Influence of wealth distribution on growth status in a rural indigenous community in Oaxaca, southern Mexico: height z-scores of children by postnuptial residence and current resident type. (Drawn from data reported in Little and Malina under review.)

height, and genetic variance in the community. Social processes can thus positively and significantly affect biological and genetic variation in the population.

13.7 SOCIAL PROCESSES AND MENARCHE

Age at menarche is perhaps the most extensively studied maturational event. Many environmental factors are associated with its timing. More recently, emphasis has shifted to the family environment in the context of life history theory as an influence on menarcheal timing. Accordingly, high-quality, warm-home developmental environments are associated with a later age at menarche, while socially-adverse home developmental environments are associated with earlier menarche (Ellis 2004; Ellis and Essex 2007). This implies, of course, an interaction between behavioural development and biological maturation. It is proposed 'that quality of parental investment, as indexed by measures such as parental supportiveness, is the most important mechanism through which young children receive information about levels of stress and support in their local environments, and that this information provides a basis for adaptively adjusting pubertal timing' (Ellis and Essex 2007, 1814). As a point of historical reference, it should be noted that Whiting (1965) suggested an association between stress early in life and age at menarche. Using cross-cultural ethnographic data from the Human Relations Area files, several practices believed to be stressful were identified, such as piercing, moulding, and several others. Stress as defined during the first two years of life was associated with earlier menarche (Whiting 1965).

It is of interest to examine sport training as a stressor in the context of the family environment hypothesis given the degree of parental investment in the sporting careers of their offspring. It is often accepted that the stress of training for sport 'delays' sexual maturation (Ellis 2004). Data generally cited for this generalisation do not meet criteria for causality. Prospective data for young athletes are limited to nine samples, four of gymnasts, and one each for swimming, tennis, athletics, rowing, and ballet (see Malina, Bouchard, et al. 2004). Status quo estimates for young athletes are also limited and estimated median ages refer to the sample and not individuals. All other data for age at menarche in athletes are retrospective and limited largely to late adolescents and adults; as such, they do not permit cause–effect conclusions. Training for sport is specific; it refers to systematic, specialised practise for a sport or sport discipline for most of the year and over several years. Most discussions of training and menarche neither define nor specify the nature of the training stimulus except in general terms; hours per week and years of training are common descriptors. Diet is not ordinarily controlled. Moreover, sport-specific selection and exclusion are not considered—for example, selective dropout in gymnastics or ballet. Nevertheless, available data indicate later mean ages at menarche in athletes in many, but not all, sports (Malina 1983; Malina, Bouchard, et al. 2004). If training for sport is related to the age at menarche, it probably interacts with, or is confounded by, other factors so that a specific effect of training per se may be difficult to extract. Two critical evaluations offer the following conclusions on training and menarche: 'although menarche occurs later in athletes than in nonathletes, it has yet to be shown that exercise delays menarche in anyone' (Loucks et al. 1992, S288), and 'the general consensus is that while menarche occurs later in athletes than in nonathletes, the relationship is not causal and is confounded by other factors' (Clapp and Little 1995, 2–3).

Other factors include the family of athletes. Mothers of athletes tend to be later maturers and mother–daughter and sister–sister correlations for age at menarche are of similar magnitude as those for nonathletes (Malina et al. 1994; Malina, Bouchard, et al. 2004). Family size is an additional factor and some data suggest athletes come from larger families (Malina et al. 1982, 1997). Larger family size is associated with later menarche. The estimated sibling effect, controlling for birth order, on menarche ranges from 0.08 to 0.19 years per additional sibling in families of nonathletes and from 0.15 to 0.22 years per additional sibling in families of athletes (Malina et al. 1997; Malina, Bouchard, et al. 2004a).

Information on the home environment of young athletes is reasonably well established. Parents are primary socialising agents for children's participation in sport and likewise are a major influence on psychosocial outcomes associated with sport (Weiss 2003; Brustad 2003). Parental involvement with elite athletes (swimmers, tennis players) is similar to that of parents of talented individuals in the arts (concert pianists, sculptors) and sciences research (mathematicians, neurologists) who attained international status at relatively young ages (Sloane 1985). Given parental investment—socially, emotionally, and economically—in the sport training and achievements of their offspring, is it possible that later menarche of athletes reflects, in part, familial correlation and the high-quality, warm-home developmental environments? Indicators of a positive developmental environment are parental approval, greater family cohesiveness, frequency of positive interactions, and so on. Unknown factors in the development of young athletes, however, are coaches and the sport system, which may function to complement or to disrupt the home environments of young athletes.

13.8 ENVIRONMENTAL POLLUTION

Environmental pollution, especially that associated with human activities, is an important agent associated with growth and maturation (Schell and Denham 2003; Schell et al. 2006). It is currently of much concern in many areas of the world. Among pollutants,

TABLE 13.5

Estimated Decrements in Linear Dimensions for Each 10 μg/dL Increase in Blood Lead Level

	Males	Females
Height	3.2 cm	4.0 cm
Leg length	2.1	2.9
Trunk length	1.2	1.1
Arm length	1.8	1.9

Source: Adapted from Ignasiak et al. (2006).

Note: Leg length is symphysion height; trunk length is estimates as height minus symphysion height; arm length is acromiale to dactylion.

lead and lead compounds are well-known toxic agents to growth, maturation, and development pre- and postnatally. In this context, the relationship between blood lead levels and the growth status of rural Polish youth 7–15 years of age was evaluated (Ignasiak et al. 2006). The children were resident in the copper basin in southwestern Poland. The mines and associated smelting and refining facilities generate large amounts of industrial wastes, including heavy metals. Weight, height, segment lengths, and limb circumferences were considered relative to blood lead levels, which ranged 2.0 to 33.9 μg/dL. Decrements in linear growth associated with blood lead levels are summarised in Table 13.5. Reduction in growth is more severe in extremity lengths than trunk length and it appears that the lower extremity is more affected than the upper extremity. The reduction in height associated with blood lead is more attributable to stunted leg growth (males, 2.1 cm, 66%; females, 2.9 cm, 73%) than to reduced growth in trunk length (males, 1.2 cm, 34%; females, 1.1 cm, 27%). The results collectively suggest that growth stunting associated with blood lead may be differentiated to a major effect on chondrocyte proliferation, hypertrophy, and matrix calcification at the epiphyseal growth plates in long bones of the extremities, and a somewhat lesser effect on the growth in height and remodelling of vertebral bodies (trunk length).

13.9 SUMMARY

Much remains to be done in understanding variation in biological growth and maturation. Growth studies, with relatively few exceptions, are largely field-based—in schools, communities, and clinics throughout the world. There is a wide range of environmental factors that influence growth and maturation, and in identifying growth within healthy parameters, one often goes from the field to the laboratory, rather than the other way around.

REFERENCES

Bandini, L.G., Schoeller, D.A. and W.H. Dietz. 1990. Energy expenditure in obese and non-obese adolescents. *Pediatr. Res.* 27:198–203.

Berkey, C.A., Rockett, H.R.H., Gillman, M.W. and G.A. Colditz. 2003. One-year changes in activity and inactivity among 10- to 15-year old boys and girls: Relationship to change in body mass index. *Pediatrics* 111:836–43.

Black, A.E., Coward, W.A., Cole, T.J. and A.M. Prentice. 1996. Human energy expenditure in affluent societies: An analysis of 574 doubly labelled water measurements. *Eur. J. Clin. Nutr.* 50:72–92.

Bodzsar, E.B. and C. Susanne (eds). 1998. *Secular Growth Changes in Europe*. Budapest: Eotvos University Press.

Bogin, B., Smith, P., Orden, A.B., Varela-Silva, M.I. and J. Loucky. 2002. Rapid change in height and body proportions of Maya American children. *Am. J. Hum. Biol.* 14:753–61.

Bray, G.A., Nielsen, S.J. and B.M. Popkin. 2004. Consumption of high-fructose corn syrup in beverages may play a role in the epidemic of obesity. *Am. J. Clin. Nutr.* 79:537–43.

Briefel, R.R. and C.L. Johnson. 2004. Secular trends in dietary intake in the United States. *Annu. Rev. Nutr.* 24:401–31.

Brustad, R.J. 2003. Parental roles and involvement in youth sport: Psychosocial outcomes for children. In *Youth Sports: Perspectives in a New Century*, ed. R.M. Malina and M.A. Clark, 127–38. Monterey, California: Coaches Choice.

Centers for Disease Control. 2005. YRBSS: Youth online—comprehensive results. Available at http://www.cdc.gov/HealthyYouth/index.htm.

Chumlea, W.C., Schubert, C.M., Roche, A.F. et al. 2003. Age at menarche and racial comparisons in U.S. girls. *Pediatrics* 111:110–3.

Clapp, J.F. and K.D. Little. 1995. The interaction between regular exercise and selected aspects of women's health. *Am. J. Obstet. Gynecol.* 173:2–9.

Cook, S., Weitzman, M., Auinger, P., Nguyen, M. and W.H. Dietz. 2003. Prevalence of a metabolic syndrome phenotype in adolescents: Findings from the third National Health and Nutrition Examination Survey, 1988–1994. *Arch. Pediatr. Adolesc. Med.* 157:821–7.

Durnin, D.V.G.A. 1992. Physical activity levels—past and present. In *Physical Activity and Health* (Society for the Study of Human Biology Symposium 34), ed. N.G. Norgan, 20–27. Cambridge: Cambridge University Press.

Eisenmann, J.C., Katzmarzyk, P.T. and M.S. Tremblay. 2004. Leisure-time physical activity levels among Canadian adolescents, 1981–1998. *J. Phys. Act. Health* 1:154–62.

Ellis, B.J. 2004. Timing of pubertal maturation in girls: An integrated life history approach. *Psychol. Bull.* 130:920–58.

Ellis, B.J. and M.J. Essex. 2007. Family environments, adrenarche, and sexual maturation: A longitudinal test of a life history model. *Child Dev.* 78:1799–817.

Flegal, K.M. and R.P. Troiano. 2000. Changes in the distribution of body mass index of adults and children in the US population. *Int. J. Obes. Relat. Metab. Disord.* 24:807–18.

Fredriks, A.M., van Buuren, S., Burgmeijer, R.J.F. et al. 2000. Continuing positive secular growth change in the Netherlands 1955–1997. *Pediatr. Res.* 47:316–23.

Gerrior, S., Bente, L. and H. Hiza. 2004. Nutrient content of the U.S. food supply, 1909–2000. Center for Nutrition and Policy Promotion, Home Economics Research Report No 56. http://www.usda.gov/cnpp (accessed 21 November 2005).

Hamill, P.V.V., Johnston, F.E. and W. Grams. 1970. Height and weight of children, United States. *Vital. Health Stat. 11* no. 104.

Hamill, P.V.V., Johnston, F.E. and S. Lemeshow. 1973. Height and weight of youths 12–17 years, United States. *Vital. Health Stat. 11*, no. 124.

Hamill, P.V.V., Drizd, T.A., Johnson, C.L., Reed, R.B. and A.F. Roche. 1977. NCHS growth curves for children birth-18 years, United States. *Vital. Health Stat. 11*, no. 165.

Himes, J.H. 2006. Examining the evidence for recent secular changes in the timing of puberty in US children in light of increases in the prevalence of obesity. *Mol. Cell. Endocrinol.* 254–5:13–21.

Ignasiak, Z., Sławinska, T., Rożek, K., Little, B.B. and R.M. Malina. 2006. Lead and growth status of schoolchildren living in the copper basin of south-western Poland: Differential effects on bone growth. *Ann. Hum. Biol.* 33:401–14.

Komlos, J. and A. Breitfelder. 2008. Height of US-born non-Hispanic children and adolescents 2–19, born 1942–2002 in the NHANES samples. *Am. J. Hum. Biol.* 20:66–71.

Komlos, J. and B.E. Lauderdale. 2007. Underperformance in affluence: The remarkable relative decline in U.S. heights in the second half of the 20th century. *Soc. Sci. Q.* 88:283–305.

Kroeber, A.L., Kluckhohn, C. and W. Unterreiner. 1952. Culture: A critical review of concepts and definitions. *Papers of the Peabody Museum of American Archaeology and Ethnology, 47/1*. Cambridge, Massachusetts: Peabody Museum, Harvard University.

Kuczmarski, R.J., Ogden, C.L., Grummer-Strawn, L.M., Kuczmarski, R.J., Ogden, C.L. and L.M. Grummer-Strawn. 2000. CDC growth charts: United States. *Advance Data from Vital and Health Statistics* no. 314, Hyattsville, Maryland: National Center for Health Statistics.

Lazzer, S., Boirie, Y., Bitar A. et al. 2003. Assessment of energy expenditure associated with physical activities in free-living obese and nonobese adolescents. *Am. J. Clin. Nutr.* 8:471–9.

Little, B.B. and R.M. Malina. (under review). Marriage patterns in a Mesoamerican peasant community are biologically adaptive.

Little, B.B., Buschang, P.H. and R.M. Malina. 1988. Socioeconomic variation in estimated growth velocity of school children from a rural, subsistence agricultural community in southern Mexico. *Am. J. Phys. Anthropol.* 76:443–8.

Loucks, A.B., Vaitukaitis, J., Cameron, J.L. et al. 1992. The reproductive system and exercise in women. *Med. Sci. Sports Exerc.* 24:S288–93.

Lowry, R., Brener, N., Lee, S., Epping, J., Fulton, J. and D. Eaton. 2004. Participation in high school physical education—United States, 1991–2003. *Morb. Mortal. Wkly. Rep.* 5336:844–7.

MacMahon, B. 1973. Age at menarche. *Vital and Health Statistics* Series 11, No 133.

Malina, R.M. 1983. Menarche in athletes: Synthesis and hypothesis. *Ann. Hum. Biol.* 10:1–24.

Malina, R.M. and P.T. Katzmarzyk. 2006. Physical activity and fitness in an international growth standard for preadolescent and adolescent children. *Food Nutr. Bull.* (Suppl 27):S295–313.

Malina, R.M. and B.B. Little. 2008. Physical activity: The present in the context of the past. *Am. J. Hum. Biol.* 20:373–91.

Malina, R.M., Bouchard, C., Shoup, R.F. and G. Lariviere. 1982. Age, family size and birth order in Montreal Olympic athletes. In *Physical Structure of Olympic Athletes:* Part I. *The Montreal Olympic Games Anthropological Project*, ed. J.E.L. Carter, 13–24. Basel: Karger.

Malina, R.M., Selby, H.A., Buschang, P.H., Aronson, W.L. and R.G.Wilkinson. 1983. Adult stature and age at menarche in Zapotec-speaking communities in the Valley of Oaxaca, Mexico, in a secular perspective. *Am. J. Phys. Anthropol.* 60:437–49.

Malina, R.M., Little, B.B., Buschang, P.H., Demoss, J. and H.A. Selby. 1985. Socioeconomic variation in the growth status of children in a subsistence agricultural community. *Am. J. Phys. Anthropol.* 68:385–91.

Malina, R.M., Ryan, R.C. and C.M. Bonci. 1994. Age at menarche in athletes and their mothers and sisters. *Ann. Hum. Biol.* 21:417–22.

Malina, R.M., Katzmarzyk, P.T., Bonci, C.M., Ryan, R.C. and R.E. Wellens. 1997. Family size and age at menarche in athletes. *Med. Sci. Sports Exerc.* 29:99–106.

Malina, R.M., Bouchard, C. and O. Bar-Or. 2004. *Growth, maturation, and physical activity.* 2nd edition. Champaign, Illinois: Human Kinetics.

Malina, R.M., Peña Reyes, M.E., Tan, S.K., Buschang, P.H., Little, B.B. and S. Koziel. 2004. Secular change in height, sitting height and leg length in rural Oaxaca, southern Mexico: 1968–2000. *Ann. Hum. Biol.* 31:615–33.

Malina, R.M., Howley, E. and B. Gutin. 2007. Body mass and composition. Report prepared for the Youth Health subcommittee, Physical Activity Guidelines Advisory Committee.

Malina, R.M., Peña Reyes, M.E. and B.B. Little. (2009). Socioeconomic variation in the growth status of urban school children 6–13 years in Oaxaca, Mexico, in 1972 and 2000. *Am. J. Hum. Biol.* 21:805–16.

Malina, R.M., Peña Reyes, M.E., Tan, S.K. et al. (2010). Secular change in heights of indigenous adults from a Zapotec-speaking community in Oaxaca, southern Mexico. *Am. J. Phys. Anthropol.* 141: in press.

McDowell, M.A., Fryar, C.D., Hirsch, R. and C.L. Ogden. 2005. Anthropometric reference data for children and adults: U.S. population, 1999–2002. *Advance Data from Vital and Health Statistics* No. 361.

Moore, L.L., Gao, D., Bradlee, M.L. et al. 2003. Does early physical activity predict body fat change throughout childhood? *Prev. Med.* 37:10–7.

Mul, D., Fredriks, A.M., van Buuren S., Oostdijk, W., Verloove-Vanhorick, S.P. and J.M. Wit. 2001. Pubertal development in the Netherlands 1965–1997. *Pediatr. Res.* 50:479–86.

Mundt, C.A., Baxter-Jones, A.D.G., Whiting, S.J., Bailey, D.A., Faulkner, R.A. and R.L. Mirwald. 2006. Relationships of activity and sugar drink intake on fat mass development in youths. *Med. Sci. Sports Exerc.* 38:1245–54.

Murphy, A.D. and A. Stepick. 1991. *Social Inequality in Oaxaca: A History of Resistance and Change.* Philadelphia: Temple University Press.

Najjar, M.F. and M. Rowland. 1987. *Anthropometric Reference Data and Prevalence of Overweight. United States.* DHHS Publication No. PHS 87–1688. Washington, D.C.: U.S. Government Printing Office.

National Center for Health Statistics. 2005. National Health and Nutrition Examination Survey, anthropometric reference data, United States, 1988–1994. http://www.cdc.gov (accessed 29 May 2006).

Nutini, H.G. 1967. A synoptic comparison of Mesoamerican marriage and family structure. *Southwest J. Anthropol.* 23:383–404.

Ogden, C.L., Flegal, K.M., Carroll, M.D. and C.L. Johnson. 2002. Prevalence and trends in overweight among U.S. children and adolescents, 1999–2000. *J. Am. Med. Assoc.* 288:1728–32.

Ogden, C.L., Carroll, M.D., Curtin, L.R., McDowell, M.A., Tabak, C.J. and K.M. Flegal. 2006. Prevalence of overweight and obesity in the United States, 1999–2004. *J. Am. Med. Assoc.* 295:1549–55.

Programa de las Naciones Unidas para el Desarrollo. 2003. *Informe Sobre Desarrollo Humano, Mexico 2002.* Mexico, DF: Mundi Prensa Mexico, SA.

Schell, L.M. and M. Denham. 2003. Environmental pollution in urban environments and human biology. *Annu. Rev. Anthropol.* 32:111–34.

Schell, L.M., Gallo, M.V., Denham, M. and J. Ravenscroft. 2006. Effects of pollution on human growth and development: An introduction. *J. Physiol. Anthropol.* 25:103–12.

Schoonover, H. and M. Muller. 2006. *Food without Thought: How U.S. Farm Policy Contributes to Obesity.* Minneapolis, Minnesota: Institute for Agriculture and Trade Policy.

Selby, H.A. 1966. Social structure and deviant behavior in a Zapotec community. PhD diss., Stanford University, Palo Alto, California.

Selby, H.A. 1974. *Zapotec Deviance: The Convergence of Folk and Modern Sociology.* Austin: University of Texas Press.

Selby, H.A. and A.D. Murphy. 1979. *The City of Oaxaca: Final Technical Report.* Washington, D.C.: Office of Urban Development, Technical Assistance Bureau.

Sloane, K.D. 1985. Home influences on talent development. In *Developing Talent in Youth*, ed. B.S. Bloom, 439–76. New York: Ballantine Books.

Sun, S.S., Schubert, C.M., Liang, R. et al. 2005. Is sexual maturity occurring earlier among U.S. children? *J. Adolesc. Health* 37:345–55.

Torun, B., Davies, P.S.W., Livingstone, M.B.E., Paolisso, M., Sackett, R. and G.B. Spurr. 1996. Energy requirements and dietary energy recommendations for children and adolescents 1 to 18 years old. *Eur. J. Clin. Nutr.* 50(Suppl 1):S37–81.

Treuth, M.S., Figueroa-Colon, R., Hunter, G.R., Weinsier, R.L., Butte, N.F. and M.I. Goran. 1998. Energy expenditure and physical fitness in overweight vs. non-overweight prepubertal girls. *Int. J. Obes.* 22:440–7.

Troiano, R.P., Briefel, R.R., Carroll, M.D. and K. Bialostosky. 2000. Energy and fat intakes of children and adolescents in the United States: Data from the National Health and Nutrition Examination Surveys. *Am. J. Clin. Nutr.* 72(Suppl):S1343–53.

Weiss, M.R. 2003. Social influences on children's psychosocial development in youth sports. In *Youth Sports: Perspectives in a New Century*, ed. R.M. Malina and M.A. Clark, 109–26. Monterey, California: Coaches Choice.

Whiting, J.M.W. 1965. Menarcheal age and infant stress in humans. In *Sex and Behavior*. ed. F.A. Beach, 221–33. New York: Wiley.

14 Bone Health and Body Composition Measurement in Older People
Challenges Imposed by Variability

Katherine Brooke-Wavell[*]
Department of Human Sciences, Loughborough
University, Leicestershire, United Kingdom

CONTENTS

---

[*] Address all correspondence to k.s.fbrooke-wavell@lboro.ac.uk.

14.1   INTRODUCTION

It is estimated that over half of women, and one in five men, will sustain osteoporotic fractures (Van Staa et al. 2001). These cause substantial morbidity and increased mortality (Van Staa et al. 2001). High fat mass and low fat-free mass have also been associated with increased mortality (Wannamethee et al. 2007; Bigaard et al. 2004) and increased risk of disability (Visser et al. 1998). Obesity is associated with increased risk of cancer, type 2 diabetes, and cardiovascular disease (Visscher and Seidell 2001). As these are conditions that become more prevalent with age, research into the aetiology and prevention of such conditions requires measures that are valid for use in older people. This chapter reviews the status of bone health and body composition measures in older people, and considers to what extent the changes that tend to be associated with ageing can confound the techniques that aim to evaluate them.

14.2   BONE HEALTH

14.2.1   Laboratory Techniques

14.2.1.1   Dual X-Ray Absorptiometry

The most widely used measure of skeletal status is dual X-ray absorptiometry (DXA) (Sambrook and Cooper 2006), alone or combined with clinical risk factors for prediction of osteoporotic fracture risk (Kanis et al. 2007). DXA measures the attenuation of two energies of X-ray. Measurement is repeated during a scan of the measurement region to produce a two-dimensional image. In the whole body, or particular regions of interest (most often clinically relevant sites such as lumbar vertebrae, various regions of the hip and distal forearm), bone mineral content (BMC, g), and bone mineral density (BMD, g/cm^2) are determined. BMC will depend largely on body, and hence bone, size. BMD offers some adjustment for size but is an areal rather than volumetric density: it is not possible to measure bone depth so this parameter will still be size dependant. BMD is related to risk of low trauma fracture (Marshall et al. 1996), so the technique is widely used for the diagnosis of osteoporosis.

Cross-sectional data demonstrate that bone mass and bone mineral density tend to rise with age to reach a peak in the third or fourth decade. There is a subsequent decline, particularly in the postmenopausal period. This is demonstrated in Figure 14.1, where BMDs of 2,728 women from Changsha in China are compared with that in a national representative sample of Japanese women and the U.S. Caucasian American reference dataset (Wu et al. 2003). However, the higher hip BMD in Caucasian than Japanese women contradicts the data on hip fracture risk, which is higher in U.S. women than Japanese women (Kanis et al. 2002). Ethnic differences in bone mass may be explained by differences in body size, as they tend to disappear after adjustment for body size (Ross et al. 1996).

Given that differences in body size may confound differences in BMD between ethnic groups, it also seems likely that secular trends in body size may confound comparisons of BMD between older and younger individuals. Determination of effects of ageing per se requires longitudinal comparisons. Melton and colleagues compared bone changes assessed longitudinally over four years with those from cross-sectional

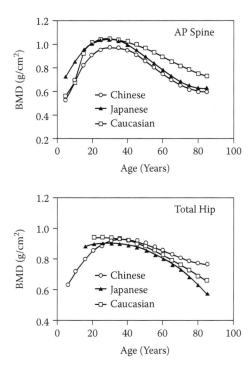

FIGURE 14.1   Bone mineral density according to age in 2,728 Chinese women compared to Japanese and U.S. reference databases. (Reprinted with permission from Wu et al. 2003.)

comparisons (Figure 14.2). Lumbar spine BMD (measured anteroposteriorly) increased rather than showing the expected decline (Melton et al. 2000), although this increase could reflect artefactual change due to an increased prevalence of aortic calcification, osteophytosis, or vertebral fracture (Figure 14.3), all of which can artefactually elevate BMD (Damilakis et al. 2007). The decline in hip BMD measured longitudinally was again smaller than that estimated from cross-sectional comparisons. Conversely, radius BMD loss was underestimated from cross-sectional comparisons. The oldest partici- pants in this cohort had smaller stature and arm span than the youngest, so it seems possible that some of the differences in BMD in different age groups are explained by secular trends in size rather than age-related bone loss.

Duan et al. (2001) took account of body size effects by performing DXA scans in both anteroposterior and lateral orientations, allowing estimation of vertebral dimen- sions and volumetric bone density. Although volumetric bone density declined with age (to a greater extent in women), this decline was compensated for by increased cross-sectional area, resulting presumably from continued periosteal apposition of bone such that the load per unit cross-sectional area declined with age (Figure 14.4 and Figure 14.5).

To allow assessment of structural parameters that may be important predictors of bone strength from DXA scans, algorithms such as the hip structural analysis have been developed. The mass profile is examined in thin regions of the hip, to estimate

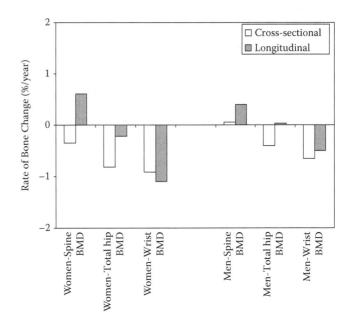

FIGURE 14.2 Comparison of annual change in bone mineral density assessed cross-sectionally and longitudinally over four years in men aged > 50 years and postmenopausal women. (Data from Melton et al. 2000.)

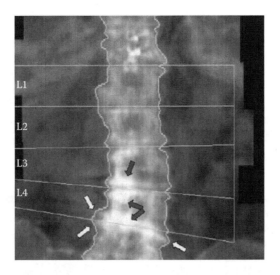

FIGURE 14.3 Antero-posterior lumbar spine scan by DXA in an eighty-year-old woman, showing artefactually increased BMD due to osteophyte formation (white arrows), end plate sclerosis (black arrows), and scoliosis. (Reprinted with permission from Damilakis et al. 2007.)

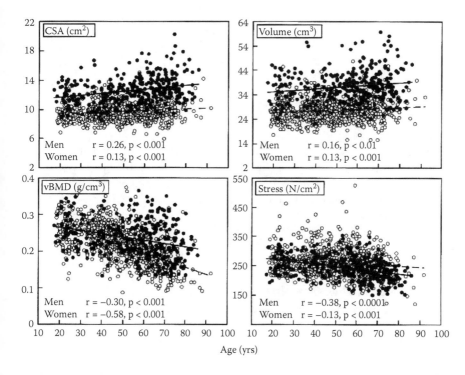

FIGURE 14.4 Vertebral body cross-sectional area (CSA), volume, volumetric bone mineral density (vBMD), and stress (load per unit CSA) against age in men (solid line) and women (dashed line). (Reprinted with permission from Duan et al. 2001.)

the cross-sectional area and cross-sectional moment of inertia from which section modulus (a measure of bending and torsional strength) can be estimated (Beck 2003). Findings using this technique reveal that the femur also shows periosteal expansion which maintains section modulus despite losses of BMD (Beck et al. 2001). Although

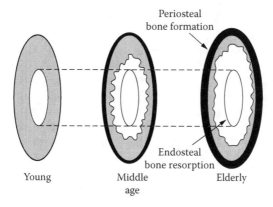

FIGURE 14.5 Diagrammatic representation of bone geometric changes with age. (Reprinted with permission from Duan et al. 2001.)

bone mineral density declined substantially with age (by 15–22% in women and 6–13% in men), the estimated stress from a fall on the greater trochanter increased by only 2–13% in women and declined in men (Beck et al. 2006).

Differences in the degree of bone mineralisation would also influence DXA measures. Examination of biopsies using backscattering electron imaging did not observe variation with age per se (Roschger et al. 2003) although mineralisation is reduced with primary hyperparathyroidism (Roschger et al. 2007), which is more common in older people.

Studies using DXA measures of bone thus demonstrate that lower bone mass with age observed cross-sectionally may be partly related to size differences between age cohorts, although loss in BMD with advancing age is also observed longitudinally. Bone loss is accompanied by increased diameter associated with periosteal expansion, which will help to maintain strength in bending in bone. However, such structural change could confound interpretation of bone density measurements. DXA has limited ability to assess bone structural parameters and is unable to discriminate trabecular and cortical bone.

14.2.1.2 Quantitative Computed Tomography

Quantitative computed tomography (QCT), like DXA, involves measuring the attenuation of an X-ray beam as it passes through the body. QCT, however, performs measurement of slices, producing a three-dimensional scan and so allowing assessment of true volumetric BMD (vBMD) and discrimination of trabecular and cortical bone (Bousson 2007). Furthermore, QCT allows assessment of structural parameters such as periosteal and endosteal circumferences and cortical thickness. QCT scanners are expensive, less widely available, and involve a higher X-ray dose than DXA, so have been less widely used for research. QCT measurements are most often conducted at vertebral and hip sites. Peripheral QCT (pQCT) technology is cheaper and involves lower radiation dose but can only provide measurement of appendicular bones such as the radius and tibia.

Whilst the continued periosteal apposition as detected by DXA may maintain strength in bending despite bone loss with age, it is argued that the concomitant thinning of the bone cortex may contribute to increased risk of fracture following a fall. Bone from fracture patients had reduced cortical but not trabecular bone (Bell et al. 1999a; Crabtree et al. 2001). Fractured hips also have higher cortical porosity (Bell et al. 1999b) and studies involving high-resolution scanning of bone samples have reported increased number and size of pores (canals and resorption cavities) in cortical bone with age (Bousson et al. 2001; Bell et al. 1999b). It may thus be important to differentiate cortical and trabecular bone and to assess cortical porosity. Whilst most in vivo techniques do not have the resolution to directly assess cortical porosity, increased porosity may be reflected in reduced cortical vBMD (Bousson 2007).

Studies using QCT have allowed examination of the changes in cortical and trabecular bone components with age. Cross-sectional studies have reported greater loss of trabecular bone with age: trabecular vBMD declined by around a half at central sites and a quarter at peripheral sites, whilst cortical vBMD declined more in women (25%) than men (18%) (Riggs et al. 2004). These QCT comparisons were consistent with DXA studies in showing increased periosteal apposition with age (~15%) (Riggs et al. 2004). Longitudinal study has demonstrated greater rates of loss of cortical than trabecular vBMD in women, particularly those who were postmenopausal (Uusi-Rasi et al. 2007).

Exercise may influence skeletal dimensions as well as bone mineral density, possibly confounding detection of a bone response (Jarvinen et al. 1999). Studies using pQCT have demonstrated increased cortical area and BMC of the distal radius, accompanied by reduced trabecular BMC in the absence of any change in BMD (Adami et al. 1999).

The location of bone loss may also be important to fracture risk. Fracture patients have greatest deficits in infero-anterior to supero-posterior axis (Bell et al. 1999a; Crabtree et al. 2001). Femurs from older people had proportionately greater loss from the superior aspect (Mayhew et al. 2005), whilst hip structural analysis of DXA data reveals that the centre of mass tends to be distributed infero-medially with age (Kaptoge et al. 2007). It is suggested that this redistribution of bone may reflect habitual loading patterns, with the activities that persist into older age (such as slower walking) loading the inferior rather than superior femoral neck (Kaptoge et al. 2007; Mayhew et al. 2005). It is thus difficult to differentiate the effect of ageing *per se* with the effects of the reduced activity that tends to be associated with ageing.

Measurements by QCT and pQCT thus overcome some of the limitations of DXA, in that they provide true volumetric densities, can distinguish trabecular and cortical bone, and can assess structural parameters and detect location of bone change. However, the limited availability of equipment and the somewhat higher X-ray dose can limit the use of QCT.

14.2.2 FIELD TECHNIQUES

Quantitative ultrasound (QUS) techniques involve transmitting pulses of ultrasound through or along bone. The measured parameters are broadband ultrasonic attenuation (BUA)—the change in ultrasonic attenuation with frequency, or speed of sound (SOS) (Langton et al. 1984). There are two major methodological approaches: transmission of ultrasound through the heel or phalanges, either with the foot in a footbath or using gel to couple transducers to skin (Gluer 1997). Axial transmission devices measure the speed of sound transmitted along the bone cortex, at phalanges, tibia, radius, or metatarsal. BUA and SOS are related to bone density and also bone structural parameters (Gluer 1997). QUS devices are relatively cheap, do not involve exposure to ionising radiation (so there are not the legislative constraints on use), and are portable, so could be appropriate for use in field situations where there is a power supply (and access to distilled water for the models incorporating a footbath).

QUS measurements are able to predict hip fracture in elderly women (Marin et al. 2006) and men (Bauer et al. 2007) with predictive power similar to that of DXA. Several different instruments are available, and differences in the technology mean that the measured BUA or SOS can vary substantially between instruments (Njeh et al. 2000). The decline in measurements with age differs in magnitude between skeletal sites and between QUS and DXA (Knapp et al. 2004; Frost et al. 2000). Calcaneal QUS may be more sensitive to effects of weight-bearing exercise as BUA increased to a greater extent than calcaneal and femoral neck BMD (Brooke-Wavell et al. 1997).

In QUS, as with DXA, differences in bone size between populations may contribute to differences in bone measurements, as calcaneal measurements are dependent upon bone width (Hans et al. 1995). Differences in body or bone marrow composition could affect measurements (Kotzki et al. 1994; Alves et al. 1996) so changes in these parameters in older people could also confound interpretation of results. Differences in bone size may affect the placement of the region of interest in some devices where transducers are at a fixed location (Brooke-Wavell et al. 1995b) and use of a scan image to identify the region of interest improved the ability to discriminate a fracture (Damilakis et al. 1998).

14.3 BODY COMPOSITION

The composition of the human body cannot be directly determined in vivo, although cadaveric study suggests that some compartments are of relatively constant composition (Ellis 2000). By making assumptions about the properties of the compartments, it is possible to estimate the composition of the whole body.

14.3.1 LABORATORY TECHNIQUES

Laboratory techniques mostly describe the body as consisting of two compartments: commonly the fat mass and the fat-free mass (the total mass of all other tissues). Differences in the properties of these compartments are used to estimate their size. However, the composition of these compartments (particularly the fat-free mass) is variable; for example, the hydration, mineral, or protein content may differ (Ellis 2000). As such, three-, four- and multicompartment models that involve separate determination of some of the more variable components are used as "reference" methods against which other techniques are validated or compared.

14.3.1.1 Hydrodensitometry

Hydrodensitometry (HD) involves the determination of body density. The participant is submerged in water, and underwater weight is determined to estimate the mass and hence calculate the volume of water displaced by the body. Correction for the volume of air in the lungs is necessary to calculate body volume. This is usually achieved by having the participant breathe down to residual volume during measurement, and then measuring residual volume (for example, by oxygen dilution or nitrogen washout techniques) either during or immediately after the measurement. An allowance is also made for the (considerably smaller) volume of gas in the gastrointestinal tract. Body density can then be calculated from mass divided by volume. The conversion of body density to fat content requires assumptions about the density of the fat mass and fat-free mass; for example, Siri's equation (%fat = 4.95/density − 4.50) assumes these to be 0.9 and 1.1 g/ml respectively (Ellis 2000).

The validity of these assumptions in older people has been questioned (Heymsfield et al. 1989), as bone loss might be expected to reduce the density of the fat-free mass (FFM), whilst muscle is also lost with age. Lower density of FFM has been reported in older women (Visser et al. 1997). Some studies have observed significantly different estimates of body fat content by hydrodensitometry and four-compartment (4-C) techniques in older people, attributing this to differences in the density of the FFM (Goran et al. 1998; Bergsma-Kadijk et al. 1996), although in one study no significant differences in estimates by these techniques were observed (Yee et al. 2001). For this reason, it has been recommended that Siri or Brozek's equations are not appropriate for use in older people (Heyward and Wagner 2004) and additional measurements to allow body fat estimations from three- or four-compartment models may be preferable.

At a practical level, some older people may have difficulties getting into some hydrostatic weighing facilities. There may be further difficulties in breathing down to the residual volume, which is higher in older people. These factors may result in higher measurement error in older than younger people (Chumlea and Baumgartner 1989).

14.3.1.2 Air Displacement Plethysmography

Air displacement plethysmography involves the participant sitting in a sealed container. The relationship between pressure and volume (as described by Boyle's law) allows estimation of body volume and hence body density (Dempster and Aitkens 1995). Body fat is then estimated by the same methods as for hydrodensitometry (Fields et al. 2002). The technique provides similar estimates of body density to HD in older people and provided similar accuracy in assessing body fat relative to a 4-C model (Yee et al. 2001). As the technique does not require submersion in water, this technique offers practical advantages over hydrodensitometry for use in older people.

14.3.1.3 Dual X-Ray Absorptiometry (DXA)

DXA measurement of body composition, like DXA measurement of bone, relies on the differential attenuation of an X-ray beam by different tissues, particularly at lower energies. Measurements of attenuation at two energies are repeated at each pixel across the body to generate a two-dimensional scan. By making assumptions about the attenuation properties of the tissues at each of the energies, it is possible to calculate the mass of fat and lean in pixels that do not include bone. In bone pixels, tissues can only be resolved into bone and soft tissue, and the soft tissue is further subdivided into lean and fat by assuming the ratio is the same as that in nearby non-bone pixels (Ellis 2000).

One advantage of DXA is that it allows measurement of body regions. It requires little effort from the participant (lying still on the scanner couch) and so is likely to be feasible for most older people. However, measurement involves radiation exposure (although the dose for a whole body scan is substantially smaller than that for a standard chest radiograph), and the limited size of the scanner bed or height of the arm may mean that very large or obese participants do not fit onto the scanner.

There are three major scanner manufacturers and a number of models. Estimates of body composition are reported to vary between scanner manufacturers and models (Genton et al. 2002). In older pencil-beam devices, DXA was found to have a greater error than HD in some (Bergsma-Kadijk et al. 1996; Goran et al. 1998) but not all (Clasey et al. 1997) comparisons with a 4-C model in older people. The pencil-beam devices, which perform a rectilinear scan, have now often been replaced by fan-beam devices, where a broader beam and multiple detector array allow faster scan times. Fan-beam devices can give substantially different body composition estimates than pencil beam models (Ellis and Shypailo 1998). One fan-beam model tended to over-estimate FFM and underestimate FM (Schoeller et al. 2005; Salamone et al. 2000; Tylavsky et al. 2003), although the relationship between methods was not influenced by age (Schoeller et al. 2005). Significant differences are also observed between different fan-beam densitometer models and scan modes (Genton et al. 2006); a nar-rower angle fan-beam device provided a better estimate of body mass. Accuracy thus seems to differ between scanner technologies. As both the technologies and software are frequently updated, ongoing evaluation of their accuracy may be needed.

14.3.1.4 Hydrometry

A dose of labelled water [such as deuterium oxide (D_2O), tritium oxide (^3H_2O), or oxygen-18 labelled water ($H_2{}^{18}O$)] is ingested. Samples of body water, such as saliva, urine, or blood, are collected before the labelled water dose, and after the dose has had three to four hours to equilibrate. The concentration of the label is compared between samples (for example, by mass spectroscopy) to calculate isotope dilution space, which is then adjusted to allow for exchange with nonaqueous hydrogen/oxy-gen which would overestimate by 1% and 4%, respectively. By assuming that total body weight (TBW) comprises 73.2% of FFM (Pace and Rathbun 1945), FFM can be estimated, and FM by subtraction of FFM from body mass. Alternatively, TBW can be incorporated in multicompartment models.

The error of TBW estimation is reported to be typically within 1 kg, which might equate to 1.4 kg FM (Ellis 2000). The assumption that water comprises 73.2% of fat-free mass has been questioned (Wang et al. 1999). Hydration status may alter with age, although findings have been inconsistent; values close to, above, and below the assumed 73.2% of FFM have been reported (Visser et al. 1997; Baumgartner et al. 1991; Clasey et al. 1999; Goran et al. 1998). HD was, however, reported to have lower error in estimating fat content from a 4-C model than HD from DXA (Goran 1998). Measurement is feasible in most older people, although conditions affecting hydra-tion status will affect measurement.

14.3.2 FIELD TECHNIQUES

The techniques mentioned above are generally expensive and/or require access to specialised laboratory equipment. Several field techniques have been developed that are, in general, relatively inexpensive and portable, and allow rapid measurement. This allows body composition measurements in field situations or large samples where the laboratory techniques might not be possible or feasible.

The field techniques generally do not make direct assumptions about the prop-erties of the body compartments but instead measure a property (for example, the thickness of a skinfold) that correlates with the size of the compartment of interest. Regression equations are then developed to predict the size of the compartment from the measured property. These prediction equations should ideally then be validated in an independent sample. Equations tend to be population specific (Norgan and Ferro-Luzzi 1985) and perform best in a population similar to that in which they were developed and validated. For studies in older people, it is thus necessary to ensure that any prediction equations have been appropriately validated in a corresponding age group.

As prediction equations are usually developed by comparison with a reference technique such as one of the laboratory techniques described above, any error in the reference technique might thus be retained in the estimate of body fat from prediction equations. Errors in body fat estimates by field techniques were found to be related to differences in hydration of the FFM in older people (Baumgartner et al. 1991). Unfortunately, as few equations suitable for use in older people have been validated using 4-C models, they are likely to be affected by the sources of error that affect two-compartment models, as described above.

14.3.2.1 Anthropometric and Skinfold Techniques

The most widely used body composition measure is the body mass index (body mass/stature^2) (BMI), although this is a measure of weight for height rather than body composition per se. Its interpretation in older people can be confounded by the height loss that occurs with age: averaging 5 cm in men and 8 cm in women between ages 30 and 80, which would introduce an artefactual increase in BMI of 1.4 and 2.6 kg/m^2 in men and women respectively (Sorkin et al. 1999). Furthermore, age-related loss in skeletal muscle (Gallagher et al. 2000) would imply a greater fat content at a given weight or BMI. The validity of BMI as a measure of body fatness may thus be reduced in older people (Zamboni et al. 2005).

The estimation of body fat from skinfold thicknesses involves measuring the thickness of skin and subcutaneous fat at several sites across the body, and incorporating these into prediction equations to estimate body composition as assessed by a reference technique. The technique makes assumptions that the skinfold thickness is a valid measure of subcutaneous adipose tissue thickness, that the thicknesses at the measured sites are proportional to the total subcutaneous fat mass, and that the subcutaneous fat mass is proportion to total fat mass. Unfortunately, fat distribution changes with age, with body fat becoming more centrally and internally distributed (Chumlea and Baumgartner 1989). Skinfold compressibility is greater, producing a different relationship between skinfold thickness and subcutaneous adipose tissue thickness.

One skinfold prediction equation has been developed to estimate the percentage of body fat (% body fat) in older people using a reference 4-C model (Williams et al. 1992). This equation and those of Jackson and Pollock (Jackson et al. 1980; Jackson and Pollock 1978) were found to overestimate % body fat in older people (Clasey et al. 1999), and prediction errors were greater than in younger people.

Body composition has also been estimated from other anthropometric measurements, such as body circumferences. This approach will presumably be less affected by any age-related change in skinfold compressibility, although age-related changes in fat distribution could potentially still affect results. Equations have been developed in men and women aged 15–79 years to estimate body composition from abdominal, iliac, and hip circumferences (Tran and Weltman 1989, 1988). Means from this technique did not differ significantly from those of a 4-C model in older people and prediction errors were smaller than those for skinfold techniques (Clasey et al. 1999).

14.3.2.2 Bioelectrical Impedance Analysis

In the bioelectrical impedance analysis (BIA) technique, a small electrical current is passed through the body. The current will follow the path of least resistance. The fat-free mass, which contains a much higher proportion of water and electrolytes, is a better conductor than the fat mass. The impedance (Z) of a conductor is related to its length (L), specific resistivity (ρ), and cross-sectional area (A) such that $Z = \rho L/A$. This equation can be rearranged to $Z = \rho L^2/V$, where V is the volume. Although the body is not a perfect cylinder and specific resistivity is not constant throughout, the FFM or TBW is proportional to height and inversely proportional to impedance (Kyle et al. 2004). The impedance (Z) is a function of resistance (R) and the reactance (X_c) introduced by potential differences across cell membranes. At high frequencies, the current penetrates cell membranes, whereas at low frequencies it passes through extracellular fluid. Multifrequency approaches have also been developed that allow separate evaluation of intracellular and extracellular water (Kyle et al. 2004).

The most widely used mode is single frequency BIA, often at 50 kHz, where the current will pass predominantly through extracellular fluid (Ellis 2000). Using the tetrapolar approach, electrodes are attached to the hand, wrist, foot, and ankle. The current is introduced between hand and foot, and the voltage drop between wrist and ankle determined. Other approaches are foot-to-foot and segmental, which involves measurement of leg, trunk, and limb to reduce error introduced by assuming the body behaves as a cylinder (Kyle et al. 2004). Regression equations have been developed to estimate body composition from impedance in a number of populations.

As the BIA technique is dependent on total body water, measures are affected by any factors that influence hydration. Any age-related changes in hydration status will thus potentially introduce error, as for hydrometry. As the technique is more dependent upon the composition of the arm (where the relatively greater length and smaller cross-sectional area contribute to higher impedance) than the trunk (which is shorter and has greater cross-section, providing less impedance), any age-related differences in tissue distributions could affect measurements. Furthermore, single frequency measures could be influenced by differences in the proportions of intracellular and extracellular fluid as may occur with age (Silva et al. 2005), although multifrequency analysis and/or inclusion of reactance may overcome this limitation (Kyle et al. 2004).

As with skinfold and anthropometric techniques, the accuracy of body composition measures varies between different prediction equations. Several prediction equations have been developed in older people (Heyward and Wagner 2004), two of which used 4-C models as the reference technique (Baumgartner et al. 1991; Williams et al. 1995). Different prediction equations have been found to produce large differences in mean FFM and error in underweight older people, particularly women (Lupoli et al. 2004). An evaluation of body fat by several BIA equations derived for use in older people found best agreement with FFM determined by DXA from the equation of Kyle et al. (2001), which incorporates reactance. The errors in body composition estimation by BIA do seem smaller than those for skinfold equations (Heyward and Wagner 2004).

14.3.2.3   Near-Infrared Interactance

Near-infrared interactance involves shining near-infrared light on the upper arm. The emitted light includes two wavelengths, one of which is absorbed to a greater extent by fat, the other to a greater extent by water (present in lean mass). By measuring the absorbance of light at both of these wavelengths, it is thus theoretically possible to have a measure of fat and lean composition at the measurement sites (Conway et al. 1984). The absorbances are included in regression equations to estimate total body fat content. The original manufacturers' equations underestimated the composition, to an increasing extent with increasing fatness (Elia et al. 1990). Some alternative equations have been proposed (Heyward and Wagner 2004), but not for older people, or using multi-component models as reference. The technique has not been evaluated in older people, but in middle-aged men updated manufacturer's equations underestimated results at higher fat content (Brooke-Wavell et al. 1995a). As the technique involves measurement of subcutaneous adipose tissue, it will be subject to the same potential sources of error as skinfold techniques, except perhaps those pertaining to skinfold compressibility. Furthermore, as measurement is made at just one site, the technique is likely to be more greatly affected by differences in fat distribution as may occur with age.

14.4   DESIGN ISSUES

Many design issues may affect interpretation of findings in older people. If the aim is to study changes associated with ageing, there may be some limitations to conclusions that can be drawn from cross-sectional study. Secular change, such as the increases in body size, might confound the determination of age-related change by comparing different age groups in cross-sectional study. This is demonstrated in the study of Ding et al. (Figure 14.6), who observed that later cohorts had greater fat mass and percentage fat, largely accounting for the lower fatness with greater age reported in cross-sectional study (Ding et al. 2007). Particular cohorts may have experienced conditions (for instance, major conflicts) that make them unrepresentative of earlier or later cohorts. At increasing age, a smaller proportion of the age cohort will survive, and it is likely that these survivors are the healthier members of the cohort, introducing a survivor bias. For instance, survival may be lower in the obese, introducing an apparent reduction in mean body fat content with age. Furthermore, changes reflect differences in other parameters; for instance, a loss in muscle and bone mass may result from the reduced activity that tends to accompany ageing in addition to the ageing process per se.

People who volunteer to participate in research studies may not be representative of the population from which they come, introducing recruitment bias. In those with cognitive impairment, there may be difficulties in following procedures for obtaining informed consent or understanding complex instructions. Those with physical limitations may have more difficulty attending, or completing, some measurements. Some medical conditions and medications can confound measurements. The presence of osteoarthritic change, aortic calcification, or the use of strontium-containing medication can influence bone measurements. Joint replacements or other prostheses will invalidate bone measurements and influence measurements by DXA and some other body composition techniques. Conditions that influence hydration status will

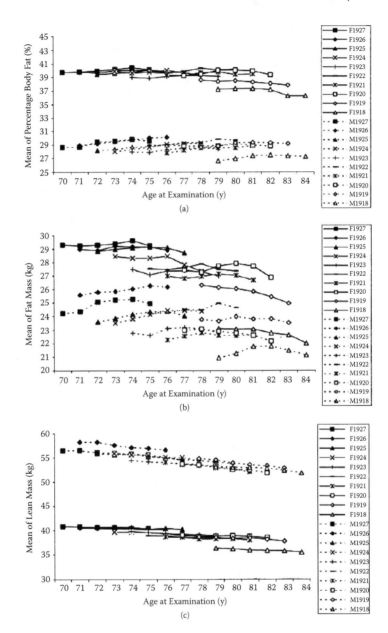

FIGURE 14.6 Mean percentage body fat (a), fat mass (b), and lean mass (c) by age at examination and birth cohorts (birth years 1918–1927) in 855 men (M) and 931 women (F) who participated in all six dual-energy X-ray absorptiometry (DXA) examinations during the five-year follow-up in the Health, Aging, and Body Composition (Health ABC) Study. (Reprinted with permission from Ding et al. 2007.)

affect body composition measurement by BIA or hydrometry. Bias may thus become increasingly marked with increasing age.

14.5   CONCLUSIONS

Whilst many changes are evident with age, some may be enhanced (or obscured) by the choice of measurement technique or experimental design. This chapter has discussed some examples of how the variability associated with ageing may confound methods that attempt to quantify age-related change. Bone mineral density declines with age, although cross-sectional studies may be confounded by secular change in body dimensions. Furthermore, changes in areal bone mineral density will be confounded by changes in bone geometry, so the true pattern of bone loss with age may be markedly different from the cross-sectional trends in areal BMD shown in Figure 14.1. QCT may offer advantages over DXA for assessing age-related change as it can assess bone geometry and true volumetric density, although instruments are less widely available.

For body composition measurement, changes in properties of body components such as hydration status or density of the FFM mean that laboratory techniques that depend on these properties may be less accurate in older people. It may thus be preferable to use multicompartment models for older people.

Body composition techniques suitable for field use, such as skinfold thicknesses, bioelectrical impedance, and near-infrared interactance are all affected by changes in fat distribution. Furthermore, any errors in reference techniques upon which regression equations are based will also produce errors in the body composition estimates produced using these equations. It is important to ensure that such methods have been validated in older people, and ideally that multicompartment models have been used as reference techniques. Evaluations so far seem to suggest that anthropometric equations and bioelectrical impedance techniques incorporating reactance may perform better than skinfold and other BIA techniques.

Some techniques may require greater physical or cognitive contribution from the participant, which may make them unfeasible, more frequently, in older people. Furthermore, recruitment of a representative sample may be an increasing challenge in older age groups.

Choosing a method that is appropriate in older people may require consideration of some more practical considerations. Is it feasible for those with physical or cognitive impairment? Is it robust across disease states? Is it feasible and available for use with the necessary size of sample? Finally, it is important to consider the validity of the technique for use in older people.

REFERENCES

Adami, S., Gatti, D., Braga, V., Bianchini, D. and M. Rossini. 1999. Site-specific effects of strength training on bone structure and geometry of ultradistal radius in postmenopausal women. *J. Bone Miner. Res.* 14(1):120–4.

Alves, J.M., Ryaby, J.T., Kaufman, J.J., Magee, F.P. and R.S. Siffert. 1996. Influence of marrow on ultrasonic velocity and attenuation in bovine trabecular bone. *Calcif. Tissue Int.* 58(5):362–7.

Bauer, D.C., Ewing, S.K., Cauley, J.A., Ensrud, K.E., Cummings, S.R. and E.S. Orwoll. 2007. Quantitative ultrasound predicts hip and non-spine fracture in men: The MrOS study. *Osteoporos. Int.* 18:771–7.

Baumgartner, R.N., Heymsfield, S.B., Lichtman, S., Wang, J. and R.N. Pierson, Jr. 1991. Body composition in elderly people: Effect of criterion estimates on predictive equations. *Am. J. Clin. Nutr.* 53(6):1345–53.

Beck, T. 2003. Measuring the structural strength of bones with dual-energy X-ray absorptiometry: Principles, technical limitations, and future possibilities. *Osteoporos. Int.* 14:S81–8.

Beck, T.J., Oreskovic, T.L., Stone, K.L. et al. 2001. Structural adaptation to changing skeletal load in the progression toward hip fragility: The study of osteoporotic fractures. *J. Bone Miner. Res.* 16(6):1108–19.

Beck, T.J., Looker, A.C., Mourtada, F., Daphtary, M.M. and C.B. Ruff. 2006. Age trends in femur stresses from a simulated fall on the hip among men and women: Evidence of homeostatic adaptation underlying the decline in hip BMD. *J. Bone Miner. Res.* 21(9):1425–32.

Bell, K.L., Loveridge, N., Power, J. et al. 1999a. Structure of the femoral neck in hip fracture: Cortical bone loss in the inferoanterior to superoposterior axis. *J. Bone Miner. Res.* 14(1):111–9.

Bell, K.L., Loveridge, N., Power, J., Garrahan, N., Meggitt, B.F. and J. Reeve. 1999b. Regional differences in cortical porosity in the fractured femoral neck. *Bone* 24(1):57–64.

Bergsma-Kadijk, J.A., Baumeister, B. and P. Deurenberg. 1996. Measurement of body fat in young and elderly women: Comparison between a four-compartment model and widely used reference methods. *Br. J. Nutr.* 75(5):649–57.

Bigaard, J., Frederiksen, K., Tjonneland, A. et al. 2004. Body fat and fat-free mass and all-cause mortality. *Obes. Res.* 12(7):1042–9.

Bousson, V. 2007. QCT, pQCT, microCT, and bone architecture. *Osteoporos. Int.* 18(6):867–871.

Bousson, V., Meunier, A., Bergot, C. et al. 2001. Distribution of intracortical porosity in human midfemoral cortex by age and gender. *J. Bone Miner. Res.* 16(7):1308–17.

Brooke-Wavell, K., Jones, P.R.M., Norgan, N.G. and A.E. Hardman. 1995a. Evaluation of near infra-red interactance for assessment of subcutaneous and total body fat. *Euro. J. Clin. Nutr.* 49(1):57–65.

Brooke-Wavell, K., Jones P.R.M. and D.W. Pye. 1995b. Ultrasound and dual x-ray absorptiometry measurement of the calcaneus—influence of region of interest location. *Calcif. Tissue Int.* 57(1):20–4.

Brooke-Wavell, K., Jones, P.R.M. and A.E. Hardman. 1997. Brisk walking reduces calcaneal bone loss in healthy women aged 60–70. *Clin. Sci.* 92(1):75–80.

Chumlea, W.C. and R.N. Baumgartner. 1989. Status of anthropometry and body-composition data in elderly subjects. *Am. J. Clin. Nutr.* 50(5):1158–66.

Clasey, J.L., Hartman, M.L., Kanaley, J. et al. 1997. Body composition by DEXA in older adults: Accuracy and influence of scan mode. *Med. Sci. Sports Exerc.* 29(4):560–7.

Clasey, J.L., Kanaley, J.A., Wideman, L. et al. 1999. Validity of methods of body composition assessment in young and older men and women. *J. Appl. Physiol.* 86(5):1728–38.

Conway, J.M., Norris, K.H. and C.E. Bodwell. 1984. A new approach for the estimation of body composition: Infrared interactance. *Am. J. Clin. Nutr.* 40:1123–30.

Crabtree, N., Loveridge, N., Parker, M. et al. 2001. Intracapsular hip fracture and the region-specific loss of cortical bone: Analysis by peripheral quantitative computed tomography. *J. Bone Miner. Res.* 16(7):1318–28.

Damilakis, J., Perisinakis, K., Vagios, E., Tsinikas D. and N. Gourtsoyiannis. 1998. Effect of region of interest location on ultrasound measurements of the calcaneus. *Calcif. Tissue Int.* 63(4):300–5.

Damilakis, J., Maris, T.G. and A.H. Karantanas. 2007. An update on the assessment of osteoporosis using radiologic techniques. *Eur. Radiol.* 17(6):1591–602.

Dempster, P. and S. Aitkens. 1995. A new air displacement method for the determination of human-body composition. *Med. Sci. Sports Exerc.* 27(12):1692–97.

Ding, J.Z., Kritchevsky, S.B., Newman, A.B. et al. 2007. Effects of birth cohort and age on body composition in a sample of community-based elderly. *Am. J. Clin. Nutr.* 85(2):405–10.

Duan, Y.B., Turner, C.H., Kim, B.T. and E. Seeman. 2001. Sexual dimorphism in vertebral fragility is more the result of gender differences in age-related bone gain than bone loss. *J. Bone Miner. Res.* 16(12):2267–75.

Elia, M., Parkinson, S.A. and E. Diaz. 1990. Evaluation of near infra-red interactance as a method for predicting body composition. *Eur. J. Clin. Nutr.* 44(2):113–21.

Ellis, K.J. 2000. Human body composition: In vivo methods. *Physiol. Rev.* 80(2):649–80.

Ellis, K.J. and R.J. Shypailo. 1998. Bone mineral and body composition measurements: Cross-calibration of pencil-beam and fan-beam dual-energy X-ray absorptiometers. *J. Bone Miner. Res.* 13(10):1613–18.

Fields, D.A., Goran, M.I. and M.A. McCrory. 2002. Body-composition assessment via air-displacement plethysmography in adults and children: A review. *Am. J. Clin. Nutr.* 75(3):453–67.

Frost, M.L., Blake, G.M. and I. Fogelman. 2000. Can the WHO criteria for diagnosing osteoporosis be applied to calcaneal quantitative ultrasound? *Osteoporos. Int.* 11(4):321–30.

Gallagher, D., Ruts, E., Visser, M. et al. 2000. Weight stability masks sarcopenia in elderly men and women. *Am. J. Physiol. Endocrinol. Metab.* 279(2):E366–75.

Genton, L., Hans, D., Kyle, U.G. and C. Pichard. 2002. Dual-energy x-ray absorptiometry and body composition: Differences between devices and comparison with reference methods. *Nutrition* 18(1):66–70.

Genton, L., Karsegard, V.L., Zawadynski, S. et al. 2006. Comparison of body weight and composition measured by two different dual energy X-ray absorptiometry devices and three acquisition modes in obese women. *Clin. Nutr.* 25(3):428–37.

Gluer, C.C. 1997. Quantitative ultrasound techniques for the assessment of osteoporosis: Expert agreement on current status. *J. Bone Miner. Res.* 12(8):1280–8.

Goran, M.I., Toth, M.J. and E.T. Poehlman. 1998. Assessment of research-based body composition techniques in healthy elderly men and women using the 4-compartment model as a criterion method. *Int. J. Obes.* 22(2):135–42.

Hans, D., Schott, A.M., Arlot, M.E., Sornay, E., Delmas, P.D. and P.J. Meunier. 1995. Influence of anthropometric parameters on ultrasound measurements of Os calcis. *Osteoporos. Int.* 5(5):371–6.

Heymsfield, S.B., Wang, J., Lichtman, S., Kamen, Y., Kehayias, J. and R.N. Pierson. 1989. Body-composition in elderly subjects—a critical-appraisal of clinical methodology. *Am. J. Clin. Nutr.* 50(5):1167–75.

Heyward, V.H. and D.R. Wagner. 2004. *Applied body composition assessment.* 2nd edition, Champaign, Illinois: Human Kinetics.

Jackson, A.S. and M.L. Pollock. 1978. Generalized equations for predicting body density of men. *Br. J. Nutr.* 40(3):497–504.

Jackson, A.S., Pollock, M.L. and A. Ward. 1980. Generalized equations for predicting body density of women. *Med. Sci. Sports Exerc.* 12(3):175–82.

Jarvinen, T.L.N., Kannus, P. and H. Sievanen. 1999. Have the DXA-based exercise studies seriously underestimated the effects of mechanical loading on bone? *J. Bone Miner. Res.* 14(9):1634–5.

Kanis, J.A., Johnell ,O., De Laet, C., Jonsson, B., Oden, A. and A.K. Ogelsby. 2002. International variations in hip fracture probabilities: Implications for risk assessment. *J. Bone Miner. Res.* 17(7):1237–44.

Kanis, J.A., Oden, A., Johnell, O. et al. 2007. The use of clinical risk factors enhances the performance of BMD in the prediction of hip and osteoporotic fractures in men and women. *Osteoporos. Int.* 18(8):1033–46.

Kaptoge, S., Jakes, R.W., Dalzell, N. et al. 2007. Effects of physical activity on evolution of proximal femur structure in a younger elderly population. *Bone* 40(2):506–15.

Knapp, K.M., Blake, G.M., Spector, T.D. and I. Fogelman. 2004. Can the WHO definition of osteoporosis be applied to multi-site axial transmission quantitative ultrasound? *Osteoporos. Int.* 15(5):367–74.

Kotzki, P.O., Buyck, D., Hans, D. et al. 1994. Influence of fat on ultrasound measurements of the os calcis. *Calcif. Tissue Int.* 54(2):91–5.

Kyle, U.G., Genton, L., Karsegard, L., Slosman, D.O. and C. Pichard. 2001. Single prediction equation for bioelectrical impedance analysis in adults aged 20–94 years. *Nutrition* 17(3):248–53.

Kyle, U., Bosaeus, G.I., De Lorenzo, A.D. et al. 2004. Bioelectrical impedance analysis principles and methods. *Clin. Nutr.* 23(5):1226–43.

Langton, C.M., Palmer, S.B. and R.W. Porter. 1984. The measurement of broadband ultrasonic attenuation in cancellous bone. *Eng. Med.* 13:89–91.

Lupoli, L., Sergi, G., Coin, A. et al. 2004. Body composition in underweight elderly subjects: Reliability of bioelectrical impedance analysis. *Clin. Nutr.* 23(6):1371–80.

Marin, F., Gonzalez-Macias, J., Diez-Perez, A., Palma, S. and M. Delgado-Rodriguez. 2006. Relationship between bone quantitative ultrasound and fractures: A meta-analysis. *J. Bone Miner. Res.* 21(7):1126–35.

Marshall, D., Johnell, O. and H. Wedel. 1996. Meta-analysis of how well measures of bone mineral density predict occurrence of osteoporotic fractures. *Br. Med. J.* 312(7041):1254–9.

Mayhew, P.M., Thomas, C.D., Clement, J.G. et al. 2005. Relation between age, femoral neck cortical stability, and hip fracture risk. *Lancet* 366(9480):129–35.

Melton, L.J., Khosla, S., Atkinson, E.J., O'Connor, M.K., O'Fallon, W.M. and B.L. Riggs. 2000. Cross-sectional versus longitudinal evaluation of bone loss in men and women. *Osteoporos. Int.*11(7):592–9.

Njeh, C.F., Hans, D., Li, J. et al. 2000. Comparison of six calcaneal quantitative ultrasound devices: Precision and hip fracture discrimination. *Osteoporos. Int.* 11(12):1051–62.

Norgan, N.G. and A. Ferro-Luzzi. 1985. The estimation of body density in men—are general equations general. *Ann. Hum. Biol.* 12(1):1–15.

Pace, N and E N. Rathbun. 1945. Studies on body composition. III. The body water and chemically combined nitrogen content in relation to fat content. *J. Biol. Chem.* 158:685–691.

Riggs, B.L., Melton, L.J., Robb, R.A. et al. 2004. Population-based study of age and sex differences in bone volumetric density, size, geometry, and structure at different skeletal sites. *J. Bone Miner. Res.* 19(12):1945–54.

Roschger, P., Gupta, H.S., Berzanovich, A. et al. 2003. Constant mineralization density distribution in cancellous human bone. *Bone* 32(3):316–23.

Roschger, P., Dempster, D.W., Zhou, H. et al. 2007. New observations on bone quality in mild primary hyperparathyroidism as determined by quantitative backscattered electron imaging. *J. Bone Miner. Res.* 22(5):717–23.

Ross, P.D., He, Y.F., Yates, A.J. et al. 1996. Body size accounts for most differences in bone density between Asian and Caucasian women. *Calcif. Tissue Int.* 59(5):339–43.

Salamone, L.M., Fuerst, T., Visser, M. et al. 2000. Measurement of fat mass using DEXA: A validation study in elderly adults. *J. Appl. Physiol.* 89(1):345–52.

Sambrook, P. and C. Cooper. 2006. Osteoporosis. *Lancet* 367(9527):2010–8.

Schoeller, D.A., Tylavsky, F.A., Baer, D.J. et al. 2005. QDR 4500A dual-energy X-ray absorptiometer underestimates fat mass in comparison with criterion methods in adults. *Am. J. Clin. Nutr.* 81(5):1018–25.

Silva, A.M., Wang, J., Pierson, R.N. et al. 2005. Extracellular water: Greater expansion with age in African Americans. *J. Appl. Physiol.* 99(1):261–7.

Sorkin, J.D., Muller, D.C. and R. Andres. 1999. Longitudinal change in height of men and women: Implications for interpretation of the body mass index—The Baltimore Longitudinal Study of Aging. *Am. J. Epidemiol.* 150(9):969–77.

Tran, Z.V. and A. Weltman. 1988. Predicting body-composition of men from girth measurements. *Hum. Biol.* 60(1):167–75.

Tran, Z.V. and A. Weltman. 1989. Generalized equation for predicting body density of women from girth measurements. *Med. Sci. Sports Exerc.* 21(1):101–4.

Tylavsky, F., Lohman, T., Blunt, B.A. et al. 2003. QDR 4500A DXA overestimates fat-free mass compared with criterion methods. *J. Appl. Physiol.* 94(3):959–65.

Uusi-Rasi, K., Sievanen, H., Pasanen, M. and P. Kannus. 2007. Age-related decline in trabecular and cortical density: A 5-year peripheral quantitative computed tomography follow-up study of pre- and postmenopausal women. *Calcif. Tissue Int.* 81(4):249–53.

Van Staa, T.P., Dennison, E.M., Leufkens, H.G.M. and C. Cooper. 2001. Epidemiology of fractures in England and Wales. *Bone* 29(6):517–22.

Visscher, T.L.S. and J.C. Seidell. 2001. The public health impact of obesity. *Annu. Rev. Public Health* 22:355–75.

Visser, M., Gallagher, D., Deurenberg, P., Wang, J., Pierson, R.N. and S.B. Heymsfield. 1997. Density of fat-free body mass: Relationship with race, age and level of body fatness. *Am. J. Physiol. Endocrinol. Metab.* 35(5):E781–7.

Visser, M., Harris, T.B., Langlois, J. et al. 1998. Body fat and skeletal muscle mass in relation to physical disability in very old men and women of the Framingham Heart Study. *J. Gerontol. A Biol. Sci. Med. Sci.* 53(3):M214–21.

Wang, Z.M., Deurenberg, P., Pietrobelli, A., Baumgartner, R.N. and S.B. Heymsfield. 1999. Hydration of fat-free body mass: Review and critique of a classic body-composition constant. *Am. J. Clin. Nutr.* 69(5):833–41.

Wannamethee, S.G., Shaper, A.G., Lennon, L. and P.H. Whincup. 2007. Decreased muscle mass and increased central adiposity are independently related to mortality in older men. *Am. J. Clin. Nutr.* 86(5):1339–46.

Williams, D.P., Going, S.B., Lohman, T.G., Hewitt, M.J. and A.E. Haber. 1992. Estimation of body-fat from skinfold thicknesses in middle-aged and older men and women—a multiple component approach. *Am. J. Hum. Biol.* 4(5):595–605.

Williams, D.P., Going, S.B., Milliken, L.A., Hall, M.C. and T.G. Lohman. 1995. Practical techniques for assessing body-composition in middle-aged and older adults. *Med. Sci. Sports Exerc.* 27(5):776–83.

Wu, X.P., Liao, E.Y., Huang, G., Dai, R.C. and H. Zhang. 2003. A comparison study of the reference curves of bone mineral density at different skeletal sites in native Chinese, Japanese, and American Caucasian women. *Calcif. Tissue Int.* 73(2):122–32.

Yee, A.J., Fuerst, T., Salamone, L. et al. 2001. Calibration and validation of an air-displacement plethysmography method for estimating percentage body fat in an elderly population: A comparison among compartmental models. *Am. J. Clin. Nutr.* 74(5):637–42.

Zamboni, M., Mazzali, G., Zoico, E. et al. 2005. Health consequences of obesity in the elderly: A review of four unresolved questions. *Int. J. Obes.* 29(9):1011–29.

15 Research Designs and Statistical Methods in Both Laboratory and Field Settings

C.G. Nicholas Mascie-Taylor*
Department of Biological Anthropology, University of Cambridge, Cambridge, United Kingdom

CONTENTS

* Address all correspondence to nmt1@cam.ac.uk.

15.1   BACKGROUND

Research designs broadly fall into two groups of descriptive (or observational) and analytical (or experimental) studies and may involve individuals or populations. Which of these research designs is used depends upon the objectives and hypotheses to be tested. Traditionally, researchers define 'exposure' and 'outcome' measures, where exposure refers to the factors to which a person, or group of people, comes into contact (Margetts and Nelson 1998), while outcome measures refer to factors which are being studied in relation to the effects of an exposure. From a statistical/analytical point of view, exposures are the independent variables while outcomes are the dependent variables.

15.2   DESCRIPTIVE STUDIES: THE 5 'W' QUESTIONS

A descriptive study is 'concerned with, and designed only, to describe the existing situation of variables without regard to causal or other hypotheses' (Last 1988). In epidemiology, descriptive studies focus either on host, agent, and environment (Lilienfield and Lilienfield 1980) or person, place, and time (Hennekens and Buring 1987). Descriptive studies can be a very important first step, since they can provide information on the frequency and possible determinants of a condition or circumstance. The early reports of Legionnaire's disease (Keys 1977) and toxic-shock syndrome (McKenna et al. 1980) are examples of descriptive studies.

So descriptive studies often represent the 'first scientific toe in the water in new areas of inquiry, and are designed to answer the 5 basic questions of who, what, why, when, and where, and a sixth: so what?' (Grimes and Schultz 2002b).

For example, **Who** has *Giardia*? [*Giardia* is the most common waterborne protozoan parasite in the world, in both developing and developed countries (Farthing 1994) and it causes intestinal infection in mammals as well as birds, reptiles, and amphibians (Adam 1991)]. The answer is: People living in unsanitary conditions exposed to contaminated water and food.

What is the disease being studied? In the case of *Giardia*, diagnosis has traditionally been through analysis of a faecal sample and the detection of *Giardia* cysts.

But infants can show a very low prevalence of *Giardia* cysts (2–3%) but very high levels of the *Giardia*-specific IgM antibody (95%), illustrating that case definition can vary enormously depending on choice of outcome (stool or blood sample) (Goto et al. 2009).

Why did the disease arise? Descriptive studies can often provide clues about cause that can be pursued using more sophisticated research designs, such as a randomised clinical trial to assess the impact of anti-*Giardia* treatment on infant nutritional status and growth.

When is *Giardia* rare or common? Time often provides important clues about many health events; in the case of *Giardia*, there are marked seasonal patterns (Goto et al. 2009).

Where does *Giardia* occur? In developed countries, *Giardia* infection can occur when people camp in the wilderness or swim in contaminated streams or lakes, especially artificial lakes formed by beaver dams (hence the popular name for giardiasis in North America, 'Beaver Fever'). *Giardia* may be ingested at camping areas, day care centres, waterborne outbreaks, and is also highly infectious to other family members once one individual is infected. Other causes can be uncooked food, contaminated wells, and failed municipal water systems. In developing countries, the poor water and sanitary conditions exacerbate the spread of the condition.

So what? From a public health perspective, *Giardia* is probably very important in contributing to gut mucosal damage and reduced food intake and thereby contributing to growth failure in infants in developing countries.

15.3 TYPES OF DESCRIPTIVE STUDIES

Figure 15.1 provides a breakdown of the types of descriptive studies with the main difference being between individual and populations studies. Cross-sectional studies are conducted in the present, case-control refer to past events, while cohort and surveillance studies mainly deal with follow-up and future events. The time frame of individual descriptive studies is summarised in Figure 15.2.

15.3.1 Cross-Sectional Studies

Cross-sectional studies are also known as prevalence studies, and the researcher has no control over the exposure of interest. For example, a one-off sample of 2,020 Bangladeshi children aged 2–6 years was used to examine the relationship between hookworm infection and haemoglobin status. The prevalence of anaemia was higher in infected children and lower in uninfected children. Since the data were cross-sectional, it would not be possible to prove that the hookworm infection was responsible for the lowered haemoglobin levels. So the main problem with this type of study is that exposure and outcome are identified at the same time and so it is not possible to say which is cause and which effect. In addition, studies may be affected by Neyman's bias; this occurs where subjects are not included in the measure of prevalence because they have died early from the disease or their symptoms have disappeared.

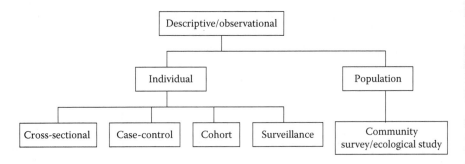

FIGURE 15.1   Types of descriptive studies.

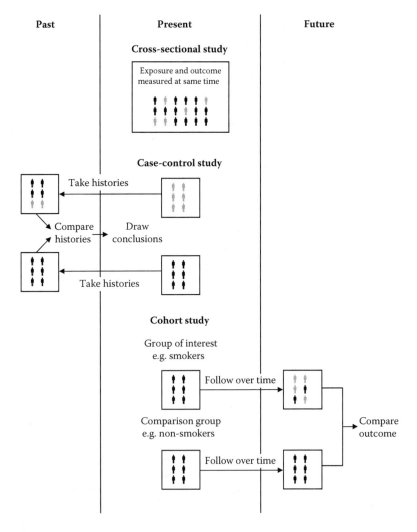

FIGURE 15.2   Summary of time frame of individual descriptive studies.

15.3.2 COHORT STUDIES

A cohort study is one in which subjects who presently have a certain condition and/ or receive a particular treatment are followed over time and compared with another group who are not affected by the condition under investigation.

For instance, a randomised trial to test the effect of smoking on health would be unethical, but a reasonable alternative would be a study that identifies two groups, a group of people who smoke and a group of people who do not, and follows them forward through time to see what health problems they develop.

The data in Table 15.1 show a cohort study of 200 smokers and 200 nonsmokers (controls) for the occurrence of myocardial infarction (MI).

The risk (incidence) of MI in smokers is 32/200 = 16% and the risk in nonsmokers is 15/200 = 7.5%. There are various ways in which the two risks or incidence rates can be compared. These are called the relative risk (RR, or risk ratio) and the attributable risk.

$$\text{The relative risk} = \frac{\frac{32}{200}}{\frac{15}{200}} = 2.13$$

$$\text{The attributable risk} = 32/200 - 15/200 = 17/200 = 8.5\%$$

The relative risk can take on any value ≥ 0, and a RR of 1 indicates no exposure–outcome association (1 = the null value). Values between 0 and 1 indicate a negative association, that is, exposure protects against the outcome; values above 1 indicate a positive association, that is, exposure increases the risk of the outcome. In this example, smoking increases the risk of myocardial infarction. The attributable risk is simply the rate of MI in smokers minus the rate of MI in nonsmokers. The attributable risk is more useful for public health purposes since it indicates the frequency with which the outcome (myocardial infarction) can be attributed to exposure (smoking) in the sample studied.

The statistical significance of a relative risk can be determined by a χ^2 test.

$$\chi^2 = \frac{(ad - bc)^2 N}{(a+b)(c+d)(a+c)(b+d)}$$

$$= \frac{[(32)(185) - (168)(15)]^2 (400)}{(200)(200)(47)(353)} = 6.968$$

(15.1)

TABLE 15.1

Cohort Study of Smokers and Nonsmokers for Occurrence of Myocardial Infarction (MI)

	MI	No MI	Total
Smokers	32	168	200
Nonsmokers	15	185	200
Total	47	353	400

which corresponds to a p value of < 0.01, so the relative risk of 2.13 is significantly greater than 1.

The confidence interval (CI) for the relative risk is

$$CI = RR^{1\pm(z_a/\chi)} \tag{15.2}$$

where z_a is the two-sided z value corresponding to the chosen width of the confidence interval (1.96 for 95% and 2.57 for 99%) and χ is the square root of the observed value of χ^2.

$$95\% \ CI = RR^{1\pm(z_a/\chi)}$$
$$= 2.13^{(1\pm1.96/\sqrt{6.968})} = 1.21 \text{ to } 3.73 \tag{15.3}$$

Since this interval excludes 1, the relative risk is statistically significant.

Cohort studies are not as reliable as randomised clinical trials (see Section 15.2.3.1) since the two groups may differ in ways other than in the variable under study. For example, if the subjects who smoked had less access to health care, that would exaggerate the difference between the two groups. The main problem with cohort studies, however, is that they can take a very long time to complete and are expensive since the researchers have to wait for the conditions of interest to develop. Also this means that large groups are needed in order to allow the appearance of the condition of interest, especially so for rare conditions, which adds expense. In addition, attrition occurs, and new and promising treatments may arise, making the study obsolete.

15.3.3 CASE CONTROL

Case-control (case-referent) studies compare a group of patients who have a condition with a group of patients who do not have the condition. Perhaps the best-known example of a case-control study is that of Sir Richard Doll and his colleagues who demonstrated a link between tobacco usage and lung cancer. However, the retrospective nature of the study meant that it was not possible to determine causality or to derive incidence of exposed and unexposed individuals.

Selection of a control group is the Achilles heel of case-control studies; controls should be as similar to cases as possible except for not having the outcome in question. Inappropriate control groups have ruined many case-control studies (Grimes and Schulz 2002a). In addition, recall bias, whereby recollection of exposures is greater in cases than controls, is a persistent difficulty in studies that rely on memory. Because case controls lack denominators, the incidence rates, relative risks, and attributable risks cannot be calculated and instead odds ratios are used. For example, Table 15.2 presents data on a case-control study relating tea drinking and renal cancer (RC).

The odds of a given event are the ratio of the probability of its occurrence to the probability of its nonoccurrence. So the odds of tea drinking among renal cancer

TABLE 15.2
Relationship between Tea Drinking and Renal Cancer (RC)

	RC	No RC	Total
Tea drinkers	400 (a)	333 (b)	733 ($a+b$)
Non–tea drinkers	100 (c)	167 (d)	267 ($c+d$)
Total	500 ($a + c$)	500 ($b + d$)	1000 ($a + b + c + d$)

cases is 400/100 (a/c), and the odds of tea drinking among controls is 333/167 (b/d). The ratio of these two odds is called the odds ratio (OR).

$$OR = \frac{\frac{400}{100}}{\frac{333}{167}} = \frac{(400)(167)}{(100)(333)} = 2.006$$

So tea drinkers are twice as likely to suffer from renal cancer than are non–tea drinkers. The statistical significance of an odds ratio is determined by a χ^2 test.

$$\chi^2 = 22.957, p < 0.001$$

$$95\% \, CI = OR^{1 \pm (z_a/\chi)}$$

$$= 2.006^{(1 \pm 1.96/\sqrt{22.937}} = 1.60 \text{ to } 2.42$$

(15.4)

Since this interval excludes 1, and the odds ratio is statistically significant.

Case-control studies are less reliable than either randomised clinical trials or cohort studies. Just because there is a statistical relationship between two conditions does not mean that one condition actually caused the other. Lung cancer rates are higher for people without a college education (who tend to smoke more), but that does not mean that someone can reduce his or her cancer risk just by getting a college education. The main advantages of case-control studies are that they can be done quickly and are generally less expensive. By asking patients about their past history, researchers can quickly discover effects that otherwise would take many years to show themselves.

15.3.4 Surveillance

In public health, surveillance is an important descriptive study. It can either be passive, with reliance on data gathered through traditional channels, or active, which searches for new cases. Surveillance is formally defined as 'the ongoing systematic collection, analysis, and interpretation of health data essential to the planning, implementation, and evaluation of public health practice, closely integrated with the timely dissemination of these data to those who need to know' (CDC 1986). Smallpox was eradicated through the combination of surveillance and containment (Foege 1998).

15.3.5   COMMUNITY SURVEYS/ECOLOGICAL STUDIES

Community surveys or ecological studies (which do not have any environmental connotation) focus on population groups rather than individuals. The group might be defined geographically, socio-demographically, or by time. Ecological studies are frequently the starting point in the construction of an epidemiological picture of, for example, the distribution of diseases among people with different risk profiles.

However, the ecological approach is limited as a source of causal inference and researchers have to be aware of the 'ecological fallacy' (Robinson 1950), which applies to errors that may result from making inferences about exposure–effect relationships at the individual level on the basis of relationships observed at the group level. Robinson computed the literacy rate and proportion of the population born outside the United States for all forty-eight U.S. states using the 1930 census, and found a strong positive correlation ($r = +0.53$), but at the individual level, the correlation was negative ($r = -0.11$). The positive correlation at state level was because immigrants tended to settle in states where the native population was more literate.

15.3.6   ADVANTAGES AND DISADVANTAGES OF DESCRIPTIVE STUDIES

Descriptive studies do have a number of uses. They are able to monitor changing patterns and trends. Health service managers and planners make use of cross-sectional surveys to assess utilisation and effectiveness of services. In addition, descriptive studies might provide the impetus for the development of hypotheses about causes.

The disadvantage of descriptive studies is that of 'post hoc ergo propter hoc' reasoning (after the thing, therefore on account of the thing), where a temporal association is incorrectly inferred to be a causal one. In the 1980s, seven women from Pasadena, California, developed functional ovarian cysts while taking one form of a multiphasic contraceptive pill. On the basis of this uncontrolled observation, a case report warned that this contraceptive pill might be detrimental to the health and well-being of women. The media printed this report and many women stopped taking the pill because they did not understand the difference between ovarian cysts and ovarian cancer. It took five years of case-control and cohort studies to show that there was no association between use of these multiphasic pills and ovarian cysts.

15.4   ANALYTICAL/EXPERIMENTAL DESIGNS

As with descriptive statistics, there is a clear delineation between individual and population designs (Figure 15.3).

15.4.1   RANDOMISED CLINICAL TRIAL—PARALLEL DESIGN

The first attempts at a controlled clinical trial can be traced back to 1753 when Lind (described by Carpenter 1986) examined the effect of different treatments on sailors with scurvy; those receiving oranges and lemons improved after six days. But the first randomised trials were not undertaken until the late 1940s by the UK Medical

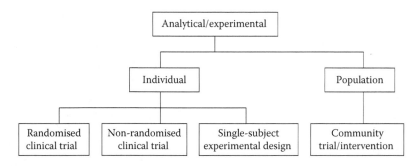

FIGURE 15.3 Types of analytical/experimental designs.

Research Council (MRC) and involved the use of streptomycin in the treatment of pulmonary tuberculosis (MRC 1948) and antihistamines for the treatment of the common cold (MRC 1950).

A clinical trial is one in which there are treatment and control regimens. The treatment group by definition receives some intervention, often in the form of medication or supplement, while the control group receives either no treatment, some standard default treatment, or a placebo. In a parallel design, subjects receive one or other regimen, that is, either treated or control. Individuals are assigned to treatment and control groups at random, which reduces the risk of bias, and provides an unbiased distribution of confounders, as well as increasing the probability that any differences between the groups can be attributed to the treatment (Figure 15.4). Confounding is defined as a situation in which the effects of two processes are not separated so that the apparent effect is not the true effect and the interpretation of the results is likely to be faulty (Clancy 2007). A good example of confounding is that death from trauma is more likely if treatment is by a consultant or senior doctor rather than a junior doctor. This association is confounded by the severity of the illness, which is associated with increased mortality and a senior doctor being involved in the patient's care. In many trials, bias is reduced by using either single-blind (in which only the researcher knows whether an individual is taking the treatment or a placebo) or a double-blind (in which neither the research staff nor the individual knows which of several therapies the individual is receiving) approach is used.

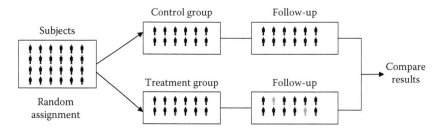

FIGURE 15.4 Randomised clinical trial.

The Consolidated Standards of Reporting Trials (CONSORT) published a twenty-one-item checklist and a flow diagram, covering title, abstract, introduction, methods, results, and comment as a means of improving the quality of reporting of randomised clinical trials (Begg et al. 1996). A CONSORT equivalent for infection control studies (ORION) has been proposed (Cooper et al. 2007).

A sequential clinical trial is one in which there is continuous comparison of two treatments as the data become available. Unlike the usual randomised trial, sequential clinical trials do not require sample size determination (see Section 15.5) prior to the start of the study. Based on the analyses, the researcher can decide either to stop the study and recommend that a specific treatment does or does not work or continue to collect data because the accumulated data are insufficient to draw a conclusion.

15.4.2 Nonrandomised Clinical Trial

Although randomisation is preferable, nonrandom assignment may be necessary where randomisation is not feasible or ethical. Nonrandom trials are sometimes referred to as 'quasi-experimental' or 'non-equivalent control group' designs. Nonrandom clinical trials might also be flawed because of selection bias. For example, there is a commonly held belief that people who practice yoga are less likely to suffer from heart disease, but if people who practice yoga ate better diets than the comparison group, then it is the diet, not the yoga, which might lead to reduction in heart disease!

15.4.3 Single-Subject Experimental Design

In single-subject experiments, the individual is the unit of analysis and serves as his or her own control. Single-subject designs involve taking measurements during the baseline phase (A), then introducing an intervention, followed by measurements during the intervention phase (B). By comparing baseline and intervention measurements the researcher hopes to detect the impact of the intervention. However, the internal validity of the A-B design is limited, because factors extraneous to the intervention that happened during the intervention phase may be the cause of the observed changes. One way to improve internal validity of the A-B design is to extend it to an A-B-A design or an A-B-A-B design, both of which contain a withdrawal phase, that is, a second A phase. A-B-A and A-B-A-B designs are most suitable for outcomes that are reversible.

Alternatively, a crossover design can be used in which each subject receives both, or all, treatments in a randomised order with sufficient gaps between treatments (wash out) and the outcome/response is compared within subjects. The advantages of the crossover design over the parallel design (see Section 15.4.1) are that (1) subjects serve as their own controls, so error variance is reduced and thus sample sizes needed are also reduced and the statistical power of the study is also greater and (2) all subjects receive treatment at least some of the time. Like parallel studies, statistical tests assuming randomisation can be used and blinding can be maintained. The disadvantages of crossover designs is that (1) all subjects receive placebo or alternative treatment at some point; (2) the washout period may be lengthy or unknown and there may be some carryover effects, and Senn (2002) has argued that it is virtually impossible to be sure that carryover effects are not present; and (3) it cannot be used for treatments with permanent effects.

15.4.4 COMMUNITY TRIALS

Community trials are where communities are allocated randomly to each study group. They are commonly used in the evaluation of lifestyle interventions that cannot be allocated to individuals. However, problems can arise due to secular trends and insufficient numbers of people who are targeted within a community.

15.5 SAMPLE SIZE DETERMINATION

15.5.1 SIMPLE RANDOM SAMPLING

A study with an overlarge sample may be costly and unnecessary while one that is too small will be unable to detect important differences. In order to calculate the desirable sample size, the significance level (α) and the power of the test ($1 - \beta$) are usually essential. The power of the test is a measure of how likely that test is to produce a statistically significant result for a population difference of a given magnitude (that is, the power of the test is defined as the probability of correctly rejecting the null hypothesis, H_0, given H_0 is false).

For example, it is planned to conduct a randomised clinical trial to test whether an iron-folate supplementation during pregnancy will increase birth weight. One group of pregnant women will receive the new supplement and the other group will receive a placebo. A pilot study showed that the standard deviation in birth weight is 500 g and it is assumed to be the same for both groups. The hypothesis of no difference is to be tested at the 5% level of significance. It is desired to have 80% power ($\beta = 0.20$) of detecting an increase of 100 g.

$$N = \frac{2\sigma^2[Z_{1-\alpha} + Z_{1-\beta}]^2}{(\mu_1 - \mu_2)^2}$$

$$= \frac{2(500)^2[1.645 + 0.842]^2}{(100)^2} = 309.26$$

(15.5)

Hence a sample of 310 subjects should be studied in each of the two groups. In this example, a one-tailed test was used, hence $Z_{1-\alpha}$ is 1.645, whereas the equivalent value for a two-tailed test is 1.96. Lemeshow et al. (1990) and Barker Bausell and Li (2002) provide details of the power/sample sizes required with different research designs; a number of computer programmes—for example, Sample Power (SPSS 2002)—are also available for determining sample sizes for proportions, means, regressions, correlations, ANOVA, logistic regression, and survival analysis.

15.5.2 Cluster Randomisation

In many studies, cluster randomisation is used and the assumption of independence is violated because two individuals within the same cluster are more likely to respond in a similar manner than are two individuals in different clusters. So sample sizes for cluster samples need to be increased to adjust for clustering. A statistical measure of the extent of clustering is the 'intracluster correlation coefficient' (ICC) which is based on the relationship of the between cluster to within-cluster variance (Donner and Koval 1980). Both the intraclass correlation coefficient and the cluster size influence the increase sample size required, and so the sample size can be very considerable if the average cluster size is large.

The loss of effectiveness by the use of cluster sampling, instead of simple random sampling, is called the design effect (DEFF) and is the ratio of the actual variance (under the sampling method actually used) to the variance computed under the assumption of simple random sampling.

The design effect is calculated as follows:

$$DEFF = 1 + ICC\,(n - 1) \tag{15.6}$$

where n is the average size of the cluster; the design effect increases as both the cluster size and ICC increase.

Hayes and Bennett (1999) provide sample size calculations for cluster-randomised trials based on unmatched and pair-matched trials. So, for example, supposing that n individuals are sampled in each cluster, then c, the number of clusters required, is given by

$$c = 1 + \frac{(Z_{1-\alpha}+Z_{1-\beta})^2\left[\left(\sigma_0^2 + \sigma_1^2\right)/n + k^2\left(\mu_0^2 - \mu_1^2\right)\right]}{(\mu_0 - \mu_1)^2} \tag{15.7}$$

where k is the coefficient of variation of true means between clusters within each group, and σ_0 and σ_1 are within-cluster standard deviations. A problem faced by researchers is that data on between-cluster variation is seldom known and so it is necessary to 'guestimate' k. Experience from a number of field trials suggests that k is often ≤ 0.25 and seldom exceeds 0.5 for most health outcomes.

If the previous example is changed from simple random sampling to cluster sampling where σ_0 and σ_1 are 500 g, $\mu_0 - \mu_1$ are 2600 g and 2500 g respectively, $k = 0.25$, and n, the number in each cluster, is 400, then

$$c = 1 + \frac{(1.645 + 0.842)^2[(500^2 + 500^2)/400 + 0.25^2(2600^2 - 2500^2)]}{(2600 - 2500)^2}$$

$c = 20.48$, so ten sets of villages per study arm would be required with a total sample size of 8,000. Increasing n to 1,000 would have little impact on c ($c = 20.02$) but overall sample size would increase to 20,000. Alternatively, keeping n at 400 and decreasing k from 0.25 to 0.1 would reduce c to 3.94, and a total sample size of only 1,600. So the magnitude of k is very important in determining the number of clusters and the overall sample size.

15.6 STATISTICAL AND MEASUREMENT ERRORS

There are four stages in the execution of an analytical study: (1) research hypothesis, (2) research design, (3) data collection, and (4) statistical analysis. Very few studies can avoid bias at some point in sample selection, conducting the study, and interpretation of results, and Sackett (1979) identified thirty-five types of potential bias. Most biases occur during data collection, often as a result of taking observations from an unrepresentative subset of the population rather than from the population as a whole. For example, during the Second World War, a group was studying planes returning from bombing Germany. They drew a rough diagram showing where the bullet holes were and recommended that those areas be reinforced. A statistician pointed out that essential data were missing from the sample they were studying—the planes that did *not* return from Germany!

The statistical tests available for use in the field and the laboratory are essentially the same. It is very common now for projects to involve both fieldwork and laboratory components. For example, height and weight of a child will be measured at the field site but the analyses of stool samples to determine parasite egg counts will be undertaken in the laboratory. Similarly, the collection of urine for five hours after giving a lactulose/mannitol sugar solution to a child would be in the field but the analyses to determine the extent of gut leakage and damage would be conducted in the laboratory.

Statistical analysis commonly involves both descriptive statistics (data summary and displays) and statistical inference, which includes hypothesis testing. Usually the researcher studies the data with respect to the null hypothesis (H_0), which implies that there is no association between the exposure and outcome in the target population. Thus, the null hypothesis is a 'straw man' that provides a reference by which the actual data are compared with data that would be expected under a null hypothesis.

The probability (p) value for rejecting the null hypothesis should be decided a priori and is conventionally set at 0.05. Conversely, the null hypothesis is accepted (not willing to reject H_0) if $p > 0.05$. But few people would drive a car or take a flight if the p value of crashing was 0.05, so the accepted α level for most science and medical literature is very high compared with real-life events where $p \ll 10^6$.

Even if $p < 0.05$, rejecting the null hypothesis may be wrong. When the null hypothesis is rejected, when it is, in fact, true, then a Type I error is made. When, however, the null hypothesis is accepted when H_0 is false, a Type II error (or β error) is made. The decision to reject or not to reject H_0 is an inference—an inference that might or might not be correct. Thus, there is a risk of committing either a Type I or a Type II error depending on which inference is made (see Table 15.3), but not both errors.

Whether the work is conducted in the field or the laboratory, the aim is to limit measurement error so that analyses reflect the 'real' variation. Measurement error is usually greater in a field situation than in the laboratory due to more human measurement error and more inaccurate instrumentation. Even so, researchers train their field staff to measure as accurately as possible. To test for accuracy of, for example, height and weight, the intra- and interobserver technical error of measurements (TEMs) can be calculated (Ulijaszek and Kerr 1999) and the coefficients of reliability determined. Reliability values of 95% and above indicate good quality control. Recently Goto and Mascie-Taylor (2007) found that the combination of human measurement

TABLE 15.3
Two Errors of Hypothesis Testing

		Truth	
		H_0 False	H_0 True
Inference	Reject H_0	Correct	Type I error
	Do not reject H_0	Type II error	Correct

error and instrument error accounted for only a very small proportion of the total variation of weight in a Bangladeshi field study.

Other errors include (1) using the same set of data to formulate hypotheses and to test them, (2) failing to draw random, representative samples, (3) measuring the wrong variables or failing to measure what had been hoped to measure, and (4) using small sample sizes.

Every statistical procedure relies on certain assumptions for correctness. Errors in testing hypotheses occur either because the assumptions underlying the chosen test are not satisfied or because the chosen test is less powerful than other competing procedures. One of the main errors is not determining whether the test is to be one-sided or two-sided before the test is performed and before the data are examined. Secondly, virtually all statistical procedures rely on the assumption that the observations are independent.

15.7 EFFECT SIZE

If the p value is less than the predetermined significance level, it does not necessarily mean that the effect detected is of practical significance. For example, in a Sudan nutrition study of 80,000 children a correlation coefficient of 0.01 is statistically significant ($p < 0.005$) but has practically little meaning. Alternatively, if the sample size is too small, the null hypothesis will be accepted. If p values can be misleading, then confidence intervals can be used.

It is often useful to know not only whether an experiment has a statistically significant effect but also the *size* of any observed effect. Effect sizes are useful because they provide an objective measure of the importance of an independent variable (that is, an effect).

Many measures of effect size have been proposed, including Pearson's correlation coefficient and Cohen's d (see below). Some researchers prefer the correlation coefficient because it has a finite range of between 0 (no effect) and 1 (perfect effect).

15.7.1 Effect Size for the Association between Two Continuous Variables

The association between two continuous characters is either measured by a correlation coefficient or by a regression coefficient/equation, if prediction of a dependent variable is required.

15.7.1.1 Example 1

In a study, 196 children had their height and weight measured and the correlation coefficient was $r = +0.844$. Does this indicate a significant association between height and weight and what is the effect size?

$$t_{(n-2)} = r\sqrt{\frac{n-2}{1-r^2}} \tag{15.8}$$

$$t_{(194)} = 0.844\sqrt{\frac{196-2}{1-0.844^2}} = 21.93, \, p < 0.001$$

So there is a very significant positive association between height and weight.

Cohen (1988, 1992) has made some widely accepted suggestions about what constitutes a small, medium, or large effect based on correlation coefficients. If $r = 0.10$, small effect, the effect explains only 1% (R^2) of the total variance. If $r = 0.30$, medium effect, the effect explains 9% (R^2) of the total variance. If $r = 0.50$, large effect, the effect explains 25% (R^2) of the total variance.

So, in this example the effect size (0.844) is high.

15.7.1.2 Example 2

A regression analysis was undertaken with height as the dependent variable and weight as the independent variable. The ANOVA and constant and regression coefficients are shown below in Table 15.4.

The ANOVA shows that F is very significant and the regression equation is Height = $41.437 + 4.024 \times$ Weight. The significance of the regression coefficient is shown as $t = 21.929, \, p < 0.001$.

TABLE 15.4

SPSS Output Showing ANOVA and Regression Equation

ANOVA[b]

Model		Sum of Squares	df	Mean Square	F	Sig.
1.	Regression	13344.49	1	13344.485	480.881	.000[a]
	Residual	5383.515	194	27.750		
	Total	18728.00	195			

Coefficients[b]

Model		Unstandardized Coefficients		Standardized Coefficients		
		B	Std. Error	Beta	T	Sig.
1	(*Constant)	41.437	2.131		19.447	.000
	WT1 Weight-1 In Kg	4.024	.184	.844	21.929	.000

[a] Predictors: (Constant), WT1 Weight-1 in kg
[b] Dependent Variable: HT1 Height-1 in cm

F and t are related (if $df = 1$), so $\sqrt{F} = t$, or $t^2 = F$. Therefore, $\sqrt{480.881} = 21.929$ or $21.929^2 = 480.881$.

F and t values can be converted to r values using the formulae

$$r = \sqrt{\frac{t^2}{t^2 + df}} \tag{15.9}$$

$$r = \sqrt{\frac{F(1, df_R)}{F(1, df_R) + df_R}} \tag{15.10}$$

$$r = \sqrt{\frac{t^2}{t^2 + df}} = \sqrt{\frac{21.929^2}{21.929^2 + 194}} = 0.844$$

$$r = \sqrt{\frac{F(1, df_R)}{F(1, df_R) + df_R}} = \sqrt{\frac{480.881}{480.881 + 194}} = 0.844$$

So the effect size is high.

Another measure of the strength of the association between two continuous variables is the coefficient of determination, which is calculated by squaring the correlation coefficient or 'r-squared' but symbolised by R^2. It is a measure of the proportion of the variance shared by the two variables, and varies from 0 to 1. In the example $R^2 = 0.844^2 = 0.713$, so 71.3% of the total variance is shared by height and weight.

It is also possible to compute an 'adjusted R^2' using Wherry's equation

$$R_a^2 = R^2 - \frac{p(1 - R^2)}{N - p - 1} \tag{15.11}$$

where p is the number of independent variables in the equation, so $R_a^2 = 0.711$ in this example.

Wherry's equation has been criticised and some authors prefer to use Stein's formula:

$$\text{adjusted } R^2 = 1 - \left[\left(\frac{n-1}{n-k-1} \right) \left(\frac{n-2}{n-k-2} \right) \left(\frac{n+1}{n} \right) \right] (1 - R^2) \tag{15.12}$$

where n is the number of cases and k is the number of predictors.

15.7.2 Effect Size for the Comparison of Two Means

The effect size is obtained from the following equation which converts a t value to an r value (correlation coefficient):

$$r = \sqrt{\frac{t^2}{t^2 + df}}$$

Example

A random sample of 50 girls and 50 boys had their haemoglobin determined. The girls mean was 11.7 g/dl and SD 1.4, and the boys mean 11.0 g/dl and SD 1.3.

$$t_{n-2} = \frac{\overline{X} - \overline{X}}{\sqrt{\dfrac{SD_1^2 + SD_2^2}{n_1 \quad n_2}}} \tag{15.13}$$

$$t_{98} = \frac{11.7 - 11.0}{\sqrt{\dfrac{1.4^2 + 1.3^2}{50 \quad 50}}} = 2.59, p < 0.02$$

Girls, on average, have a 0.7 g/dl significantly higher mean haemoglobin than boys. So

$$r = \sqrt{\frac{t^2}{t^2 + df}} = \sqrt{\frac{2.59^2}{2.59^2 + 98}} = 0.25$$

Other ways of determining the effect size in the context of a t-test of means are by calculating Cohen's d or Hedges' $\hat{g}$:

$$d = \frac{\overline{X}_1 - \overline{X}_2}{\sqrt{(SD_1^2 + SD_2^2)/2}} \tag{15.14}$$

$$d = \frac{11.7 - 11.0}{\sqrt{(1.4^2 + 1.3^2)/2}} = 0.52$$

Cohen suggests that $d = 0.2$ is a small effect, 0.5 medium, and ≥ 0.8 a large effect, so the effect size using Cohen's d and the correlation coefficient is very similar.

It is possible to convert Cohen's d to r:

$$r = \frac{d}{[(d^2 + 4)^{0.5}]} \tag{15.15}$$

Suppose $d = 0.52$:

$$r = \frac{0.52}{[(0.52^2 + 4)^{0.5}]} = 0.25$$

If the total sample size is small or group sizes are very different, a slightly more precise formula should be used:

$$r = \frac{d}{[(d^2 + [(N^2 - 2N)/(n_1 n_2)]^{0.5}]} \tag{15.16}$$

It is also possible to convert r to Cohen's d:

$$d = \left[\frac{2(r)}{(1-r^2)^{0.5}} \right] \tag{15.17}$$

Suppose $r = 0.25$:

$$d = \left[\frac{2(0.25)}{(1-0.25^2)^{0.5}} \right]$$

$$d = 0.52$$

Hedges and Olkin (1985) noted that the problem with Cohen's d is that the outcome is heavily influenced by the denominator in the equation:

$$g = \frac{\overline{X} - \overline{X}}{\sqrt{\dfrac{(n_1-1)SD_1^2 + (n_2-1)SD_2^2}{(N_{total}-2)}}} \times \left(1 - \frac{3}{4(n_1+n_2)-9} \right) \tag{15.18}$$

So in this example $\hat{g} = 0.51$, but one would not expect much change since the sample sizes are the same and the variances are also similar.

15.7.3 EFFECT SIZE FOR THE COMPARISON OF THREE OR MORE GROUPS, ANOVA

The effect size can be determined from the coefficient of determination:

$$R^2 = \frac{SS_B}{SS_T} \tag{15.19}$$

where SS_B refers to between-group sums of squares and SS_T the total sums of squares.

Example

Table 15.5 shows the mean heights of a sample of Sudanese children in relation to their mother's education level (classified as none, primary, or secondary) as well as the output from the one-way ANOVA.

$$R^2 = \frac{SS_B}{SS_T} = \frac{6187.2}{18728.0} = 0.330$$

So, $r = \sqrt{0.330} = 0.57$, and a high effect size.

Alternatively using Cohen's f^2:

$$f^2 = \frac{R^2}{1-R^2} \tag{15.20}$$

$$f^2 = \frac{R^2}{1-R^2} = \frac{0.33}{1-0.33} = 0.49$$

TABLE 15.5

SPSS Output Showing Heights of Sudanese Children by Mother's Education

Descriptives

	N	Mean	Std. Deviation	Std. Error	95% Confidence Interval for Mean Lower Bound	95% Confidence Interval for Mean Upper Bound	Minimum	Maximum
0 none	96	83.00	7.835	.800	81.41	84.59	67	99
1 primary	60	87.60	7.792	1.006	85.59	89.61	77	104
2 Secondary	40	97.80	8.953	1.416	94.94	100.66	76	106
Total	196	87.43	9.800	.700	86.05	88.81	67	106

ANOVA

	Sum of Square	df	Mean Square	F	Sig.
Between Groups	6187.200	2	3093.600	47.610	.000
Within Groups	12540.80	193	64.978		
Total	18728.00	195			

HT1 Height-1 in cm

by convention f^2 of 0.02, 0.15, and 0.35 are considered small, medium, and large effects. If the adjusted R^2 had been used (0.323), then $r = 0.57$ while f^2 falls very slightly to 0.48.

15.7.4 Effect Size for Categorical Variables

The best measure of association for a 2×2 chi-square test is phi (φ) or Cramer's phi or V. Phi may not lie between 0 and 1 because the chi-square can exceed the sample size. Pearson suggested the use of the contingency coefficient (which does lie between 0 and 1), but it seldom reaches its upper limit; for this reason, Cramer devised Cramer's V. This measure should be used when variables have more than two levels.

$$\phi = \sqrt{\frac{\chi^2}{N}} \tag{15.21}$$

Cramer's V

$$\phi_c = \sqrt{\frac{\chi^2}{N(k-1)}} \tag{15.22}$$

where k is the smaller of the number of rows and columns.

Using the χ^2 from the data comparing smokers and nonsmokers in relation to myocardial infarction (in Section 1.3.2), where $\chi^2 = 6.968$

$$\phi = \sqrt{\frac{\chi^2}{N}} = \sqrt{\frac{6.968}{400}} = 0.13$$

and

$$\phi_c = \sqrt{\frac{\chi^2}{N(k-1)}} = \sqrt{\frac{6.968}{400(2-1)}} = 0.13$$

Cramer's V uses the same cut-offs as the correlation coefficient of 0.1, 0.3, and 0.5, for small, medium, and large effects, so here there is a small effect size, with phi and Cramer's V of 0.13.

The odds ratio is another useful measure of effect size (see Section 15.3.3 for details).

15.8 OUTLIERS

Some statistical tests assume that the data approximate to a normal distribution. Departures from normality result from either outliers (one or more observations that are extreme in value compared with the other data in the sample) or some degree of skewness or kurtosis. The first thing to do with suspicious values is to check that the outlier is not the result of incorrect recording of data. A box plot provides a graphical representation of dispersion of the data. The graphic represents the lower quartile (Q1) and upper quartile (Q3) along with the median. The upper and lower fences usually are set a fixed distance from the interquartile range (Q3 – Q1). Any observation outside these fences is considered a potential outlier. Even when data are not normally distributed, a box plot can be used because it depends on the median and not the mean of the data.

There are lots of different methods for outlier detection such as Grubbs (Grubbs 1969; Stefansky 1972), Dixon, and Nalimov tests (see Iglewicz and Hoaglin 1993). These tests are flexible enough to allow for specific observations to be tested and they perform well with small sample sizes. Because they are based on ordered statistics, there is no need to assume normality of the data. Depending on the number of suspected outliers, different ratios are used to identify potential outliers.

Outliers in regression analysis are usually detected with graphical methods such as residual plots including deleted residuals. A common statistical test is Cook's distance measure (which is a measure of 'influence', combining both discrepancy and leverage) and which provides an overall measure of the impact of an observation on the estimated regression coefficient. However, just because the residual plot or Cook's distance measure test identifies an observation as an outlier does not mean that the point should automatically be eliminated (see next paragraph). The regression equation should be fitted with and without the suspect point and the coefficients of the model compared as well as the mean-squared error and R^2 from the two models.

Assuming that the outlying observation is correct, it is not appropriate to discard data simply because they appear to be unreasonably extreme or unlikely. Logarithmic or square root transformations are one way to reduce the impact of outliers since larger values shrink to a much greater extent than they shrink smaller values. However, transforming the data may not fit into the theory of the model or they may affect its interpretation. Taking the log of a variable does more than make a distribution less skewed; it changes the relationship between the original variable and the other variables in the model. In addition, most commonly used transformations require non-negative data or data that is greater than zero, so they are not always the answer. Working effectively with outliers in numerical data can be a rather difficult and frustrating experience. One solution is to report the results both with and without outliers to see how much they change. Another is to use robust statistical methods such as weighted least-squares regression, which minimise the effect of an outlier observation, or nonparametric statistical methods. In small samples, outlying values may have a very large influence on the results; for example, a regression line will be pulled towards outlying values.

15.9   NORMAL DISTRIBUTION

An easy way of ascertaining whether the distribution is showing some skewness is by comparing the mean and median values. For perfectly normal distributions, the mean and median are numerically identical. As the distribution becomes more skewed, so the difference between then mean and median increases. A number of goodness-of-fit procedures can test the null hypothesis that a sample came from a normal distribution. These include the Kolmogorov-Smironov, D'Agostino and Pearson, and Shapiro and Wilk tests, which assess normality using symmetry and kurtosis measures.

If the null hypothesis is rejected, then it may be desired to determine whether the non-normality is due to departure from symmetry or mesokurtosis or both. The Cox test (coefficient of skewness/standard error of skewness) provides a simple method of determining the extent of skewness, and Zar (1999) provides tables of critical values of both symmetry and kurtosis. Nevertheless, significant skewness and/or kurtosis may occur with large samples even though the magnitude of the effect(s) is small. Graphical assessment of normality is also recommended (Zar 1999).

15.10   PARAMETRIC AND NONPARAMETRIC TESTS

Statistical tests fall into two groups of parametric and nonparametric; parametric tests assume that the data conform to a normal distribution, whereas nonparametric tests are distribution-free tests. Nonparametric tests should be used when the data are ranked or ordinal.

The choice of test depends on the number of dependent (DV) and independent (IV) variables and whether they are continuous (interval), ordinal, or categorical. Table 15.6 provides a summary of the main statistical tests and when they should be used. For more complex analyses involving one dependent variable with one or more continuous independent variables and one or more categorical variables, there

TABLE 15.6
Summary of Commonly Used Statistical Tests

No. of DVs	No. of IVs	Type of IV	Type of DVs	Tests
	0 (1 sample)	None	Categorical (2 categories)	Binomial
				Chi-square
			Categorical	One-sample t-test
			Continuous & normal	One-sample median
			Ordinal or continuous non-normal	
	1	2 categories	Categorical	Chi-square or Fisher's exact
			Continuous & normal	Independent sample t-test
			Ordinal or continuous non-normal	Wilcoxon or Mann-Whitney
	1	3 or more categories	Categorical	Chi-square
			Continuous & normal	One way ANOVA
			Ordinal or continuous non-normal	Kruskal Wallis
1	0	None (Paired or matched)	Categorical	McNemar
			Continuous & normal	Paired t-test
			Ordinal or continuous non-normal	Wilcoxon signed rank
	0	None (3 or more paired or matched)	Categorical	Repeated measures logistic
			Continuous & normal	Repeated measures ANOVA
			Ordinal or continuous non-normal	Friedman
	2 or more	2 or more categories	Categorical	Factorial logistic regression
			Continuous & normal	Factorial ANOVA
	1	Continuous	Categorical	Logistic regression
			Continuous & normal	Regression or correlation
			Ordinal or continuous non-normal	Nonparametric correlation
	2 or more	1 or more continuous &/ or 1 or more categorical	Categorical	Multiple logistic regression or discriminant function
			Continuous & normal	Multiple regression or ANCOVA
2 or more	1	2 or more categories	Continuous & normal	One-way MANOVA
2 or more	2	2 or more categories	Continuous & normal	Multivariate multiple regression
2 sets of 2 or more	0	None	Continuous & normal	Canonical correlation
2 or more	0	None	Continuous & normal	Factor analysis

are no equivalent nonparametric tests. The same constraint occurs when two or more dependent variables are analysed.

With survival analysis, the Kaplan Meier curve is used. To compare survival curves for two unpaired groups, either the log-rank test or Mantel-Haenszel is used. For comparisons of survival curves of paired data and three or more unmatched groups, the conditional proportional hazards regression is used.

15.11   SOME LIMITATIONS OF STATISTICAL TESTS

For a long time many statisticians, including Fisher (1925), advised that the chi-square analysis of contingency tables should only be used if the expected frequencies were at least 5.0. However, empirical analysis by Roscoe and Byars (1971) provided more useful guidelines, and a secure procedure is to have an average expected frequency of at least 6.0 when testing with $\alpha = 0.05$ and 10.0 for $\alpha = 0.01$. If a 2×2 table or $N \times N$ table has insufficiently large frequencies for a chi-square analysis, then an exact test can be used.

The t-test for comparing two means is very robust and is affected very little by skewness provided a two-tailed test is used. Some authors recommend using Bessel's correction when one of the sample sizes falls below thirty ($N < 30$). The comparison of two means from a normal population without assuming equal variances is known as the 'Behrens-Fisher problem'. One of the easiest and most reliable solutions is known as 'Welch's approximate t' in which the normal test statistic is computed but degrees of freedom are corrected (downwards). The degrees of freedom thus computed are usually not integers, in which case the next smaller integer should be used.

For multivariate statistics, it is important to avoid multicollinearity (correlation between independent variables) and one of the key questions is how many subjects are required since the ratio of number of cases to independent variables (IV) has to be substantial, otherwise the solution will be perfect and yet meaningless. For example, in a multiple regression analysis, if there are more independent variables than cases, a solution can be found which predicts the dependent variable for each case, but only as an artefact of the ratio of cases to independent variables. Fewer than five cases per independent variable are generally considered unacceptable, even for exploratory research.

According to Tabachnick and Fidell (2001), a rule of thumb for testing regression coefficients is $N \geq 50 + 8m$, where m is the number of independent variables. Another popular rule of thumb is that there must be at least 20 times as many cases as independent variables. A higher cases-to-IVs is needed when the dependent variable is skewed or when a small effect size is anticipated or substantial measurement error is expected from unreliable variables. A more complex rule that takes into account size effects is $N \geq (8/f^2)$, where $f^2 = 0.01, 0.15,$ and 0.35 for small, medium, and large effects, respectively. More precisely estimated effect sizes can be calculated by $f^2 = R^2/(1 - R^2)$, where R^2 is the coefficient of determination. If stepwise multiple regression is being used, yet more cases are needed and a case-to-IV ratio of 40 to 1 is reasonable.

15.12   MULTIPLE TESTS

15.12.1   How Many False Positives?

If n independent tests are undertaken, each with a preset Type I error of α, then the number of false positives follows from the binomial distribution, with α the probability of 'success' (a false positive) and n the number of trials. Hence, the probability of k such false positives is

$$\text{Pr}(k \text{ false positives}) = \frac{n!}{(n-k)! \, k!}(1-\alpha)^{n-k}\alpha^k \qquad (15.23)$$

For n large and α small, this is closely approximated by the Poisson, with Poisson parameter $n\alpha$ (the expected number of false positives):

$$\text{Pr}(k \text{ false positives}) \approx \frac{(n\alpha)^k e^{-n\alpha}}{k!} \qquad (15.24)$$

Example

Suppose 250 independent tests are performed and a false positive probability of $\alpha = 0.025$ is chosen for each test, and 12 significant tests are observed by this criterion. Is this number of significant tests greater than expected by chance?

Here, $n\alpha = 250 \times 0.025 = 6.25$, so expect 6.25 significant tests by chance alone. The probability of observing 12 (or more) significant tests is

$$\sum_{k=12}^{250} \text{Pr}(k \text{ false positives}) = \sum_{k=12}^{250} \frac{250!}{(250-k)! \, k!}(1-0.025)^{250-k}0.025^k$$

So one would only expect to get this number of significant values 2.5% by chance, and some of these 'significant' values are truly significant, not false positives.

15.12.2   Bonferroni Correction

The Bonferroni correction is a method of correcting for significance based on the number of tests undertaken. If an experimenter is testing n independent tests on a set of data, then the statistical significance level that should be used for each hypothesis separately is $1/n$ times what it would be if only one hypothesis were tested. So, for example, to test two independent hypotheses on the same data at 0.05 significance level, instead of using a p value threshold value of 0.05, one would use a stricter threshold of 0.025.

Bonferroni correction, $\alpha = \pi/n$, where α is the new stricter threshold and π, the p value threshold value (usually 0.05).

TABLE 15.7

Significance with and without Bonferroni Correction, an Example

Test (i)	1	2	3	4	5	6	7	8	9	10
P	0.002	0.0045	0.006	0.008	0.0085	0.009	0.0175	0.025	0.1055	0.535
$\pi/n - i + 1$ (i.e., .05/n $- i +$ 1)	0.005	0.0056	0.0063	0.0071	0.0083	0.0100	0.0125	0.0167	0.0250	0.05

Example

Suppose ten tests were undertaken and the p values in order are as shown in Table 15.7

Without correction, eight of the ten tests would be significant, but with the Bonferroni correction, $\alpha = \pi/n = 0.05/10 = 0.005$, only the first two tests, 1 and 2, would be significant.

But when a hypothesis is rejected, there remain one fewer tests, and the multiple comparison correction should take this into account, resulting in so-called sequential Bonferroni corrections. Such corrections have increased power.

The Bonferroni sequential correction uses the formula $\pi/n - i + 1$.

With the Holm's method, one works from the smallest p towards the largest p value. The first test is significant; for the second test, $0.05/9 - 1 + 1 = 0.0056$ (as shown in the table), and again is significant; the third test is just significant (0.006 cf. 0.0063); but all the remaining tests are not significant. So with this method, Tests 1 to 3 are significant.

With the Simes-Hochberg method, one works from the largest p values first, still using the same formulation of $\pi/n - i + 1$. So 10, 9, 8, and 7 are all not significant, but 6 is significant (0.009 cf. 0.0100), and so Tests 6 to 1 are significant (that is, reject the null hypothesis).

With Hommel's method, all hypotheses whose p values are less than or equal to π/k are rejected where

$$k = \max_i p(n - i + j) > \pi \frac{j}{i} \qquad (15.25)$$

Start with $i = 1$. Here ($i = 1, j = 1$), $p(10) = 0.530 > \pi = 0.05$.

Now with $i = 2$, giving (for $j = 1,2$), $p(9) = 0.1055 > \pi(1/2) = 0.025$ and $p(10) > \pi$.

For $i = 3$ and $j = 1,2,3$, $p(8) = 0.025 > \pi(1/3) = 0.0167$, $p(9) > \pi(2/3) = 0.033$, $p(10) > \pi$

For $i = 4, j = 1,2,3,4$, $p(7) = 0.0175 > \pi(1/4) = 0.0125$, but ($i = 4, j = 2$) $p(8) = 0.025 = \pi(1/2)$.

Hence $k = 3$, and all hypotheses are rejected whose p values are $\leq 0.05/3 = 0.0167$, which are Tests 1 to 6.

So the strict Bonferroni declared the fewest of the hypotheses to be significant, and the Simes-Hochberg and Hommel's found the most significant.

TABLE 15.8

P values for Significance of All Pair-Wise Comparisons (Independent Sample t-Tests)

	Group 1 Mean = 34.5	Group 2 Mean = 37.8	Group 3 Mean = 41.0	Group 4 Mean = 39.3
Group 1		0.05[†]	0.0001[†*]	0.0103[†]
Group 2			0.0846	0.4444
Group 3				0.2820
Group 4				

[†] Significant before multiple comparison test
[*] Remained significant after Dunn-Šidák α' test applied

Sokal (2004) highlighted the failure to employ multiple comparison methods using examples drawn from papers published in recent issues of the *American Journal of Human Biology*, *American Journal of Physical Anthropology*, *Human Biology*, and *Proceedings of the National Academy of Sciences*.

In one paper, the authors compared four group means using all six possible combinations of t-tests (see below) and the *p* values indicated that three of the comparisons were significant (Group 1 versus Groups 2, 3, and 4). However when the Dunn-Šidák α' test was applied [$\alpha' = 1 - (1 - \alpha)^{1/k}$, where α is the desired experiment-wise Type I error, and k is the number of tests undertaken] based on α of 0.05, $\alpha' = 0.0085$, and only one test, between Groups 1 and 3, remained significant (Table 15.8).

Another example concerned an author who constructed a correlation matrix for seven variables and wanted to compute significance for each correlation coefficient.

Of the twenty-one coefficients, twelve were nominally significant ($p < 0.05$) by the conventional test, but when the Dunn-Šidák α' test was applied, the adjusted critical value of $\alpha' = 0.0024$, three additional coefficients lose significance (Table 15.9).

Multiple comparison tests have come in for criticism by some researchers in the field of behavioural ecology and animal behaviour (because of the reduction in statistical power) and Nakagawa (2004) recommended reporting effect size instead.

15.13   COMBINING *P* VALUES OVER INDEPENDENT TESTS

An interesting example of multiple comparisons is when the same hypothesis is independently tested. If the raw data are available, then they can be combined into a single dataset. However, often this is not possible, either because the data are not fully reported, or the experiments are such that different variables are being followed, and hence the raw data cannot be easily combined.

Fisher (1925) offered a simple, yet powerful, way around this based on the *p* values for each independent test. If *k* independent tests—usually different studies

TABLE 15.9
P values for Significance of All Correlation Coefficients

Variable	1	2	3	4	5	6	7
1	–	<0.001[†*]	<0.001[†*]	<0.001[†*]	0.157	0.400	<0.001[†*]
2		–	<0.001[†*]	<0.001[†*]	0.101	0.225	<0.001[†*]
3			–	<0.001[†*]	0.843	0.064	0.038[†]
4				–	0.050[†]	0.282	<0.001[†*]
5					–	0.085	0.444
6						–	0.010[†]
7							

[†] Significant before multiple comparison test
[*] Remained significant after Dunn-Šidák α' test applied

from different groups—are performed, and the p value from each test is natural log transformed, then the sum of the transformed values multiplied by -2 follows a chi-squared distribution with df equal to $2k$.

$$\chi^2_{(2k)} = -2 \sum_{i=1}^{k} \ln(p_i) \tag{15.26}$$

Fisher's method started the field of meta-analysis, wherein one searches the literature to find a number of tests of a particular hypothesis, and then combines these tests into a single test. An import caveat to keep in mind during a literature search is the bias of reporting p values that are close to significant and not reporting p values that are far from significant.

Example

Suppose five different groups collected data to test the same hypothesis and the reported p values were 0.10, 0.06, 0.15, 0.08, and 0.07. Note that none of these individual tests are significant, but the trend is clearly that all are close to being significant (Table 15.10).

$$\chi^2_{(10)} = -2 \sum_{i=1}^{k} \ln(p_i) = -2(-12.1981) = 24.3962$$

$$\chi^2_{(10)} = 24.3962 \; p < 0.01$$

Hence, taken together, the five tests show a highly significant p value.

TABLE 15.10
Combining *p* Values over
Independent Tests

p Value	Natural Log (ln) *p*
0.10	−2.3026
0.06	−2.8134
0.15	−1.8971
0.08	−2.5257
0.07	−2.6593
Sum	−12.1981

15.14 USING INAPPROPRIATE OR INEFFICIENT STATISTICAL METHODS

McKinney et al. (1989) reviewed over seventy articles that appeared in six medical journals and found that in over half of these articles, Fisher's exact test was applied improperly. Wang and Zhang (1998) reviewed all original articles published in five leading Chinese medical journals in 1985 ($N = 640$) and in 1995 ($N = 954$). Compared with 1985, significant improvement was seen and the percentage using appropriate methods increased from 22% to 46%. In both years, the most commonly used statistical methods were t-tests and contingency tables. The most common errors were presentation of *p* values without specifying the test used, use of multiple t-tests instead of analysis of variance, and use of unpaired t-tests when paired tests were required. In clinical research, Scales et al. (2005) examined eighty-three studies and found that sixty-nine (71%) providing statistical comparisons had at least one statistical error, including using the wrong test for the data type (28%), inappropriate use of a parametric test (22%), and failure to account for multiple comparisons (65%). In studies applying multivariate analysis (29%), overfitting the model with too many variables was the most common statistical flaw (39%).

15.15 CONCLUSIONS

Understanding what kind of study to undertake and the advantages and disadvantages of different types of study that can be used to answer a specific hypothesis is a prerequisite to research. Then the sample size required needs to be justified. All statistical tests have limitations, and hypothesis testing involves one of two types of potential errors. Multiple tests are likely to increase the risk of making a Type I error. Testing null hypotheses at arbitrary *p* values of 0.05 can be misleading and the effect size provides a more objective measure of the importance of an independent variable.

Finally one should pay heed to an admonition from George Dyke (1997), who wrote,

> The availability of 'user-friendly' statistical software has caused authors to become increasingly careless about the logic of interpreting their results, and to rely uncritically on computer output, often using the 'default option' when something a little different (usually, but not always, a little more complicated) is correct, or at least more appropriate.

REFERENCES

Adam, R.D. 1991. The biology of *Giardia* spp. *Clin. Microbiol. Rev.* 55, 706–32.

Barker Bausell, R. and Y-F. Li. 2002. *Power Analysis for Experimental Research.* Cambridge: Cambridge University Press.

Begg, C., Cho, M., Eastwood, S. et al. 1996. Improving the quality of reporting of randomized controlled trials: The CONSORT statement. *JAMA* 276, 637–9.

Carpenter, K. 1986. *The History of Scurvy and Vitamin C.* Cambridge: Cambridge University Press.

Centers for Disease Control (CDC). 1986. *Comprehensive Plan for Epidemiological Surveillance.* Atlanta, Georgia: Centers for Disease Control.

Clancy, M.J. 2007. Overview of research designs. *Emergency Med J.* http://www.emjonline.com.

Cohen, J. 1988. *Statistical Power Analysis for the Behavioural Sciences (2nd Edition).* New York: Academic Press.

Cohen, J. 1992. A power primer. *Psychol. Bull.* 112:155-9.

Cooper, B.S., Cookson, B.D., Davey, P.G., and S.P. Stone. 2007. Introducing the ORION statement: A CONSORT equivalent for infection control studies. *J. Hosp. Infect.* 65:85–7.

Donner, A. and J.J. Koval. 1980. The estimation of intraclass correlation in the analysis of family data. *Biometrics* 36(1):19–25.

Dyke, G. 1997. How to avoid bad statistics. *Field Crops Res.* 51:165–97.

Farthing, M.J.G. 1994. Giardiasis as a disease. In *Giardia: From Molecules to Disease,* ed. R.C.A. Thompson, J.A. Reynoldson, and A.J. Lymbery. Wallingford, Oxon: CAB International.

Fisher, R.A. 1925. *Statistical Methods for Research Workers.* 1st edition. Edinburgh, Scotland: Oliver and Boyd.

Foege, W.H. 1998. Smallpox eradication in West and Central Africa revisited. *Bull. World Health Organ.* 76:233–5.

Goto, R. and C.G.N. Mascie-Taylor. 2009. Precision of measurement as a component of human variation. *J. Physiol. Anthrop.* 26:253–6.

Goto, R. Mascie-Taylor, C.G.N., and P.G. Lunn. 2009. Impact of anti-*Giardia* and antihelmintic treatment on infant growth and intestinal permeability in rural Bangladesh. *Trans. R. Soc. Trop. Med.* 103(5):520–9.

Grubbs, F. 1969. Procedures for detecting outlying observations in samples. *Technometrics.* 11(1):1–21.

Grimes, D.A. and K.F. Schulz. 2002a. An overview of clinical research: The lay of the land. *Lancet* 359:57–61.

Grimes, D.A. and K.F. Schulz. 2002b. Descriptive statistics: What they can and cannot do. *Lancet* 359:145–9.

Hayes, R.J. and S. Bennett. 1999. Simple sample size calculation for cluster-randomized trials. *Int. J. Epidemiol.* 28:319–26.

Hedges, L.V. and I. Olkin.1985. *Statistical Methods for Meta-Analysis.* Orlando: Academic Press.

Hennekens, C.H. and J.E. Buring. 1987. *Epidemiology in Medicine Little.* Boston: Brown and Company.

Iglewicz, B. and D. Hoaglin. 1993. *How to Detect and Handle Outliers.* Milwaukee, Wisconsin: ASQ Quality Press.

Keys, T.F. 1977. A sporadic case of pneumonia due to legionnaires disease. *Mayo Clin. Proc.* 52:657–60.

Last, J.M. 1988. *A Dictionary of Epidemiology.* 2nd edition. New York: Oxford University Press.

Lemeshow, S., Hosmer, D.W., Klar, J., and S.K. Lwanga. 1990. *Adequacy of Sample Size in Health Studies.* Chichester: John Wiley & Sons.

Lilienfeld, A.M., and D.E. Lilienfeld. 1980. *Foundations of Epidemiology.* 2nd edition. New York: Oxford University Press.

Margetts, B.M., and M. Nelson. 1997. *Design Concepts in Nutritional Epidemiology*. Oxford: Oxford University Press.

McKenna, U.G., Meadows, J.A.III., Brewer, N.S., Wilson, W.R., and J. Perrault. 1980. Toxic shock syndrome, a newly recognized entity: Report of 11 cases. *Mayo Clin. Proc.* 55:663–72.

McKinney, W.P., Young, M.J., Hartz, A., and M.B. Lee. 1989 The inexact use of Fisher's Exact Test in six major medical journals. *JAMA* 261:3430–3.

Medical Research Council (MRC). 1948. Streptomycin treatment of pulmonary tuberculosis *Br. Med. J.* 2:769–82.

Medical Research Council (MRC). 1950. Clinical trials of antihistamine drugs in the prevention and treatment of the common cold. *Br. Med. J.* 2:425–9.

Nakagawa, S. 2004. A farewell to Bonferroni: The problems of low statistical power and publication bias. *Behav. Ecol.* 15:1044–5.

Robinson, W.S. 1950. Ecological correlations and the behavior of individuals. *Am. Sociol. Rev.* 15:351–7.

Roscoe, J.T. and A. Byars. 1971. Sample size restraints commonly imposed on the use of the chi-square statistic. *J. Am. Stat. Assoc.* 6:755–9.

Sackett, D.L. 1979. Bias in analytic research. *J. Chronic Dis.* 32:51–63.

Scales, C. Jr., Norris, R., Peterson, B., Preminger, G., and P. Dahm. 2005. Clinical research and statistical methods in the urology literature. *J. Urol.* 174:1374–9.

Senn, S. 2002. *Cross-over Trials in Clinical Research*. 2nd edition. Chichester: John Wiley.

Sokal, R.R. 2004. Raymond Pearl's Legacy: The proper measure of Man. *Am. J. Hum. Biol.* 16:113–24.

SPSS. 2002. *Sample Power 2*. Chicago: SPSS Inc.

Stefansky, W. 1972. Rejecting outliers in factorial designs. *Technometrics* 14:469–79.

Tabachnick, B.G. and L.S. Fidell. 2001. *Using Multivariate Statistics*. 4th edition. Needham Heights, Massachusetts: Allyn and Bacon.

Ulijaszek, S.J. and D.A. Kerr. 1999. Anthropometric measurement error and assessment of nutritional status. *Br. J. Nutr.* 82:165–77.

Wang, Q. and B. Zhang. 1998. Research design and statistical methods in Chinese medical journals. *JAMA*, 280:283–5.

Zar, J.H. 1999. *Biostatistical Analysis*. 4th edition. London: Prentice-Hall International (UK) Limited.

16 Impact of Natural Environmental Stressors on Physiological and Morphological Processes
Methodological Approaches in the Field and Laboratory

Ralph M. Garruto[*][1] *and Charles A. Weitz*[2]
[1]Department of Anthropology, Binghamton University,
State University of New York, Binghamton, New York
[2]Department of Anthropology, Temple
University, Philadelphia, Pennsylvania

CONTENTS

16.1 INTRODUCTION

In the 1960s and 1970s, Professor Paul T. Baker from The Pennsylvania State University (Figure 16.1), one of several human population biologists of his generation, wondered why more satisfactory answers had not been produced at a time when there was such

---

[*] Address all correspondence to rgarruto@binghamton.edu.

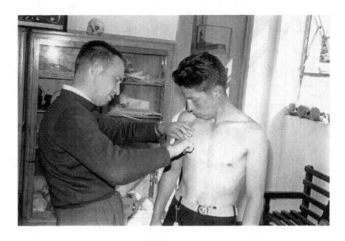

FIGURE 16.1 Professor Paul T. Baker in the Andes (Chinchero) outside of Cuzco, Peru, conducting anthropological (anthropometric) studies in 1962.

a strong professional interest in the question of how the physical environment affects human populations. He, of course, realised the level of complexity of the question he had asked. Baker raised essentially one of the same issues in a paper he published in the 1970s (Baker 1975)—the limitations of the methodological approaches used in field versus laboratory settings. At the time, he argued that human population field biologists could improve the scientific value of their results by employing some of the research design tools of the experimental (laboratory) scientist. The simplest design involved controlling as many variables as possible, except for the one under study—a one-stress–one-variable model. This design could then be followed by one stress, two variables, one stress, three variables, and so on, eventually evolving into a complex research design with multiple stresses and multiple variables.

In field studies as well as laboratory studies, increases in design complexity tend to produce results that vary in quality. In part, this variability may be the result of the level of biological organisation at which one seeks answers. Figure 16.2 shows

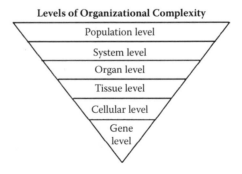

FIGURE 16.2 An example of increasing levels of biological complexity from the gene to the population with increasing interaction and epigenetic phenomena.

different levels of organisational complexity, ranging from the gene level to that of the population. Laboratory studies are directed primarily towards investigating mechanisms at the lower levels (that is, gene, cell, tissue) and therefore tend to be reductionist in approach. The major advantage of laboratory (experimental) studies is that they allow for increased specificity and sensitivity of measurement in a controlled environment, often with more technologically sophisticated equipment than is available for field use. However, laboratory studies are conducted in artificial environments, usually with small sample sizes; and because they lack a population perspective, they are unable to address evolutionary questions.

The perspective of human population field biologists tends to be more holistic. Most conduct field studies of natural populations within an evolutionary framework, with larger sample sizes under natural conditions. By using a comparative approach to investigate variation at the population level, field population biologists can eliminate some sources of variation and identify others as likely candidates for further examination at lower levels of causality. The advantages of a comparative approach to the study of physiological or morphological variation within and between populations include testing hypotheses about

- Genetic versus environmental causality or gene–environment interactions
- The influence of growth, development, or ageing
- The role of specific environmental factors, such as chronic or infectious diseases, nutrition, etc.
- Evolutionary processes, specifically those related to the operation of natural selection and adaptation

Familial-based studies within populations provide a particularly powerful tool for disentangling genetic causes of variation from those due to environmental influences, including nutrition and disease. Longitudinal data, reflecting growth/ageing, maternal health/nutrition, and foetal outcome, are highly prized but difficult to obtain since problems of follow-up are considerable in populations where individuals are prone to migration into and out of the area of study.

Offsetting the advantages, there are some important drawbacks to field studies that should be noted. First, sampling may be problematic. In population-based studies, sample structure is delineated as part of the research design but also reflects such issues as diurnal variation and standardisation of procedures and equipment (see Table 16.1). All the components listed in Table 16.1 can lead to potential sources of error since, as noted previously, the greater the number of variables, the more difficult it becomes to conduct unambiguous tests of evolutionary hypotheses. Furthermore, opportunistic sampling may not represent population structure accurately. Second, field conditions may limit the use of sophisticated instrumentation. The standard for field studies continues to be noninvasive measurements. While field instruments designed to make noninvasive measurements of physiological function or internal features of morphology have greatly improved over the past decade or two, the level of sophistication and detail does not approach laboratory study standards. Third, field studies, especially in earlier years, tended to be more descriptive than experimental or hypothesis-driven studies. The use of the comparative approach has reduced this

TABLE 16.1

Basic Components of a Research Design Necessary to Address the Impact of Environmental Stressors on Physiological and Morphological Processes

- Selection of stress (heat, cold, altitude, disease, multiple stressors)
- Selection of population (naturalistic model, single or multiple populations, other models)
- Diurnal variation (work–home environment, awake–sleep, at rest–workload)
- Sample characteristics (genetic differences, cultural differences, nutritional/disease pattern differences, environmental exposure differences, sample size, need for subpopulations, inclusion/ exclusion of healthy or all individuals, convenience vs. random samples, life-cycle approaches to physiological functioning)
- Standardization (procedures/methods/asymmetry, equipment)

Note: All components above have potential sources of error

tendency, but it still continues in many modern studies. Finally, field studies necessarily involve concerns about ethnic sensitivity. When working with non-Western populations, issues of informed consent become considerably more problematic. Not only is it necessary to explain the purpose of the study in culturally relevant ways, but it is frequently the case that certain types of studies are simply inappropriate for societies that may find particular measurements unacceptable. Thus, receiving informed consent requires the flexibility to incorporate the sensibilities of the study population to change the design or even, in rare cases, the willingness to walk away from the study if no agreement can be reached.

16.2 PROBLEMS IN THE STUDY OF ALTITUDE HYPOXIA

Historically, from the authors' perspective, there were four basic methodological problems in the study of altitude hypoxia:

- The first problem was that laboratory-based experimentalists were unselective in the sampling strategy they applied to field studies. Often under time constraints, their studies tended to involve small samples, individuals who were not part of indigenous populations, and sojourners who had lived at altitude for varying periods of time. This produced results that were not representative of natural human populations, leading early on to confusing and sometimes wrong interpretations about population-based issues of adaptation.
- The second problem was that studies lacked standardisation of methods and equipment. This resulted in different investigators generating different results for the same question.
- The third problem was that experimentalists and field scientists were applying the results from one population to genetically different populations living in different high-altitude regions of the world.
- The fourth problem was that studies of phenotypic variation were conducted with little control over genetic versus environmental causality.

Summarised below are two examples that represent methodological strategies employed by the authors to answer questions related to blood physiological changes among two very different high-altitude populations living in different regions of the world: Quechua Indians from the Andes of Peru and Golak Tibetans living on the Qinghai-Tibetan Plateau. These examples reflect issues that were relevant at the time the studies were undertaken, as well as changes incorporated into the study goals, sampling strategies, and technology.

16.2.1 EARLY PROBLEMS IN STUDIES OF BLOOD PHYSIOLOGICAL CHANGES IN THE ANDES

An example of methodological problems during the early years of high-altitude research in the Peruvian Andes concerned the question of whether or not secondary or altitude polycythemia occurred in traditional high-altitude Quechua Indians and, if it did occur, whether it was adaptive. Problems in early Andean studies from the 1930s–1960s were characterised by poor field designs, inappropriate sample selection, small sample sizes, rural vs. urban issues, and traditional vs. nontraditional participant occupations (Ballew et al. 1989). These problems were reflected in the variation in haemoglobin concentration and haematocrit that was reported in studies of adult Andean Quechua, where, as the analysis presented in Figure 16.3 demonstrates, patterns of traditional occupations and urbanism clearly affect reported values for these variables.

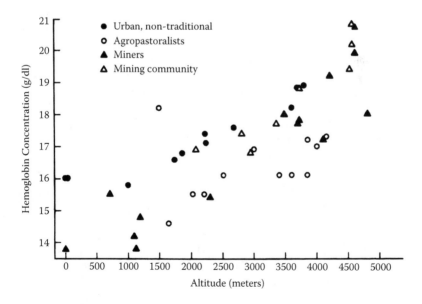

FIGURE 16.3 Mean haemoglobin concentration values by altitude of residence in the Peruvian Andes. Note the higher haemoglobin concentrations with altitude among urbanites and miners compared to traditional indigenous agropastoralists. (For details, see Ballew et al. 1989.)

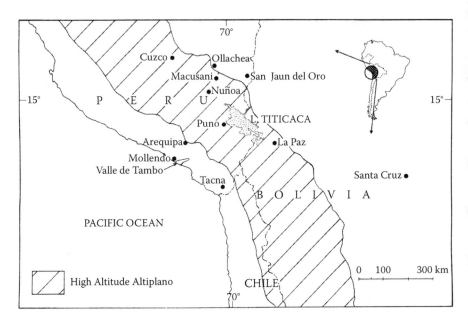

FIGURE 16.4   Map of Peru showing the location of study villages by Nuñoa (4,000 m) and Macusani (4,400 m) for the 1969 and 1971 field expeditions.

In the late 1960s and early 1970s, field studies were conducted in the southern Peruvian Andes among two traditional and geographically isolated communities at the time—Nuñoa (4,000 m) and Macusani (4,400 m) (Garruto 1976a, 1976b; Garruto and Dutt 1983) (Figure 16.4). Blood samples were collected from 303 Quechua males ranging in age from six to fifty-seven years. At the time, it was culturally more acceptable to work only with males in a society that did not easily allow for research to be conducted on females.

A specific protocol for evaluating potential participants was established to minimise the effects of factors other than altitude on blood physiological changes. Specifically, ancestry, health status, migration history, and lifestyle of the participants were closely examined. Only individuals of known Quechua heritage who had lived all of their lives at high altitude as traditional pastoralists and horticulturalists and who were in good physical health were included in the study (Figure 16.5). Evaluation of Quechua ancestry was based on a number of genetic markers: ABO and Rh blood group systems, dermatoglyphic patterns, and phenylthiocarbamide (PTC) taste sensitivity (Garruto 1976b). In addition, several cultural criteria were utilised, including an evaluation of parental surnames, place of birth, and socioeconomic status. Demographic information, including date and place of birth, current and past residences, and a detailed work and travel history, were collected on all individuals. Only individuals who were born in the highlands and had not travelled to lower altitudes were considered. From a total of 340 individuals screened, 303 met the above criteria and were included at the study.

FIGURE 16.5   Traditional indigenous Quechua male nearby the study site in the town of Nuñoa, Peru, at 4,000 m altitude. Note traditional dress and use of sandals in the snow. Children usually walked barefoot even in the snow. (Photo by Paul T. Baker in 1964.)

In this study, the microtechnique of finger puncture was utilised to collect all blood samples (Figure 16.6). With participants in a sitting position, a deep puncture was made into the finger to ensure a free flow of blood with as little admixture of tissue fluid as possible. Participants were normally active, normally hydrated, and nonfasting prior to sampling. Venipuncture was not used as there was strong reluctance on the part of the Quechua at that time to have their blood drawn. However,

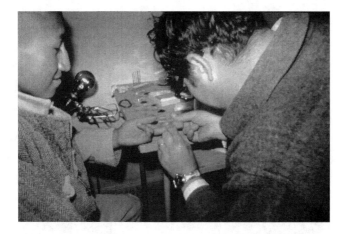

FIGURE 16.6   The microtechnique of finger puncture used in the 1969 and 1971 field expeditions to Peru. The same technique was used by the authors in the Qinghai high-altitude project in the 1990s in western China.

no significant differences in the results from the two methods (finger puncture and venipuncture) had been reported (Fahey et al. 1977; Moe 1970; Stuart et al. 1974).

Erythrocyte and leucocyte counts and haemoglobinometry were performed using two different methods. In 1969, blood counts (red and white) for the Macusani sample were performed using a Hausser haemocytometer. The same individual read all cell counts. A Lumetron haemoglobinometer, Model 15, which employed the oxyhaemoglobin method, was used to measure haemoglobin concentrations. During the 1971 field expedition, erythrocyte and leucocyte counts for the Nuñoa sample were determined with a Haemacount MK-2 electronic cell counter, with haemoglobin measured by a Haemacount MK-9 haemoglobinometer which employed cyanmethaemoglobin. Both instruments were calibrated against the Colter Counter Model-S (a widely used standard laboratory instrumentation at the time) before leaving for and after returning from Peru (Figure 16.7). The variance in accuracy of the MK-2 was about ±2% while the MK-9 readings were consistently 5–6% higher than those of the Colter Counter Model-S. Haematocrit values during both field expeditions were determined using the Adams microhaematocrit centrifuge. Duplicate readings were obtained for all values.

Reticulocyte and differential leucocyte counts were obtained using the same methods during both field sessions and blood smears were read by the same individual. Blood smears for reticulocyte counts were made at the time using a modification of the Osgood-Wilhelm technique (Davidsohn and Wells 1966). Proportional amounts of blood and brilliant cresyl blue dye were mixed together on one corner of the glass slides and allowed to stand for six minutes (rather than the usual three minutes) in a moist, enclosed chamber to prevent the blood-dye mixture from drying before the smears could be made. Reticulocytes were extremely difficult to stain at high altitude using regular procedures; lysing and dye crystallisation often occurred, and data were lost before the techniques could be adequately modified. Staining of

FIGURE 16.7 The Haemacount MK-2 electronic cell counter with associated MK-9 Haemoglobinometer (not shown) used in the 1971 field expedition to Peru.

differential counts for leucocyte analysis and microscopic examination of these cells were conducted according to prescribed methods, with one major modification: the staining time, like that for the reticulocytes, was doubled from three to six minutes and the buffering time was increased from five to ten minutes.

Although this Andean polycythemia study was problem driven, it was also descriptive. It did include several important improvements in field design and sample selection compared to earlier studies from the Andean region (Garruto 1976a). Virtually no other data existed at the time on children and adolescents. Also, at the time, most investigations failed to adequately characterise their sample group and frequently did not control for factors that might affect their results. For example, ethnic heritage was often not determined or reported, and nutritional and cultural data on the population, as well as information on the biotic environment, were meagre at best. The effects of short-term or intermittent exposure to hypoxic stress were almost never controlled for since place of birth, residence history, and work and travel history were not considered. Furthermore, the sample sizes in many studies were small and in numerous cases sample groups included participants who no longer practiced a traditional way of life (that is, pastoralism) (Figure 16.8). Volunteers were often selected from mining communities and hospital clinics or were migrants from lower altitudes. Such mining communities in most cases were located at altitudes higher than the highest traditional native communities, and a sample of miners probably included men who already had some degree of silicosis or at least those who had frequently inhaled airborne particulate matter from the mines. Those who were not miners but who worked in a mining community were, like miners, usually not involved in traditional occupational practices. In one study of highland Andean

FIGURE 16.8 Typical Quechua pastoralist with llama pack animals in the Peruvian Andes in 1969. The use of horses occurred only with wealthier Quechua, as most walked behind the pack animals.

TABLE 16.2
Percent Change in High-Altitude Blood Values from Sea-Level Norm (Male Adults)

Research Study	RBC (10⁶/mm³)	[Hb] (g/dl)	HCT (%)
Nuñoa Quechua (Garruto 1975)	9.8	12.1	10.3
Macusani Quechua (Garruto 1975)	7.8	7.0	13.9
Nuñoa Mestizos (Garruto 1975)	15.6	16.6	15.0
High-altitude residents (Hurtado 1964)	25.4	28.8	28.2
High-altitude residents (Reynafarje 1966)	27.4	29.4	28.7

Source: Data are based on the 1969 and 1971 field studies of altitude polycythemia in the Peruvian Andes compared to other major Andean studies at that time.

miners, a strong association was found between silicosis and polycythemia, the latter of which increased with an increase in severity of disease as measured by a decrease in the arterial oxygen saturation of the blood (Hurtado et al. 1945). Furthermore, mining as an economic activity, although possibly not much different in terms of total work output, is culturally a new lifestyle to which traditional Andean natives may not have developed efficient work patterns (Garruto 1972). Design flaws, such as those presented above, are much more likely to have had a major confounding effect under conditions of chronic hypoxic stress than at sea level and led, in many cases, to conflicting results and misinterpretations of the data.

The reported 20–30% increase in the red cell mass (Table 16.2), as a 'normal' long-term response to chronic hypoxic stress, did not appear to be an efficient or beneficial adaptive mechanism because of the high physiologic cost of such an increase in erythrocyte concentration (Garruto 1976a; Garruto and Dutt 1983). A polycythemic response increases blood viscosity, blood flow resistance, and workload on the heart (Figure 16.9), and tends to negate the beneficial effects of an increase in arterial oxygen saturation brought about by an increase in red cell mass and haemoglobin concentration. Clearly a change in one part of a coordinated biological system produces changes in other parts of the system, and one needs to know the consequences of the downstream changes that are produced. In this case, an increase in erythropoiesis and subsequent increase in the red cell mass, leading to increased blood viscosity, increased peripheral resistance, and an increased workload on the heart (Figure 16.9).

Data from the late 1960s and early 1970s revealed a lack of a prominent compensatory polycythemia in males of all ages between six and fifty-seven years (Table 16.2) (Garruto 1976a). Between 61 and 81% of the adult Quechua in the study fell within the Lima sea-level normal range. The results were very controversial at the time because they challenged the accepted dogma that 'normative' blood physiological values for high-altitude Andean populations should be different from low-altitude groups. Thus, the argument in the 1970s remained whether individuals

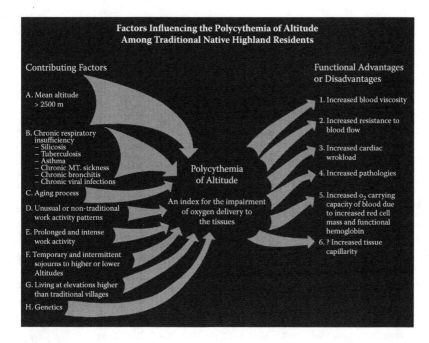

Factors Influencing the Polycythemia of Altitude Among Traditional Native Highland Residents

Contributing Factors

A. Mean altitude
   > 2500 m

B. Chronic respiratory insufficiency
   – Silicosis
   – Tuberculosis
   – Asthma
   – Chronic MT. sickness
   – Chronic bronchitis
   – Chronic viral infections

C. Aging process

D. Unusual or non-traditional work activity patterns

E. Prolonged and intense work activity

F. Temporary and intermittent sojourns to higher or lower Altitudes

G. Living at elevations higher than traditional villages

H. Genetics

Polycythemia of Altitude

An index for the impairment of oxygen delivery to the tissues

Functional Advantages or Disadvantages

1. Increased blood viscosity

2. Increased resistance to blood flow

3. Increased cardiac wrokload

4. Increased pathologies

5. Increased o_2 carrying capacity of blood due to increased red cell mass and functional hemoglobin

6. ? Increased tissue capillarity

FIGURE 16.9 A schematic that represents the factors (left) influencing the development of secondary (altitude) polycythemia in native Andean Quechua. On the right are some of the functional advantages and disadvantages associated with the development of altitude polycythemia.

whose red cell count, haemoglobin concentration, and haematocrit were closer to sea-level norms were better adapted to chronic hypoxic stress than those whose values were much higher and consistent with a secondary polycythemia. It was stressed at the time (Garruto 1976a; Garruto and Dutt 1983) that known morphological and physiological compensations other than polycythemia in a coordinated biological system were adequate to meet the demands of the body's tissues for oxygen. Thus, it was suspected that when polycythemia did occur among native Andean residents living above 3,000 m, it was most likely the result of a combination of altitude (≈10%) and some forms of chronic respiratory insufficiency. Therefore, a significant secondary polycythemia response was pathological rather than adaptational.

16.2.2 LATER PROBLEMS IN STUDIES OF BLOOD PHYSIOLOGICAL CHANGES IN QINGHAI

The second example addresses another issue of blood physiological changes at high altitude, arterial oxygen saturation. In the 1990s, in a series of six field expeditions, attempts at a more hypothesis-driven approach to the general problem of genetic versus environmental causes of Asian high-altitude adaptation was made. Of particular interest was a comparison of the morphological and physiological characteristics of indigenous Tibetan populations with those of Han who had been born and raised at the same altitudes under the same set of environmental conditions. As part of this larger research programme, the specific hypothesis that Tibetans had higher resting arterial oxygen saturation values as determined by pulse oximetry (that is, SpO_2 values) than Han populations born and raised at the same altitude in Qinghai was tested. It was reasoned that Tibetans, who had lived at high altitude for millennia, were more likely to possess genetic adaptations to hypoxia and thus would exhibit higher SpO_2 values than Han—even Han who had been born and raised at high altitude. Thus, in a study by Weitz and Garruto (2007), Tibetan and Han males and females between the ages of five and fifty-one years who were born and raised between 3,200 and 4,300 m in the high-altitude province of Qinghai in western China were evaluated (Figure 16.10). The results of this study were then compared to previous published results on SpO_2 in high-altitude Asian populations.

Even though there were a few Han residents at the 3,200 m study site (Guinan) as early as the 1920s, Han migration to high-altitude areas of Qinghai was negligible until the Cultural Revolution period between 1966 and 1976. During that period and continuing through the present, the number of Han migrants increased significantly. Nevertheless, Han and Tibetan communities still remained largely distinct culturally and genetically during the period of this study in the 1990s (Weitz and Garruto 2007). Surveys conducted at that time indicated that there were a small number of mixed Han–Tibetan marriages at the 3,200 m study site, and only a very few such marriages in the study communities at 3,800 m (Maquin) or at 4,300 m (Mado) (Figure 16.10). Thus, determining ethnicity was simply a matter of asking each study participant to identify the ethnicities and birthplaces of both parents. These statements were verified by checking whether parents had Tibetan or Han names, where the parents had been born, and by consulting with individuals in the community (such as teachers and health-care workers) who were familiar with the participants and their families. Individuals were classified as Han only if both parents could be identified unambiguously as ethnic Han, or as Tibetans only if both parents could be identified unambiguously as ethnic Tibetans. Individuals with one Han and one Tibetan parent were not included in the analysis. Unlike earlier high-altitude studies in the Andes, biomarker phenotypes were not used to discriminate between Han and Tibetans.

Birthplaces and birth dates of participants were determined from school records or through other official documents. In addition, all individuals were asked if they had made visits to lower elevations in the year preceding our study and the duration of that visit. Among younger children, this determination was made by questioning the parents directly. Slightly more than 89% of the study participants remained in the study villages throughout the year. About 60% of those who went to lower altitudes stayed for less than one month, and the remainder were absent for no longer than three months. All absentees had returned to their towns at least three months prior to participation in the study. At all study sites, some Han women chose to give birth in Xining, the provincial capital (2,300 m), or in other medium-altitude communities. Han children born at these altitudes were included in the study if they returned to the study town before their first birthday and if their parents were residents of the study town. Individuals born at lower elevations who migrated to the study town after the age of one year were not included.

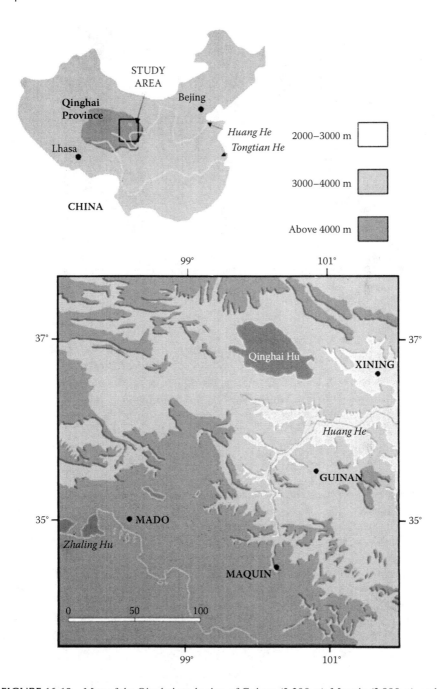

FIGURE 16.10 Map of the Qinghai study sites of Guinan (3,200 m), Maquin (3,800 m), and Mado (4,300 m) during the 1990s field expeditions to western China.

FIGURE 16.11   The battery-powered Nonen Model 8500 hand-held pulse oximeter used in the Qinghai studies in the 1990s to determine arterial oxygen saturation under field conditions.

Measuring arterial oxygen saturation under field conditions was made possible by the development of noninvasive, hand-held pulse oximeters which first became available in the 1980s. These instruments measure arterial oxygen saturation as the ratio of absorbencies of light emitted in the red and infrared wavelengths. Previously, determination of arterial oxygen saturation required arterial phlebotomy, a procedure that could only be conducted safely under hospital conditions (see, for example, Weil et al. 1968), or it required expensive and cumbersome equipment that could not be easily transported from site to site (Cerretelli 1976) and was prone to failure (Pugh 1964). Thus, the first study to report arterial oxygen saturation values on Tibetan-derived populations was conducted on a small number ($n = 9$) of adult male Sherpas in 1980 (Hackett et al. 1980). However, by the time of the authors' study, dozens of different instruments were on the market, making it possible to generate noninvasive measurements of oxygen saturation on large numbers of individuals. Thus a Nonin hand-held pulse oximeter (Nonin Medical Inc., model 8500, Plymouth, Minnesota) with a finger sensor was used (Figure 16.11). While seated in a warm environment, the left index finger of each subject was cleaned with alcohol and dried and the sensor was slipped over the finger. The SpO_2 readouts were allowed to stabilise over a period of at least two minutes and then monitored for several more minutes until the SpO_2 value remained unchanged for a minimum of ten seconds. This entire process commonly took between five and ten minutes. This value was recorded as the subject's SpO_2.

The study by Weitz and Garruto (2007), which was the first (and to date the only one) to compare Tibetan and Han males and females of the same ages who had been born at high altitude, produced unexpected results. Across the age range, no statistically significant differences in SpO_2 values between Han and Tibetans were found. Based on these results, it was concluded that indigenous Tibetans do not possess superior arterial saturation while awake and at rest compared to lowlanders who had been born and raised at high altitude.

The authors believed they effectively controlled the most significant sources of error in conducting this study and reaching this conclusion (Weitz and Garruto 2007). First, by including only Han who had been born and raised at high altitude, they eliminated individuals with varying degrees of exposure to hypoxia. Comparisons were made between Han and Tibetans of the same age, with the same lifelong exposure to high-altitude hypoxia and who lived in the same towns. Second, the individuals measured in the study were in good health. Prior to inclusion in the study, paediatricians and other health-care professionals on the research team from collaborating Chinese institutions assessed the general health status of each participant. Individuals who were identified as suffering from chronic illnesses were not included in the study. Nevertheless, the authors did include some individuals who were suffering from transient respiratory infections or related subclinical problems, but assessed the impact of including these individuals by comparing them to individuals without evidence of pulmonary obstruction or anaemia and with normal white cell and differential counts (see Garruto et al. 2003 for a description of the protocol used). In no case were the differences in arterial oxygen saturation greater than 0.2%, regardless of age group, ethnicity, or altitude. Thus, the impact of including a few individuals with mild symptoms had no effect on the results. Third, the authors took into account the effect of smoking, which increases carboxyhaemoglobin concentration (Seppanen 1977; Glass et al. 1996). Since oximeters cannot distinguish carboxyhaemoglobin from oxyhaemoglobin, SpO_2 values in smokers may be higher than their actual arterial oxygen saturation (Buckley et al. 1994). Smoking was very common among both Tibetans and Han males, but nearly absent among females in the study. Among males, there was no particular tendency for one group to more likely smoke than the other. About 50% of both Han and Tibetan boys over the age of sixteen smoked, and about two thirds of the Han and Tibetan adult males made the same claim. The average SpO_2 values between smokers and nonsmokers by age group and ethnicity were compared and found not to be statistically different. Similar results have been obtained in other studies of smoking and nonsmoking Tibetans and Han (Zhuang et al. 1993; Curran et al. 1997, 1998). Thus, the inclusion of smokers by Weitz and Garruto (2007) in their study did not bias the results.

Although it was impossible to compare the specific results of the Weitz and Garruto (2007) study with any previous study, as no others included Han born and raised at high altitude, the authors did compare the Tibetans in their study with previously published results on Tibetans. As was the case with red cell mass among the Quechua discussed earlier, there exists considerable variability in the arterial oxygen saturation data reported in the literature for Tibetans (see Figure 16.12). Some of this variability is probably a consequence of different sampling strategies. In several studies (Zhuang et al. 1993; Niermeyer et al. 1995; Zhuang et al. 1996; Chen et al. 1997; Curran et al. 1997, 1998), values are reported for relatively small numbers of individuals. Between the ages of five and nineteen years, the Weitz and Garruto (2007) Tibetan sample from Qinghai (245 Tibetans at 3,200 m, 305 Tibetans at 3,800 m, 112 Tibetans at 4,300 m, total $n = 662$) is comparable to the sample size in Beall's (2000) study of Tibetans ($n = 1063$ for a larger age range (zero to nineteen

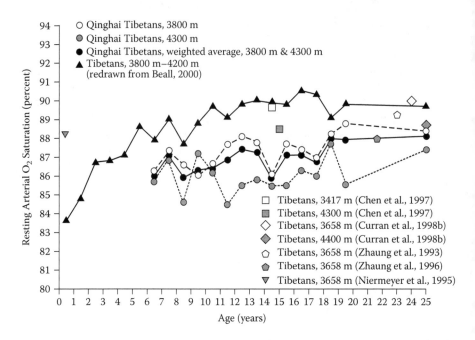

FIGURE 16.12   Comparison of SpO$_2$ values among Qinghai Tibetans (authors' study) and the values reported for Tibetans in other studies. For Qinghai Tibetans and central Tibetans, the ages of 20–29 are represented as a single value at age 25. (Reprinted with permission from Weitz and Garruto 2007.)

years). So, the consistently higher SpO$_2$ values for Tibetans reported by Beall (2000) for children, adolescents, and young adults are unlikely to be the result of differences in sample size. Also, the absence of smokers in Beall's (2000) study is unlikely to have caused the arterial oxygen saturation values to be higher, since the error introduced by smoking should result in lower—not higher—SpO$_2$ values in nonsmokers. Finally, it is unlikely that any differences in the health of individuals in the Weitz and Garruto (2007) sample resulted in their generally lower values compared to the sample collected by Beall (2000), as the former had only a few individuals with mild respiratory symptoms who tended to be concentrated among younger children. Even if mild respiratory symptoms modestly reduce SpO$_2$ values among five- to ten-year-olds, this would not explain the consistently lower values Beall (2000) reported among adolescents and young adults.

An analysis of the above discrepancies suggested that the use of different instruments was an important source of between-study differences in the arterial oxygen consumption values reported for Tibetans. Even though all pulse oximeters operate on the same general principle (Jubran 1999), there is no commonly accepted in vivo calibration method (Hornberger et al. 2002) or internal algorithm linking direct measurements of arterial oxygen saturation with the ratio of absorbencies of light emitted in the red and infrared wavelengths (Thilo et al. 1993). Consequently, studies comparing pulse oximeters manufactured by different companies commonly report within and between instrument variation (see Jensen et al. 1998 for a review of seventy-six studies conducted between 1976 and 1994). Furthermore, the variation in instrument response has been reported to increase substantially under (1) hypoxic compared to normoxic conditions (Trivedi et al. 1997; Jubran 1999), (2) conditions of poor perfusion (Morris et al. 1989; Gehring et al. 2002), (3) probe location (Clayton et al. 1991; Hanning and Alexander-Williams 1995; Hamber et al. 1999), (4) conditions when movement during measurement occurs (Gehring et al. 2002), and (5) possibly conditions relative to the degree of skin pigmentation and coarseness in both normoxic and hypoxic conditions (see Jubran 1999 vs. Bothma et al. 1996). While test conditions and subject selection may address some of these issues (for example, poor perfusion, motion artefacts), it is difficult to assess the specific bias of different instruments under the sorts of conditions commonly encountered when doing field-work at high altitude.

Finally, a variety of different protocols has been used to convert the continuous LCD readout produced by pulse oximeters to a single arterial oxygen saturation value for each individual. While the consequences of this source of error have not been assessed, future studies might benefit from the use of instruments that can produce a twenty-four-hour continuous recording of SpO_2 values, similar to the recordings made using twenty-four-hour ambulatory blood pressure monitors (James and Pickering 1991). Using this technology, which downloads data into a computer, would provide a standardised protocol and thousands of SpO_2 values for determining measures of central tendency (or the basis of a cumulative frequency analysis) and also might be important for understanding diurnal variations in saturation (sleep vs. awake) and for assessing the effects of different activity levels (at rest vs. work).

16.3 CONCLUDING REMARKS: USE OF NATURAL EXPERIMENTS

Examples of the ways field studies can be used to characterise the biology of populations living in stressful environments and to test hypotheses regarding environmental versus genetic control of differences between populations have been presented and discussed. These types of studies represent natural experiments that

- Take advantage of a population or special circumstance that naturally isolates a particular problem; that is, a population will be identified as particularly appropriate for addressing an existing scientific problem
- Use populations that have unique attributes or biomedical phenomena that direct inquiry and relevance at the local, regional, and global levels; that is, a population with unique attributes will identify a new scientific problem

Both cases are opportunistic; in the first case, the problem dictates the population, and in the second case, the problem arises from the population.

The human population biology field scientist can greatly enhance the scientific power of field studies by seeking out such natural experimental phenomena as a basis for setting up both simple and complex field designs. However, the question is not so much whether variation can and does occur, but why and how it occurs. In natural experiments, contrasts are made between populations that have a similar genetic structure and a similar culture and lifestyle and/or those that have differences in these variables but a common environmental stressor or peculiar disease pattern (Garruto 1981; Garruto et al. 1999). Thus, conceptually, natural experimental models in human populations are the reverse of the biological reductionism models discussed earlier that purport to explain living systems at all levels of organisation (Strohman 1997; Wilkins 1996). There now exists a new generation of field studies that is increasingly concerned with detecting the genetic causality of phenotypic variation (Weiss and Bigham 2007). In many cases, new technologies and new analytical strategies have created the potential for field biologists to use the same techniques that were once used only in laboratories. Human geneticists have been interested in applying quantitative trait loci analysis to human populations for over a decade (Blangero et al. 2000), and it has become technologically and methodologically possible to refine whole genome searches for ever-more-precise location of genes associated with particular phenotypes (Elston and Spence 2006). Commercially available probes and microarray panels have helped to reduce the expense associated with this type of research. However, large-scale studies of human populations, such as those searching for the genetic contribution to major diseases, are still too expensive to be used by most field biologists. Nevertheless, a few studies of indigenous populations have been conducted using more intensive genetic methodologies. Examples at high altitude include a search for candidate genes linked to chronic mountain sickness in indigenous high-altitude Andean populations (Mejia et al. 2005), and the sequencing of the HIF-1α gene among Sherpas (Suzuki et al. 2003). As the cost of conducting these types of studies declines, an exciting new era unlocking the genetic causes of variation within and between human populations can be anticipated.

In summary, the field scientist has come a long way towards approaching the level of experimental sophistication that the authors' mentors sought decades ago when working with human population groups. Today, the field scientist is in many ways much more equivalent to the experimental scientist as described by Professor Baker thirty years ago (Baker 1975). Even so, in the not-too-distant future, it is conceivable that the human population biology field scientist may be able to set up on site a portable prefabricated field laboratory outfitted with small, highly portable, and durable microelectrical mechanical system (MEMS) technology field equipment of 'experimental quality' or even nanotechnology ideally suited for field studies of the future.

ACKNOWLEDGEMENTS

The authors would like to thank Professor Nicholas Mascie-Taylor, the Japan Society for the Study of Physiological Anthropology, and the Society for the Study of Human Biology for their kind support of one of us (RMG) to attend the symposium in Cambridge upon which this paper is based. In addition, special thanks to Professor Michael Little who stepped in to deliver this lecture at the symposium when one of us (RMG) was called away on a family emergency. This research was supported by funding from the National Institutes of Health (RMG), Temple University (CAW), and the Wenner Gren Foundation for Anthropological Research (RMG).

DEDICATION

This chapter is in memory of Paul T. Baker—scholar, mentor, and friend.

REFERENCES

Baker, P.T. 1975. Research strategies in population biology and environmental stress. In *The Measures of Man: Methodologies in Biological Anthropology*, ed. E. Giles, and J. S. Friedlaender, 230–59. Cambridge, Massachusetts: Peabody Museum Press.

Ballew, C., Garruto, R.M. and J.D. Haas. 1989. High-altitude hematology: Paradigm or enigma? In *Human Population Biology: A Transdisciplinary Science*, ed. M. A. Little, and J. D. Haas, 239–62. New York: Oxford University Press.

Barker, S.J. 2002. 'Motion-resistant' pulse oximetry: A comparison of new and old models. *Anesth. Analg.* 95:967–72.

Beall, C.M. 2000. Oxygen saturation increases during childhood and decreases during adulthood among high altitude native Tibetans residing at 3,800–4,200m. *High Alt. Med. Biol.* 1:25–32.

Blangero, J., Williams, J.T. and L. Almasy. 2000. Quantitative trait locus mapping using human pedigrees. *Hum. Biol.* 72:35–62.

Bothma, P.A., Joynt, G.M., Lipman, J. et al. 1996. Accuracy of pulse oximetry in pigmented patients. *S. Afr. Med. J.* 86:594–6.

Brutsaert, T.D., Araoz, M., Soria, R., Spielvogel, H. and J.D. Haas. 2000. Higher arterial oxygen saturation during submaximal exercise in Bolivian Aymara compared to European sojourners and Europeans born and raised at high altitude. *Am. J. Phys. Anthropol.* 113:169–81.

Buckley, R.G., Aks, S.E., Eshom, J.L., Rydman, R., Schaider, J. and P. Shayne. 1994. The pulse oximetry gap in carbon monoxide intoxication. *Ann. Emerg. Med.* 24:252–5.

Cerretelli, P. 1976. Limiting factors to oxygen transport on Mount Everest. *J. Appl. Physiol.* 40:658–67.

Chen, Q.H., Ge, R.L., Wang, X.Z. et al. 1997. Exercise performance of Tibetan and Han adolescents at altitudes of 3,417 and 4,300 m. *J. Appl. Physiol.* 83:661–7.

Clayton, D.G., Webb, R.K., Ralston, A.C., Duthie D. and W.B. Runciman. 1991. Pulse oximeter probes. A comparison between finger, nose, ear and forehead probes under conditions of poor perfusion. *Anaesthesia* 46:260–5.

Curran, L.S., Zhuang, J., Sun, S.F. and L.G. Moore. 1997. Ventilation and hypoxic ventilatory responsiveness in Chinese-Tibetan residents at 3,658 m. *J. Appl. Physiol.* 83:2098–104.

Curran, L.S., Zhuang, J., Droma, T. and L.G. Moore. 1998. Superior exercise performance in lifelong Tibetan residents of 4,400 m compared with Tibetan residents of 3,658 m. *Am. J. Phys. Anthropol.* 105:21–31.

Davidsohn, I. and B. Wells. 1966. *Todd-Sanford Clinical Diagnosis by Laboratory Methods*. Philadelphia: W.B. Saunders Co.

Elston, R.C. and M.A. Spence. 2006. Advances in statistical human genetics over the last 25 years. *Stat. Med.* 25:3049–80.

Fahey, T.D., Kerr, T., Ohelson, G. and R. Schroeder. 1977. Substitution of fingertip blood for venous blood in the measurement of hematocrit and hemoglobin following exercise. *Res. Q.* 48:293–8.

Garruto, R.M. 1972. Hematological response to strenuous work activity patterns at high altitude. Abstract. *Am. J. Phys. Anthropol.* 37:437.

Garruto, R.M. 1976a. Hematology. In *Man in the Andes: A Multidisciplinary Study of High-Altitude Quechua*, ed. P.T. Baker and M.A. Little, 261–82. Stroudsburg, Pennsylvania: Dowden, Hutchinson and Ross.

Garruto, R.M. 1976b. Genetic history and affinities. In *Man in the Andes: A Multidisciplinary Study of High-Altitude Quechua*, ed. Baker, P.T. and M.A. Little, 98–114. Stroudsburg, Pennsylvania: Dowden, Hutchinson and Ross.

Garruto, R.M. 1981. Disease patterns of isolated groups. In *Biocultural Aspects of Disease*, ed. H. Rothschild, 557–97. New York: Academic Press.

Garruto, R.M. and J.S. Dutt. 1983. Lack of prominent compensatory polycythemia in traditional native Andeans living at 4,200 meters. *Am. J. Phys. Anthropol.* 61:355–66.

Garruto, R.M., Little, M.A., James, G.D. and D.E. Brown. 1999. Natural experimental models: The global search for biomedical paradigms among traditional, modernizing, and modern populations. *Proc. Natl. Acad. Sci. U.S.A.* 96:10536–43.

Garruto, R.M., Chin, C.T., Weitz, C.A., Liu, J.C., Liu, R.L. and X. He. 2003. Hematological differences during growth among Tibetans and Han Chinese born and raised at high altitude in Qinghai, China. *Am. J. Phys. Anthropol.* 122:171–83.

Gehring, H., Hornberger, C., Matz, H., Konecny, E. and P. Schmucker. 2002. The effects of motion artifact and low perfusion on the performance of a new generation of pulse oximeters in volunteers undergoing hypoxemia. *Respir. Care* 47:48–60.

Glass, K.L., Dillard, T.A., Phillips, Y.Y., Torrington, K.G., and J.C. Thompson. 1996. Pulse oximetry correction for smoking exposure. *Mil. Med.* 161:273–6

Hackett, P.H., Reeves, J.T., Reeves, C.D., Grover, R.F. and D. Rennie. 1980. Control of breathing in Sherpas at low and high altitude. *J. Appl. Physiol.* 49:374–9.

Hamber, E.A., Bailey, P.L., James, S.W., Wells, D.T., Lu, J.K. and N.L. Pace. 1999. Delays in the detection of hypoxemia due to site of pulse oximetry probe placement. *J. Clin. Anesth.* 11:113–8.

Hanning, C.D. and J.M. Alexander-Williams. 1995. Pulse oximetry: A practical review. *Br. Med. J.* 311:367–70.

Hornberger, C., Knoop, P., Matz, H. et al. 2002. A prototype device for standardized calibration of pulse oximeters II. *J. Clin. Monit. Comput.* 17:203–9.

Hurtado, A. 1964. Animals in high altitudes: Resident man. In *Handbook of Physiology.* Section 4. *Adaptation to the environment,* ed. D.B. Dill, E.F. Adolph, and C.G. Wilbur, 843–60. Washington, D.C.: American Physiological Society.

Hurtado, A., C. Merino and D. Delgado. 1945. Influence of anoxemia on the hemopoietic activity. *Arch. Intern. Med.* 75:284–323.

James, G.D. and T.G. Pickering. 1991. Ambulatory blood pressure monitoring: Assessing the diurnal variation of blood pressure. *Am. J. Phys. Anthropol.* 84:343–9.

Jensen, L.A., Onyskiw, J.E. and N.G. Prasad. 1998. Meta-analysis of arterial oxygen saturation monitoring by pulse oximetry in adults. *Heart Lung* 27:387–408.

Jubran, A. 1999. Pulse oximetry. *Crit. Care* 3(2):R11–7.

Mejia, O.M., Prchal, J.T., Leon-Velarde, F., Hurtado, A. and D.W. Stockton. 2005. Genetic association analysis of chronic mountain sickness in an Andean high-altitude population. *Haematologica* 90:13–9.

Meyts, I., Reempts, P.V. and K.D. Boeck. 2002. Monitoring of haemoglobin oxygen saturation in healthy infants using a new generation pulse oximeter which takes motion artifacts into account. *Eur. J. Pediatr.* 161:653–5.

Moe, P.J. 1970. Hemoglobin, hematocrit and red blood cell count in 'capillary' (skin-prick) blood compared to venous blood in children. *Acta. Paediatr. Scand.* 59:49–51.

Morris, R.W., Nairn, M. and T.A. Torda. 1989. A comparison of fifteen pulse oximeters. Part I: A clinical comparison; Part II: A test of performance under conditions of poor perfusion. *Anaesth. Intensive Care* 17:62–73.

Niermeyer, S., Yang, P., Shanmina, Drolkar, Zhuang, J. and L. G. Moore. 1995. Arterial oxygen saturation in Tibetan and Han infants born in Lhasa, Tibet. *N. Engl. J. Med.* 333:1248–52.

Pugh, L.G.C.E. 1964. Cardiac output in muscular exercise at 5,800 m (19,000 ft). *J. Appl. Physiol.* 19:441–7.

Reynafarje, C. 1966. Physoiological patterns: Hematological aspects. In *Life at High Altitudes*, 32–5. Washington D.C.: Pan American Health Organization. PAHO Scientific Publications no. 140.

Seppanen, A. 1977. Smoking in closed space and its effect on carboxyhaemoglobin saturation of smoking and nonsmoking subjects. *Ann. Clin. Res.* 9:281–3.

Strohman, R.C. 1997. The coming Kuhnian revolution in biology. *Nat. Biotechnol.* 15:194–200.

Stuart, J., Barrett, B.A. and D.R. Prangnell. 1974. Capillary blood collection in haematology. *J. Clin. Pathol.* 27:869–74.

Suzuki, K., Kizaki, T., Hitomi, Y. et al. 2003. Genetic variation in hypoxia-inducible factor 1 alpha and its possible association with high altitude adaptation in Sherpas. *Med. Hypotheses* 61:385–9.

Thilo, E.H., Andersen, D., Wasserstein, M.L., Schmidt, J. and D. Luckey. 1993. Saturation by pulse oximetry: Comparison of the results obtained by instruments of different brands. *J. Pediatr.* 122:620–6.

Trivedi, N.S., Ghouri, A.F., Lai, E., Shah, N.K. and S.J. Barker. 1997. Pulse oximeter performance during desaturation and resaturation: A comparison of seven models. *J. Clin. Anesth.* 9:184–8.

Weil, J.V., Jamieson, G., Brown, D.W. and R.F. Grover. 1968. The red cell mass-arterial oxygen relationship in normal man. Application to patients with chronic obstructive airway disease. *J. Clin. Invest.* 47:1627–39.

Weiss, K.M. and A.W. Bigham. 2007. 'So mortal and so strange a pang': A tribute to Paul T. Baker. *Evol. Anthropol.* 16:164–71.

Weitz, C.A. and R.M. Garruto. 2007. A comparative analysis of arterial oxygen saturation among Tibetans and Han born and raised at high altitude. *High Alt. Med. Biol.* 8:13–26.

Wilkins, A.S. 1996. Antibiotic resistance: Origins, evolution and spread. Ciba Foundation Symposium, 16–18 July 1996, London. *Bioessays* 18:847–8.

Zhuang, J., Droma, T., Sun, S. et al. 1993. Hypoxic ventilatory responsiveness in Tibetan compared with Han residents of 3,658 m. *J. Appl. Physiol.* 74:303–11.

Zhuang, J., Droma, T., Sutton, J.R. et al. 1996. Smaller alveolar-arterial 02 gradients in Tibetan than Han residents of Lhasa (3658 m). *Respir. Physiol.* 103:75–82.

Index